Progress in Nonlinear Differential Equations and Their Applications
Volume 48

S.N. Antontsev
J.I. Díaz
S. Shmarev

Energy Methods for Free Boundary Problems

Applications to Nonlinear PDEs and Fluid Mechanics

Birkhäuser
Boston • Basel • Berlin

S. N. Antontsev
Departamento de Matemática
Universidade da Beira Interior
6201-001 Covilhã, Portugal

J. I. Díaz
Departamento de Matemática Aplicada
Universidad Complutenese de Madrid
28040 Madrid, Spain

S. Shmarev
Departamento de Matemáticas
Universidad de Oviedo
33007 Oviedo, Spain

Library of Congress Cataloging-in-Publication Data

A CIP catalogue record for this book is available from the Library of Congress, Washington D.C., USA.

AMS Subject Classifications: 35B05, 35B30, 35J25, 35J60, 35K20, 35K55, 35K57, 35Q30, 35Q35, 35Q72, 35R35, 76A05, 76B10, 76Nxx, 76S05, 76Txx, 80A20

Printed on acid-free paper

Birkhäuser ®

ISBN 0-8176-4123-8 SPIN 10716475
ISBN 3-7643-4123-8

Reformatted from authors' files in LaTeX2ε by John Spiegelman, Philadelphia, PA
Printed and bound by Hamilton Printing Company, Rensselaer, NY
Printed in the United States of America

9 8 7 6 5 4 3 2 1

Contents

Preface

For the past several decades, the study of free boundary problems has been a very active subject of research occurring in a variety of applied sciences. What these problems have in common is their formulation in terms of suitably posed initial and boundary value problems for nonlinear partial differential equations. Such problems arise, for example, in the mathematical treatment of the processes of heat conduction, filtration through porous media, flows of non-Newtonian fluids, boundary layers, chemical reactions, semiconductors, and so on.

The growing interest in these problems is reflected by the series of meetings held under the title "Free Boundary Problems: Theory and Applications" (Oxford 1974, Pavia 1979, Durham 1978, Montecatini 1981, Maubuisson 1984, Irsee 1987, Montreal 1990, Toledo 1993, Zakopane 1995, Crete 1997, Chiba 1999). From the proceedings of these meetings, we can learn about the different kinds of mathematical areas that fall within the scope of free boundary problems. It is worth mentioning that the European Science Foundation supported a vast research project on free boundary problems from 1993 until 1999. The recent creation of the specialized journal *Interfaces and Free Boundaries: Modeling, Analysis and Computation* gives us an idea of the vitality of the subject and its present state of development.

This book is a result of collaboration among the authors over the last 15 years. The collaboration centered on using different energy methods to get conditions on the structure of a PDE (or system of PDEs) which yields the formation of a free boundary; in other words, this means that the support of a solution is localized in the space–time domain.

Energy methods are of special interest in those situations in which traditional methods based on comparison principles have failed. A typical example of such a situation is either a higher-order equation or a system of PDEs.[1] We note that the energy methods are well suited in the study of PDE systems which include equations of different types frequently arising in the mathematical models of continuum mechanics. In such systems, the various unknowns (velocity, density, saturation, etc.) may satisfy equations of different types and need not even be defined on the same domain. Moreover, even when the comparison principle holds, it may be extremely difficult to construct suitable sub- or supersolutions if, for instance, the equation under study contains transport terms and has either variable or unbounded coefficients or the right-hand side.

The main idea of the energy methods consists in deriving and studying suitable ordinary differential inequalities for various types of energy. In typical situations, these inequalities follow from the conservation and balance laws of continuum mechanics. In the simplest situations, the energy functions defined through a formal procedure coincide with the kinetic and potential energy.

The book is structured as follows. The first three chapters begin with a systematic explanation of the energy methods in applications to nonlinear PDEs. A thorough exposition of the application of the methods is developed for problems in fluid mechanics and is given in Chapter 4. For the convenience of the reader, we collect in the appendix some useful facts from the theory of Sobolev spaces, such as embedding and interpolation inequalities.

In each of the first three chapters, we follow the same way of presenting the material, that is, we begin with a heuristic explanation avoiding any technical detail. Next, we consider the class of general quasilinear second-order equations under minimal assumptions on the data. The results obtained then provide an analytic framework for further generalizations. Finally, we consider systems of equations and higher-order equations. The study of each of these objects requires ad hoc modifications of the basic arguments. Each chapter concludes with a section devoted to bibliographical comments and open problems.

Fluid mechanics is one of the most natural fields for the application of energy methods. This is because the fundamental conservation and balance laws employed for its description suggest an adequate choice of energy functions which, in turn, produces the formation of a free boundary. Because of the variety of problems included in this chapter, we have found it more convenient to incorporate the bibliographical comments within the sections themselves.

This book contains a selection of results obtained via the energy methods; for this reason, the presentation is not exhaustive. On the other hand, the number of possible applications is too large to be included in only one volume. This is why we have tried to keep a unified presentation, sometimes avoiding specific features of individual problems. We only briefly mention possible applications of the methods to the study of blowup regimes and to the study of the asymptotic behavior of the solutions (where a free boundary need not appear).

---

[1] For the use of comparison principles in free boundary problems, see, for example, the pioneering papers by O. Oleinik, A. Kalashnikov, and Chzhou Yui-Lin [264] on a degenerate parabolic equation published in 1958 and by H. Brezis [96, 95] on the stationary obstacle problem published in 1973–1974.

The book contains some new unpublished results. Among others, these include the phenomenon of the one-directional localization in solutions of anisotropic elliptic and parabolic equations discussed in Subsection 4.2 in Chapter 1 and Subsection 4.5 of Chapter 3, a fine balance between the diffusion and absorption in a parabolic equation, which leads to the finite time extinction of solutions given in Subsection 2.3 of Chapter 2 and the sonic motion of a gas jet under suitable nonhomogeneous boundary conditions in Subsection 3.3 of Chapter 4. Sections, formulas, figures and assertions are numbered within each chapter. When cross-referencing, we additionally indicate the chapter number.

Certain parts of the material have been used by the authors in postgraduate courses at the Novosibirsk State University, the University Beira Interior of Covilhã, the University Complutense of Madrid, and the University of Oviedo.

It is our pleasure to express our deep gratitude to many colleagues who contributed valuable discussion and critical remarks. We cannot underestimate the enriching discussions with Ph. Benilan, F. Bernis, M. Chipot, G. Galiano, G. Gagneux, A. S. Kalashnikov, A. M. Meirmanov, M. Peletier, V. V. Pukhnachov, J. F. Rodrigues, V. A. Solonnikov, I. Stakgold, and L. Veron, among others.

We especially thank H. Brezis who encouraged us during the too long period of preparation of the manuscript and the staff of Birkhäuser, especially A. Kostant, for their infinite patience and kindness.

S. Antonsev

J. I. Diaz

S. Shmarev

July 2001

1

Localized Solutions of Nonlinear Stationary Problems

1 Introduction

In this chapter, we introduce and develop the energy method as a tool for the study of nonlinear stationary problems which give rise to free boundaries. These free boundaries are usually defined as the boundaries of the supports of solutions and, as we shall indicate later on, are of great relevance in many applications.

First of all, let us define the concept of localized solution referred to in the title of the chapter.

Definition 1.1. Let Ω be an open subset of $\mathbb{R}^N$, $N \geq 1$, and let $u : \Omega \mapsto \mathbb{R}$ be a function satisfying (at least in a weak sense) a given stationary partial differential equation in Ω. We say that u is a *localized solution* if it vanishes on an open part of Ω, i.e., the set $(\operatorname{supp} u) \cap \Omega$ is strictly contained in Ω.

A special class of localized solutions, widely studied in the literature (see the references in Section 7), corresponds to the case where Ω is unbounded and the supports of solutions are bounded (and therefore compact).

Localized solutions occur in problems where the influence of data (such as, for instance, boundary conditions and/or some source terms) on the behavior of solutions is restricted to the points of Ω close enough to the support of the data (the boundary and/or the supports of the source terms). Numerous examples of equations of this sort are furnished by mathematical models of fluid mechanics. However, we postpone application of the method to problems of fluid mechanics until after a formal exhibition of its use as a tool of study of solutions to nonlinear stationary equations and systems of such equations regardless of their physical sense. It is worth noting that all the equations studied at this stage occur in applications and that in each of the cases under study the existence of localized solutions is caused by interaction of various nonlinear terms of the equation that model the phenomena of diffusion and absorption or diffusion and convection.

1.1 A heuristic explanation of the method. The main ideas of the energy method developed in this chapter can be explained via consideration (in a heuristic way) of very simple nonlinear stationary problems.

Consider the one-dimensional equation

$$Lu \equiv -\frac{d}{dx}\left(|u_x|^{p-2}u_x\right) + |u|^{q-1}u = f(x) \quad \text{in } \Omega, \tag{1.1}$$

where

$$\Omega = (L_1, L_2), \qquad -\infty \leq L_1 < L_2 \leq \infty. \tag{1.2}$$

The exponents in (1.1) are assumed to satisfy the inequalities $1 < p < \infty$ and $q > 0$. Note that in the special case $p = 2$ and $q = 1$ the corresponding equation becomes linear. For the sake of simplicity we assume about the "source" term $f(x)$ that

$$f \in L^\infty(\Omega) \cap L^1(\Omega).$$

More general assumptions are discussed in Section 2.

The occurrence of localized solutions of (1.1) is due to the key assumption on the structure of equation (1.1)

$$q < p - 1. \tag{1.3}$$

The simple form of equation (1.1) makes it possible to construct explicit solutions. These solutions are easy to derive if, for instance, equation (1.1) is endowed with the Dirichlet boundary conditions and has no source term: $f \equiv 0$. We consider this type of solution in Section 2 and use them to illustrate some complicated estimates. The explicit solutions are not as easy to obtain however if the source term $f(x)$ is neither radially symmetric, nor zero, or if the boundary conditions are not of Dirichlet type. One of the merits of the energy method (in contrast to other methods usually employed for the study of localized solutions; see, e.g., the monograph [128]), is that the associated boundary conditions play a secondary role and are only needed to obtain global estimates on the solutions.

To begin with, we consider one of the simplest cases corresponding to the situation when Ω is bounded and $f \equiv 0$. To be precise, we assume that

$$\Omega = (-L, L) \quad \text{with some } L > 0, \qquad f \equiv 0. \tag{1.4}$$

Let u be an arbitrary regular solution of (1.1) (which we assume to be regular just to simplify matters; we postpone a more rigorous approach until the consequent sections).

The energy method *starts* by defining what we call *local energy functions*. Of course, these functions happen to be very simple in our present one-dimensional case, but in the next sections we extend this idea to the solutions of multidimensional equations which will allow us to deal with scalar functions of one variable. We obtain the local energy functions, multiplying the equation by the unknown u. In the special case of equation (1.1), and under assumption (1.4), a natural choice of these auxiliary energy functions is the following: given $\rho \in (0, L)$ we define

$$E(\rho) = \int_{-\rho}^{\rho} |u_x(x)|^p dx \quad \text{and} \quad b(\rho) = \int_{-\rho}^{\rho} |u(x)|^{q+1} dx. \tag{1.5}$$

These functions can be viewed as the diffusion and absorption energies generated by the solution u in the interval $(-\rho, \rho)$. The *total energy function* is then defined by

$$T(\rho) = E(\rho) + b(\rho), \qquad \rho \in (0, L). \tag{1.6}$$

Note that the domain of integration in the definition of the local energy functions, which we call the *local energy set*, is determined by assumption (1.4) on Ω. The choice of the local energy set plays an important role. In fact, different choices of this set lead to completely different final estimates on the supports of the localized solutions.

The *second step* of the energy method consists in deriving a differential inequality for some of the energy functions. This can be formally done by multiplying equation (1.1) by u and integrating the result by parts over the local energy set. In our case, making use of (1.4) and (1.5), we obtain from (1.1)

$$T(\rho) = E(\rho) + b(\rho) = |u_x(x)|^{p-2} u_x(x) u(x) \Big|_{x=-\rho}^{x=\rho} := I(\rho). \tag{1.7}$$

The information on the boundary behavior of the local energy set will be expressed in terms of the local (and global) energy functions. First of all, we *assume* that the *global energy* is bounded, i.e.,

$$T(L) < \infty. \tag{1.8}$$

Since

$$\begin{cases} \dfrac{dE(\rho)}{d\rho} = |u_x(\rho)|^p + |u_x(-\rho)|^p, \\[2ex] \dfrac{db(\rho)}{d\rho} = |u(\rho)|^{q+1} + |u(-\rho)|^{q+1}, \end{cases} \tag{1.9}$$

then

$$|I(\rho)| \leq 2^{1/p} \left(\frac{dE(\rho)}{d\rho} \right)^{(p-1)/p} (|u(\rho)| + |u(-\rho)|). \tag{1.10}$$

There are now two possible ways of estimating the boundary values $|u(\pm\rho)|$ in (1.10). The first one is typical for the one-dimensional case. It consists in using the second of equalities (1.9). We have

$$
\begin{aligned}
|I(\rho)| &\le 2^{1/p}\left(\frac{dE(\rho)}{d\rho}\right)^{(p-1)/p}\left(\frac{db(\rho)}{d\rho}\right)^{1/(q+1)} \\
&\le 2^{1/p+q/(q+1)}\left(\frac{dT(\rho)}{d\rho}\right)^{1/(q+1)+(p-1)/p}.
\end{aligned}
\tag{1.11}
$$

The second way of estimating $I(\rho)$ is more general and can also be applied in the multidimensional case. It is based on the nontrivial *interpolation-trace inequality*

$$
|u(x)| \le C\left(\|u_x\|_{L^p(B_\rho)} + \rho^{-\delta}\|u\|_{L^{q+1}(B_\rho)}\right)^{\theta}\|u\|^{(1-\theta)}_{L^{q+1}(B_\rho)},
\tag{1.12}
$$

where $x \in B_\rho = (-\rho, \rho)$, $C = C(p, q)$,

$$
\|u\|_{L^p(B_\rho)} = \left(\int_{B_\rho} |u|^p dx\right)^{1/p},
$$

and

$$
\begin{aligned}
\theta &= \frac{p}{p + (p-1)(q+1)} \in (0, 1), \\
\delta &= \frac{1}{q+1} + \frac{p-1}{p} = \frac{1}{\theta(q+1)}.
\end{aligned}
\tag{1.13}
$$

In the one-dimensional case inequality (1.12) merely follows from the formula

$$
|u(z)|^{(1-\theta)/\theta}u(z) = \frac{1}{\theta}\int_0^z |u(y)|^{(1-2\theta)/\theta}u(y)u_x(y)dy + C_0
\tag{1.14}
$$

with

$$
C_0 = |u(0)|^{(1-\theta)/\theta}u(0).
$$

This formula is a byproduct of the fundamental theorem of calculus. In order to get (1.12), we integrate (1.14) with respect to z over the interval $(-\rho, \rho)$, whence

$$
C_0 = \frac{1}{2\rho}\left(\int_{-\rho}^{\rho} |u|^{(1-\theta)/\theta}u dz - \frac{1}{\theta}\int_{-\rho}^{\rho}\left(\int_0^z |u|^{(1-2\theta)/\theta}uu_x dy\right)dz\right).
$$

Substituting this equality into (1.14) and applying Hölder's and Young's inequalities we come to (1.12). Let us return to estimating $|I(\rho)|$ in (1.10). Using (1.9) and (1.12) we get the inequality

$$
\begin{aligned}
|I(\rho)| &\le C\left(\frac{dE}{d\rho}\right)^{(p-1)/p}\left(E^{1/p} + \rho^{-\delta}b^{1/(q+1)}\right)^{\theta} b^{(1-\theta)/(q+1)} \\
&\le C_1\rho^{-\delta\theta}\left(\frac{dE}{d\rho}\right)^{(p-1)/p}(E+b)^{\theta/p}b^{(1-\theta)/(1+q)},
\end{aligned}
\tag{1.15}
$$

where

$$C_1 = C(p,r) \max_{\rho \in (0,L)} \left(1, \rho^{\delta\theta}\right) \max_{\rho \in (0,L)} \left(1, b(\rho)^{(p-q-1)\theta/p(q+1)}\right) \le C_2 \tag{1.16}$$

with some $C_2 \equiv C_2(p, q, L, T(L))$.

Let us now apply estimates (1.11) or (1.15) to derive a differential inequality for the local energy functions. In the case of (1.11) we easily obtain from (1.7) that

$$CT^\nu(\rho) \le \frac{dT(\rho)}{d\rho} \tag{1.17}$$

with an appropriate constant $C = C(p, q)$ and the exponent

$$\nu = \frac{p(q+1)}{p + (q+1)(p-1)}. \tag{1.18}$$

If we use (1.15), we have from (1.7) that

$$E(\rho) + b(\rho) \le C_2 \rho^{-\delta\theta} \left(\frac{dE}{d\rho}\right)^{(p-1)/p} (E+b)^{(1-\theta)/(q+1)+\theta/p}$$

and then

$$E^\nu(\rho) \le (E(\rho) + b(\rho))^\nu \le C_3 \rho^{-\alpha} \left(\frac{dE(\rho)}{d\rho}\right), \tag{1.19}$$

where now

$$\begin{aligned} C_3 &= C_2^{p/(p-1)}, \qquad \alpha = \frac{\delta\theta p}{p-1} = \frac{p}{(q+1)(p-1)}, \\ \nu &= \frac{p}{p-1}\left(1 - \frac{\theta}{p} - \frac{1-\theta}{q+1}\right) = \frac{p(q+1)}{p+(q+1)(p-1)}. \end{aligned} \tag{1.20}$$

The *third and last step* of the energy method is to apply the derived differential inequalities to the study of the localization property of the solution u. We use here the fundamental assumption (1.3) which implies that the exponent ν, given by (1.18), satisfies the inclusion $\nu \in (0, 1)$. Then *formally* rewriting (1.17) (resp., (1.19)) as

$$C d\rho \le \frac{dT}{T^\nu} \quad \left(\text{resp., } \rho^\alpha d\rho \le C_3 \frac{dE}{E^\nu}\right),$$

and integrating over an interval $(\rho, \rho_1) \subset [0, L]$, we get

$$T^{1-\nu}(\rho) \le T^{1-\nu}(\rho_1) - \frac{1-\nu}{C}(\rho_1 - \rho) \tag{1.21}$$

and, correspondingly,

$$E^{1-\nu}(\rho) \le E^{1-\nu}(\rho_1) - \frac{1-\nu}{C_3(1+\alpha)}\left(\rho_1^{1+\alpha} - \rho^{1+\alpha}\right). \tag{1.22}$$

Since the energy functions are nonnegative, we conclude that

$$T(\rho) = 0 \quad (\text{resp., } E(\rho) = 0) \quad \text{and so } u(x) = 0 \text{ for } |x| \le \rho \tag{1.23}$$

with an arbitrary ρ such that

$$\rho \le \rho_1 - \frac{C}{1-\nu} T^{1-\nu}(\rho_1). \tag{1.24}$$

Correspondingly,

$$\rho^{1+\alpha} \le \rho_1^{1+\alpha} - \frac{c_3(1+\alpha)}{1-\nu} E^{1-\nu}(\rho_1). \tag{1.25}$$

A rigorous justification of these operations is given in Subsection 2.1.

In order to get a nonempty conclusion, we have to be sure of the positiveness of the right-hand side of inequalities (1.24), (1.25). This leads to a suitable balance between the size of the domain Ω and the global energy. In particular, this is true if (1.8) is replaced by the stronger assumption

$$L \ge \frac{C}{1-\nu} T^{1-\nu}(L) \qquad \left(\text{resp., } L^{1+\alpha} \ge \frac{C_3(1+\alpha)}{1-\nu} E^{1-\nu}(L)\right). \tag{1.26}$$

The above arguments prove the following assertion.

Theorem 1.1. *Assume that conditions* (1.4), (1.3), *and* (1.26) *are fulfilled. Let* $u(x)$ *be a regular solution of equation* (1.1). *Then necessarily* $u(x) = 0$ *for any* $x \in [-L_0, L_0]$, *where* L_0 *satisfies either the inequality*

$$L_0 \le L - \frac{C}{1-\nu} T(L)^{1-\nu} \tag{1.27}$$

or

$$L_0^{1+\alpha} \le L^{1+\alpha} - \frac{C_3(1+\alpha)}{1-\nu} E^{1-\nu}(L). \tag{1.28}$$

It is important to stress that the above result can be used locally and applies to study more complicated situations. Say, if we consider a nonhomogeneous equation with $f \ne 0$ and assume, for instance, that instead of (1.4)

$$\Omega = (L_1, L_2) \quad \text{and} \quad \operatorname{supp} f = (a, b) \tag{1.29}$$

with $L_1 < a < b < L_2$, then any solution of (1.1) satisfies the homogeneous equation on the set $(L_1, a) \cup (b, L_2)$. Choosing

$$x_0 = \frac{b + L_2}{2}, \qquad L = \frac{L_2 - b}{2} \tag{1.30}$$

and defining

$$v(y) = u(x_0 + y), \quad y \in (-L, L),$$

we easily check that Theorem 1.1 applies to the function v and conclude that

$$u(x) = 0 \quad \text{for any } x \in \left[\frac{b + L_2}{2} - L_0, \frac{b + L_2}{2} + L_0 \right] \tag{1.31}$$

with

$$L_0 \le L - \frac{C}{1 - \nu} \tilde{T}^{1-\nu}(L) \quad \text{or} \quad L_0^{1+\alpha} \le L^{1+\alpha} - C_3 \frac{1 + \alpha}{1 - \nu} \tilde{E}^{1-\nu}(L). \tag{1.32}$$

The energy functions $\tilde{T}$ and $\tilde{E}$ coincide with T and E but the local energy set $(-\rho, \rho)$ is replaced by $(x_0 - \rho, x_0 + \rho)$. In the same way, we deduce that

$$u(x) = 0 \quad \text{for any } x \in \left[\frac{a + L_1}{2} - L_1, \frac{a + L_1}{2} + L_1 \right] \tag{1.33}$$

with a suitable $L_1 > 0$ such that either

$$L_1 \le \widehat{L} - \frac{C}{1 - \nu} \widehat{T}^{1-\nu}(\widehat{L}), \tag{1.34}$$

or

$$L^{1+\alpha} \le \widehat{L}^{1+\alpha} - C_3 \frac{1 + \alpha}{1 - \nu} \widehat{E}^{1-\nu}(\widehat{L}). \tag{1.35}$$

Here $\widehat{L} = (a - L_1)/2$ and again $\widehat{T}$, $\widehat{E}$ coincide with T, E but the energy set $(-\rho, \rho)$ is replaced by $((a + L_1)/2 - \rho, (a + L_1)/2) + \rho)$.

The reader may also wonder how to estimate the values of the energies $\tilde{T}(L)$ and $\widehat{T}(\widehat{L})$. To make (1.31), (1.33) more effective, a constructive way of estimating these quantities is required. In many boundary-value problems associated with equations of the type (1.1) such estimates can be easily derived through the prescribed data. For example, let us consider the Dirichlet problem for equation (1.1):

$$u(L_1) = u(L_2) = 0. \tag{1.36}$$

Multiplying (1.1) by $u(x)$, integrating by parts, and applying Young's inequality we arrive at the estimate

$$\tilde{T}(L) = \int_{x_0 - L}^{x_0 + L} \left(|u_x|^{p+1} + |u|^{q+1} \right) dx \le T^*.$$

The same is true for $\widehat{T}(\widehat{L})$, if we define

$$T^* = \int_{L_1}^{L_2} |f(x)|^{(q+1)/q} dx. \tag{1.37}$$

Corollary 1.1. *Let condition* (1.29) *be true. Assume that* $f \in L^{(q+1)/q}(\Omega)$, *and let* $u(x)$ *be a solution of the problem*

$$\begin{cases} -\dfrac{d}{dx}\left(|u_x|^{p-2}u_x\right) + |u|^{q-1}u = f(x) & \text{in } \Omega = (L_1, L_2), \\ u(L_1) = u(L_2) = 0. \end{cases}$$

Then hypothesis (1.3) *implies that*

$$u(x) = 0 \quad \text{in } [L_1, a - l_a] \cup [b + l_b, L_2],$$

where

$$l_a = \frac{L_2 - b}{2} - \frac{C}{1-\nu}\left(T^*\right)^{1-\nu}, \qquad l_a = \frac{a - L_1}{2} - \frac{C}{1-\nu}\left(T^*\right)^{1-\nu},$$

and T^* *is given by* (1.37).

Proof. We only have to verify that

$$u(x) = 0 \quad \text{on } [(b + L_2)/2 + L_0, L_2] \cup [L_1, (a + L_1)/2 - L_1].$$

In both cases this is a byproduct of the energy identity similar to (1.9) (but now with $I \equiv 0$), which follows by multiplying the equation by u and integrating by parts over these intervals. □

1.2 A few examples of localized solutions. Let us present several simple localized solutions of equation (1.1). It is not difficult to check that the function

$$u(x) = u_0\left(1 - \frac{x}{x_0}\right)_+^{p/(p-q-1)} \qquad (v_+ = \max(0, v)), \tag{1.38}$$

where

$$x_0 = u_0^{(p-q-1)/p}\frac{p}{p-q-1}\left(\frac{(p-1)r}{p-q-1}\right)^{1/p} \tag{1.39}$$

is a solution of equation (1.1) in the domain $\Omega = (0, \infty)$ with $f \equiv 0$. Observe that $\operatorname{supp} u = (0, x_0)$. Due to (1.39), the size of $\operatorname{supp} u$ depends on the quantity of u_0 and the localization effect is absent if $p = q + 1$. We also point out that the function $u(x)$ defined by (1.38) is a solution of the following free-boundary problem: to find a nonnegative function $u(x)$ and a finite number x_0, $0 < x_0 < \infty$, such that u satisfies (1.1) on $(0, x_0)$ and

$$u(0) = u_0, \qquad u(x_0) = 0, \qquad |u_x(x_0)|^p u_x(x_0) = 0.$$

If we consider the boundary conditions associated with (1.1) in the bounded domain $\Omega = (0, 1)$,

$$u(0) = 0, \qquad u(1) = u_1, \tag{1.40}$$

then the condition

$$1 - x_0 = u_1^{(p-q-1)/p} \frac{p}{p-q-1} \left(\frac{(p-1)(q+1)}{p} \right)^{1/p} < 1 \tag{1.41}$$

implies that the function

$$u(x) = u_1 \left(\frac{x - x_0}{1 - x_0} \right)_+^{\frac{p}{(p-q-1)}}, \ (\operatorname{supp} u \in (x_0, 1)), \tag{1.42}$$

is a solution of problem (1.1), (1.40) localized in the interval $[0, x_0]$. Let us note once again that if either $x_0 = 1$, or $p = q + 1$, the localization effect fails. Due to (1.41), it is easy to see that here the localization effect only appears for small values of u_1.

Let us consider now the boundary conditions of the form

$$u(0) = u_0, \qquad u(\infty) = 0 \tag{1.43}$$

and equation (1.1) with "the source"

$$f(x) = \varepsilon \left(1 - \frac{x}{x_0} \right)_+^{\frac{qp}{p-q-1}}, \quad 0 < x_0 < \infty. \tag{1.44}$$

(ε is termed the "intensity" of the source $f(x)$). Let us try to find a solution of problem (1.43), (1.44) in the form (1.38). A simple computation shows that such a solution does exist if the constants u_0, x_0, ε are subject to the relation

$$u_0^q \left(1 - u_0^{p-q-1} x_0^{-p} \left(\frac{p}{p-q-1} \right)^p \frac{(p-1)(q+1)}{p} \right) = \varepsilon. \tag{1.45}$$

The following assertions hold:

(i) given $x_0 \in (0, \infty)$, there exist (ε, u_0), $(\varepsilon > 0, u_0 > 0)$ such that (1.45) is true;

(ii) given $u_0 > 0$, there exist (x_0, ε), $(x_0 > 0, \varepsilon > 0)$ such that (1.45) holds;

(iii) given (ε, u_1) such that $\varepsilon / u_0^q < 1$, there exists $x_0 \in (0, \infty)$ satisfying (1.45).

Thus, under an appropriate choice of u_0, ε and x_0 the solutions to problem (1.43), (1.44) vanish once the "source" $f(x)$ is "switched off."

Similar solutions can be constructed for equation (1.1) with the source terms of the type

$$f(x) = \varepsilon \left(1 + \frac{x - 1}{x_0} \right)_+^{\frac{qp}{p-q-1}}, \quad 0 < x_0 < 1.$$

Let us consider the N-dimensional equation

$$-\operatorname{div} \left(|\nabla u|^{(p-2)} \nabla u \right) + \lambda |u|^{(q-1)} u = 0. \tag{1.46}$$

The differential operator of equation (1.46) will appear in other chapters of the book. That it why it is convenient to use for it the already classical notation

$$\Delta_p u := \operatorname{div}\left(|\nabla u|^{p-2}\nabla u\right).$$

Under the structural assumption (1.3) this equation possesses the explicit solution

$$u(x) = C|x - x_0|^{\frac{p}{p-q-1}}$$

with the constant $C \equiv C(N, \lambda)$ given by

$$C(N,\lambda) = \left[\frac{\lambda(p-q-1)^p}{p^{p-1}(pq + N(p-q-1))}\right]^{\frac{1}{p-q-1}}.$$

We refer the reader to the monograph (Diaz [128]) for an exhaustive discussion of possible applications (via comparison principles) of explicit solutions of this type to the study of free-boundary problems for equation (1.46).

2 Second-order elliptic equations

2.1 General local theorems: The diffusion/absorption balance. Let us consider the class of nonlinear second-order elliptic equations of the form

$$-\operatorname{div}\mathbf{A}(x,u,\nabla u) + B(x,u,\nabla u) + C(x,u) = f(x), \tag{2.1}$$

where

$$\nabla u = \left(\frac{\partial u}{\partial x_1}, \dots, \frac{\partial u}{\partial x_N}\right), \qquad \mathbf{A} = (A_1, \dots, A_N),$$

$$\operatorname{div}\mathbf{A} = \sum_{i=1}^{N}\frac{d}{dx_i}A_i(x,\,u,\,\nabla u) = \sum_{i=1}^{N}\left(\frac{\partial A_i}{\partial x_i} + \frac{\partial A_i}{\partial u}\frac{\partial u}{\partial x_i} + \frac{\partial A_i}{\partial u_{x_k}}\frac{\partial^2 u}{\partial x_k x_i}\right).$$

We will consider equation (2.1) in an open domain $\Omega \subseteq \mathbb{R}^N$, $N \geq 1$. To simplify matters, we always assume that Ω is connected.

The vector-valued function $\mathbf{A}$ and the scalar functions B, C are assumed to satisfy the following structural conditions:

$$|\mathbf{A}(x,r,\mathbf{q})| \leq C_1|\mathbf{q}|^{p-1}, \tag{2.2}$$

$$C_2|\mathbf{q}|^p \leq \mathbf{A}(x,r,\mathbf{q})\cdot\mathbf{q}, \tag{2.3}$$

$$|B(x,r,\mathbf{q})| \leq C_3|r|^{\alpha}|\mathbf{q}|^{\beta}, \tag{2.4}$$

$$C_4|r|^{\sigma+1} \leq C(x,r)r. \tag{2.5}$$

The conditions on $(x,r,\mathbf{q}) \in \Omega \times \mathbb{R} \times \mathbb{R}^N$ and the positive constants p, α, β, σ, $C_1 - C_4$ will be specified later on.

Equation (2.1) occurs in a number of problems of fluid mechanics. In Chapter 4 we discuss the modelling and fields of application of some equations of this kind. The following model equation also falls into this class:

$$-\operatorname{div}\left(|\nabla u|^{p-2}\nabla u\right) + |u|^{\sigma-1}u = f(x). \tag{2.6}$$

There already exists a wide literature devoted to the study of the problem of existence of weak solutions to equations of the class (2.1) (see, for example, [128, 151, 165, 235, 243]). Here we have bound ourselves to the study of qualitative properties of weak solutions. Of course, we always assume that such solutions exist, however we do not make any special extra assumptions on the structure of nonlinear terms which would guarantee the possibility to apply any of the existence results in the literature. In other words, the existence of solutions with the required properties is assumed "by default" (which in most cases is true).

Definition 2.1. Let $f(x) \in L^1_{\mathrm{loc}}(\Omega)$. A locally integrable function $u(x)$ is said to be a weak solution of (2.1) if

(i) $u \in W^{1,p}_{\mathrm{loc}}(\Omega)$;

(ii) $B(\cdot, u, \nabla u), A_i(\cdot, u, \nabla u) \in L^1_{\mathrm{loc}}(\Omega),\ i = 1, \dots, N$;

(iii) $C(\cdot, u) \in L^1_{\mathrm{loc}}(\Omega)$;

(iv) for any test function $\varphi \in C_0^\infty(\Omega)$, the equality

$$\int_\Omega \{\mathbf{A}(x, u, \nabla u) \cdot \nabla\varphi + B(x, u, \nabla u)\varphi + C(x, u)\varphi\}\, dx = \int_\Omega f\varphi dx \tag{2.7}$$

holds.

Here and throughout the book, $L^p(\Omega)$ denotes the space of functions integrable with the power p in Ω, and $W^{1,p}(\Omega)$ is the Sobolev space of functions from $L^p(\Omega)$ whose first weak derivatives belong to $L^p(\Omega)$. The local versions of these spaces are defined as the restrictions to any open bounded subset. For instance, $f \in L^1_{\mathrm{loc}}(\Omega)$ if $f \in L^1(\sigma)$ for any open bounded set $\sigma \subset \Omega$. (More information about the functional spaces that we use is given in the appendix).

Let us introduce the notation

$$B_\rho(x_0) = \left\{x \in \mathbb{R}^N : |x - x_0| < \rho\right\}, \qquad S_\rho(x_0) = \partial B_\rho(x_0),$$

$$E(\rho, u) \equiv E(\rho) = \int_{B_\rho(x_0)} \mathbf{A}(x, u, \nabla u) \cdot \nabla u dx,$$

$$b(\rho, u) \equiv b(\rho) = \int_{B_\rho(x_0)} |u|^{\sigma+1} dx = \|u\|^{\sigma+1}_{B_\rho}.$$

Then

$$C_2\|\nabla u\|^p_{L^p(B_\rho)} \le E \le C_1\|\nabla u\|^p_{L^p(B_\rho)}.$$

As before, we call E and b the *energy functions* associated with the solution $u(x)$ of equation (2.1).

Passing to the spherical coordinates (r, ω) with origin at the point x_0, and assuming that $E(\rho, u) < \infty$, we get the equality

$$E(\rho) = \int_0^\rho \int_{B_1} \mathbf{A}(x_0 + r\omega, u, \nabla u) \cdot \nabla u r^{N-1} d\omega dr = \int_0^\rho \left(\int_{S_\rho} \mathbf{A} \cdot \nabla u dS \right) d\rho.$$

It follows from the last equality and condition (2.3) that $E(\rho)$ is a monotone nondecreasing function. Hence, it has the weak derivative $E'(\rho) = \frac{dE(\rho)}{d\rho}$ and, due to (2.2), (2.3), possesses the following properties:

$$\begin{gathered} E'(\rho) = \int_{S_\rho} \mathbf{A} \cdot \nabla u \, dS \, d\rho \quad \text{for a.e. } \rho \in (0, \rho_0), \\ C_2\|\nabla u\|^p_{L^p(S_\rho)} \le E'(\rho) \le C_1\|\nabla u\|^p_{L^p(S_\rho)}, \end{gathered} \tag{2.8}$$

where $\rho_0 = \operatorname{dist}(x_0, \partial\Omega)$, and $E' \in L^1(0, \rho_0)$. We are now in position to formulate the main result on the local vanishing property of weak solutions to equation (2.1).

Theorem 2.1. *Let $u(x)$ be a weak solution of equation* (2.1). *Assume that conditions* (2.2)–(2.4) *hold with*

$$C_2 > 0, \quad C_4 > 0, \quad 0 \le \sigma < p - 1, \quad \alpha = \sigma - \beta\frac{1+\sigma}{p}$$

and, additionally, that either

$$C_3 < C_4 \quad \textit{if } \beta = 0(\qquad \textit{respectively, } C_3 < C_2 \textit{ if } \beta = p),$$

or

$$C_3 < \left(C_4\frac{p}{p-\beta}\right)^{(p-\beta)/p} \left(C_2\frac{p}{\beta}\right)^{\beta/p} \quad \textit{if } 0 < \beta < p. \tag{2.9}$$

Assume that $f(x) \equiv 0$ in Ω. Given an arbitrary point $x_0 \in \Omega$,

$$u(x) \equiv 0 \quad \textit{a.e. in } B_{\rho_1}(x_0)$$

with ρ_1 given by the expression

$$\begin{gathered} \rho_1^\nu = \left(\rho_0^\nu - C \min_{\frac{\sigma+1}{p} < \tau \le 1} \left(E^\gamma(\rho_0) G(\rho_0) \right) \right)_+, \\ G = \frac{1}{\tau p - \sigma - 1} \max\left(1, \rho_0^{\nu-1}\right) \max\left(b^\mu(\rho_0), b^\eta(\rho_0)\right). \end{gathered} \tag{2.10}$$

Here $\rho_0 = \text{dist}\,(x_0, \partial\Omega)$, $C = C(C_1, C_2, C_3, N, p, \sigma, \beta)$ *is a constant, and*

$$\nu = \frac{(p-1)(\sigma+1)}{\kappa}, \qquad \gamma = \frac{\tau p - 1 - \sigma}{\kappa}, \qquad \mu = \frac{p(1-\tau)}{\kappa},$$
$$\eta = \left(\frac{p-\sigma-1}{(p-1)(1+\sigma)} + \frac{\tau p - 1 - \sigma}{\kappa}\right), \qquad \kappa = N(p-\sigma-1) + p(1+\sigma) > 0.$$

Remark 2.1. If $\rho_1 = 0$, then the above statement offers no information on the vanishing set of $u(x)$. Nonetheless, if the total energy $E(\rho_0) + b(\rho_0)$ is small enough, then always $\rho_1 > 0$ and the vanishing set of $u(x)$ is not empty.

Proof. The proof of Theorem 2.1 is split into several steps.

Lemma 2.1 (Step 1). *Under the above assumptions,*

$$\Big(\mathbf{A}(\cdot, u, \nabla u)\cdot \nabla u;\ B(\cdot, u, \nabla u)u;\ |u|^{\sigma+1};\ |\mathbf{A}(\cdot, u, \nabla u)|;\ |u|\Big) \in L^1(B_{\rho_0}(x_0)),$$

and for almost all $\rho \in (0, \rho_0)$ *the inequality*

$$\begin{aligned} &\int_{B_\rho(x_0)} [\mathbf{A}(x, u, \nabla u)\cdot\nabla u + B(x, u, \nabla u)u]\,dx + C_4 \int_{B_\rho(x_0)} |u|^{\sigma+1} dx \\ &\quad \le -\int_{S_\rho(x_0)} \mathbf{A}(x, u, \nabla u)u \cdot \mathbf{n}\, dS := I(\rho) \end{aligned} \tag{2.11}$$

holds, where $\mathbf{n} = (n_1, \dots, n_N)$ *is the unit outer normal vector to* $S_\rho(x_0)$. *Moreover,* $I(\rho) \in L^1(0, \rho_0)$.

Proof. Inequality (2.11) and the inclusion $I(\rho) \in L^1(0, \rho_0)$ follow from (2.2)–(2.4), Definition 2.1, and the embedding $W^{1,p}(\Omega) \subset L^m(\Omega)$ with some m. More precisely (see the appendix), we have that

$$\|u\|_{L^m(\Omega)} \le C\|u\|_{W^{1,p}(\Omega)} \tag{2.12}$$

for every $m \ge 1$ satisfying the conditions

$$m \le \frac{Np}{N-p} \quad \text{if } N > p, \qquad m \ge 1 \text{ arbitrary} \quad \text{if } N = p, \qquad m = \infty \quad \text{if } p > N.$$

Unless specially indicated, here and in what follows we denote with C any constant which depends on N and constants in conditions (2.2)–(2.5). Applying Hölder's inequality, inequality (2.12), and conditions (2.2)–(2.5), we have:

$$\begin{aligned} \|\mathbf{A}\cdot\nabla u\|_{L^1} &\le C_1\|\nabla u\|^p_{L^p} \le C_1\|u\|^p_{W^{1,p}}, \\ \|u\|_{L^{\sigma+1}} &\le C\|u\|_{W^{1,p}}, \\ \|B\cdot u\|_{L^1} &\le C_3\|\nabla u\|^\beta_{L^p}\cdot\|u\|^{\alpha+1}_{L^p} \le C\|u\|^{1+\beta+\alpha}_{W^{1,p}}, \\ \|\mathbf{A}u\|_{L^1} &\le C_1\|\nabla u\|^{p-1}_{L^p}\|u\|_{L^p} \le C_1\|u\|^p_{W^{1,p}}, \\ \|I\|_{L^1(0,\rho_0)} &\le \int_{B_\rho} |\mathbf{A}||u|\,dx \le C_1\|u\|^p_{W^{1,p}(B_{\rho_0})}. \end{aligned} \tag{2.13}$$

Let us introduce the cutoff functions $T_k(u) = \min(k, |u|)\,\mathrm{sign}\, u$ with $k \in \mathbb{N}$, and

$$\psi_n(r) = \begin{cases} 1 & \text{if } r \in [0, \rho - \frac{1}{n}], \\ n(\rho - r) & \text{if } r \in [\rho - \frac{1}{n}, \rho], \\ 0 & \text{if } r \in [\rho, \rho_0], \quad n \in \mathbb{N}. \end{cases}$$

According to results of [235] the cutoff function T_k belongs to $W^{1,p}(B_{\rho_0}) \cap L^\infty(B_{\rho_0})$ and possesses the following properties:

$$\begin{aligned} &\nabla T_k(u) = \begin{cases} \nabla u & \text{if } x \in B^k_{\rho_0} = \{x : x \in B_{\rho_0}, |u| < k\}, \\ 0 & \text{if } x \in B_{\rho_0} \setminus B^k_{\rho_0}, \end{cases} \\ &\text{meas}\left\{B_{\rho_0} \setminus B^k_{\rho_0}\right\} \to 0 \quad \text{as } k \to \infty, \\ &\|T_k(u)\|_{L^m} \le \|u\|_{L^m} \quad \text{if } m \le \frac{Np}{N-p}, \\ &\|T_k(u) - u\|_{W^{1,p}} \to 0 \quad \text{as } k \to \infty. \end{aligned} \tag{2.14}$$

According to the structural condition (2.5) we have the inequality

$$C_3 \min(k, |u|)\,|u|^\sigma \le C(x,u) T_k(u). \tag{2.15}$$

It is easy to see now that the function

$$\varphi_{n,k}(x) \equiv \psi_n(|x - x_0|)\, T_k(u(x))$$

belongs to $W^{1,p}_0(B_\rho) \cap L^\infty(B_\rho)$. The set $C_0^\infty(B_\rho)$ is dense in $W^{1,p}_0(B_\rho) \cap L^\infty(B_\rho)$. Hence, $\varphi_{n,k}(x)$ can be taken for the test function in (2.7). Letting in (2.7) $\varphi = \varphi_{n,k}$, we have

$$\begin{aligned} &\int_{B_{\rho_0}} \psi_n \{\mathbf{A}(x,u,\nabla u) \cdot \nabla T_k(u) + B(x,u,\nabla u) T_k(u) + C(x,u) T_k(u)\}\, dx \\ &\quad = \int_{B_{\rho_0}} \mathbf{A}(x,u,\nabla u) T_k(u) \cdot \nabla \psi_n dx. \end{aligned}$$

Passing to the limit as $k \to \infty$ and taking into account (2.13), (2.14), (2.15) we arrive at the inequality

$$\begin{aligned} &\int_{B_{\rho_0}} \psi_n \left\{\mathbf{A}(x,u,\nabla u) \cdot \nabla u + B(x,u,\nabla u) u + C_4 |u|^{\sigma+1}\right\} dx \\ &\quad \le - \int_{B_{\rho_0}} \mathbf{A}(x,u,\nabla u) u \cdot \nabla \psi_n dx. \end{aligned}$$

Introducing the spherical coordinates (r, ω) and recalling the properties of ψ_n, we have

$$\begin{aligned} \int_{B_{\rho_0}} u\mathbf{A} \cdot \nabla \psi_n dx &= n \int_{\rho - \frac{1}{n} < |x - x_0| < \rho} u\mathbf{A}(x,u,\nabla u) \cdot \frac{x - x_0}{|x - x_0|} dx \\ &= n \int_{\rho - \frac{1}{n}}^{\rho} \left(\int_{S_r} \mathbf{A}(r\omega, u, \nabla u) u \cdot \mathbf{n} r^{N-1} d\omega \right) dr \equiv n \int_{\rho - \frac{1}{n}}^{\rho} I(r) dr. \end{aligned}$$

Since $I \in L^1(0, \rho_0)$, it follows from the Lebesgue theorem that for a.e. $\rho \in (0, \rho_0)$

$$\lim_{n\to\infty} \int_{B_{\rho_0}} u\mathbf{A} \cdot \nabla\psi_n dx = \int_{S_\rho} u\mathbf{A}(x, u, \nabla u) \cdot \mathbf{n} dS, \tag{2.16}$$

which proves (2.11). □

Step 2. Let us prove the existence of a constant $C_5 = C_5(C_2, C_3, C_4, p, \sigma, \beta) > 0$ such that

$$C_5 \left(E(\rho) + b(\rho)\right) \leq E(\rho) + C_4 b(\rho) + \int_{B_\rho} B\left(x, u, \nabla u\right) u dx. \tag{2.17}$$

First of all, if $\beta = 0$, we have that

$$\int_{B_\rho} |Bu| dx \leq C_3 \int_{B_\rho} |u|^{\sigma+1} dx = C_3 b(\rho) \quad \text{and} \quad C_5 = \min\left(1, C_4 - C_3\right).$$

If $\beta = p$, then

$$\int_{B_\rho} |Bu| dx \leq C_3 \int_{B_\rho} |\nabla u|^p dx \leq \frac{C_2}{C_3} E(\rho) \quad \text{and} \quad C_5 = \min\left(C_4, 1 - \frac{C_3}{C_2}\right) > 0.$$

Next, let us assume that $0 < \beta < p$. Applying the Young inequality,

$$\begin{cases} ab \leq \dfrac{\varepsilon}{\tau} a^\tau + \dfrac{\tau - 1}{\tau} \varepsilon^{-1/(\tau+1)} b^{\tau/(\tau-1)} \\ \forall a, b \geq 0, \ \varepsilon > 0, \ \tau > 1, \end{cases}$$

we get

$$|u|^{\alpha+1} |\nabla u|^\beta \leq \frac{\varepsilon}{\tau} |u|^{\tau(\alpha+1)} + \frac{\tau - 1}{\tau} \varepsilon^{-1/(\tau+1)} |\nabla u|^{\beta\tau/(\tau+1)}.$$

Letting here $\tau = (\sigma + 1)/(\alpha + 1)$ and, respectively, $\beta\tau = p(\tau + 1)$, we arrive at the estimate

$$\left| \int_{B_\rho} B(x, u, \nabla u) u dx \right| \leq \varepsilon C_3 \frac{p - \beta}{p} b(\rho) + \frac{\beta C_3}{C_2 p} \varepsilon^{-\frac{(p-\beta)}{\beta}} E(\rho). \tag{2.18}$$

Since C_3 satisfies (2.9), there exists $\varepsilon > 0$ depending on p, β, C_2, C_4 such that

$$\varepsilon C_3 \frac{p - \beta}{p} < C_4 \quad \text{and} \quad \frac{\beta}{C_2} C_3 p \varepsilon^{-(p-\beta)/\beta} < 1.$$

If we now set

$$C_5 = \min\left(C_4 - \varepsilon C_2 \frac{p - \beta}{p}, 1 - \frac{\beta C_3}{C_2 p} \varepsilon^{-(p-\beta)/\beta}\right),$$

(2.17) becomes a byproduct of (2.18).

Step 3. It follows from (2.11) and (2.17) that

$$C_5\left(E(\rho)+b(\rho)\right) \leq -\int_{S_\rho} \mathbf{A}(x,u,\nabla u)u\cdot \mathbf{n}dS := -I. \tag{2.19}$$

With the help of (2.2)–(2.4) and (2.8), the right-hand side of (2.19) can be estimated in the following way:

$$\begin{aligned} |I| &\leq C_1\left(\int_{S_\rho} |\nabla u|^p dS\right)^{(p-1)/p}\cdot\left(\int_{S_\rho}|u|^p dS\right)^{1/p} \\ &\leq C_1 C_2^{(1-p)/p}\left(\frac{dE}{d\rho}\right)^{(p-1)/p} \|u\|_{L^p(S_\rho)}. \end{aligned} \tag{2.20}$$

We now apply the *interpolation-trace inequality*

$$\|u\|_{L^p(S_\rho)} \leq C\left(\|\nabla u\|_{L^p(B_\rho)}+\rho^\delta\|u\|_{L^{1+\sigma}(B_\rho)}\right)^\theta \|u\|^{1-\theta}_{L^{1+\sigma}(B_\rho)}, \tag{2.21}$$

where

$$\delta = -\frac{k}{p(1+\sigma)}, \qquad \theta = \frac{N(p-\sigma-1)+\sigma+1}{k} \in (0,1),$$
$$k = N(p-\sigma-1)+p(\sigma+1), \qquad C = C(N,\sigma,p).$$

Inequality (2.21) is true for any $u \in W^{1,p}(B_\rho)$ (see [147]). In terms of the energy functions $E(\rho)$ and $b(\rho)$ inequality (2.21) takes on the form

$$\|u\|_{L^p(S_\rho)} \leq \overline{C}\left(E^{1/p}+\rho^\delta b^{1/(\sigma+1)}\right)^\theta b^{(1-\theta)/(1-\sigma)} \tag{2.22}$$

with $\overline{C}=\overline{C}(N,\sigma,p,C_2,C_4)$. Gathering (2.19), (2.20), (2.22), we get the inequality

$$(E+b) \leq K\left(\frac{dE}{d\rho}\right)^{(p-1)/p}\left(E^{1/p}+\rho^\delta b^{1/(\sigma+1)}\right)^\theta\cdot b^{(1-\theta)/(\sigma+1)} \tag{2.23}$$

with $K = K(C_1,C_2,C_5,N,\sigma,p,\beta)$. It is easy to verify that for $0\leq\tau\leq 1$, $\rho\leq\rho_0$ the relations

$$\begin{aligned} &E^{1/p}b^{(1-\theta)/\theta(1+\sigma)}+\rho^\delta b^{1/\theta(1+\sigma)} \\ &\quad = E^{1/p}b^{\tau(1-\theta)\kappa}b^{(1-\tau)(1-\theta)\kappa}+\rho^\delta b^{1/p+\tau(1-\theta)\kappa}\cdot b^{\kappa-\tau(1-\theta)\kappa/p} \\ &\quad \leq 2\rho^\delta\max\left(1,\rho_0^{-\delta}\right)K_0^{1/\theta}\left(E+b\right)^{1/p+\tau(1-\theta)\kappa}, \end{aligned} \tag{2.24}$$

hold, where

$$K_0 = \left(\max\left(b^\mu(\rho_0),b^\eta(\rho_0)\right)\right)^{(p-1)/p}, \qquad \kappa = \frac{1}{\theta(1+\sigma)}.$$

It follows from (2.23), (2.24) that

$$E+b \le \left(K_1\rho^{(1-\nu)}\left(\frac{dE}{d\rho}\right)\right)^{(p-1)/p}(E+b)^{1-(p-1)(1-\gamma)/p}, \tag{2.25}$$

where

$$K_1 = \left(2KK_0\max\left(1,\rho_0^{-\delta\theta}\right)\right)^{p/(p-1)},$$

$$\nu = \left(1-\frac{\delta\theta}{p}p-1\right) > 1,$$

$$\gamma = \frac{(1-\theta)(p\tau-1-\sigma)}{(p-1)(1+\sigma)} < 1.$$

The constants γ and ν coincide with those which were defined in (2.10). Inequality (2.25) leads to the ordinary differential inequality

$$E^{1-\gamma}(\rho) \le K_1\rho^{1-\nu}\frac{dE}{d\rho}. \tag{2.26}$$

Let us establish the following useful auxiliary result.

Lemma 2.2. *Let $E \in W^{1,1}_{\text{loc}}(0,R)$, $E \ge 0$. Assume that the inequality*

$$E'(\rho) \ge a(\rho)\varphi(E(\rho)), \quad \textit{a.e. } \rho \in (0,R),$$

holds, where φ is a continuous nondecreasing function such that $\varphi(0)=0$,

$$\int_{0+}\frac{ds}{\varphi(s)} < \infty,$$

and $a \in L^1(0,R)$, $a \ge 0$. Let there exist some $R_0 \in (0,R)$ such that

$$\int_0^{E(R)}\frac{ds}{\varphi(s)} - \int_{R_0}^R a(s)ds = 0.$$

Then $E(\rho)=0$ for any $\rho \in (0,R_0)$.

Proof. Define the functions

$$\theta(s) := \int_0^s \frac{dt}{\varphi(t)} \quad \text{and} \quad \eta(s) := \theta^{-1}(s)$$

(notice that $\theta(s)$ is a strictly increasing function). Let us also define

$$g(\rho) := \int_0^{E(R)}\frac{ds}{\varphi(s)} - \int_\rho^R a(s)ds.$$

It is easy to check that the function $y(\rho) := \eta(g(\rho))$ satisfies the conditions

$$y'(\rho) = a(\rho)\varphi(y(\rho)) \quad \text{for a.e. } \rho \in (0, R) \text{ and } y(R) = E(R).$$

It follows from the comparison principle for ordinary differential equations that

$$0 \le E(\rho) \le y(\rho) \quad \text{for any } \rho \in (0, R).$$

Finally, from the assumptions on E we infer that $E(R_0) = 0$, and since $E'(\rho) \ge 0$ and $E \ge 0$, the conclusion follows. □

2.2 Completion of the proof of Theorem 2.1. Applying Lemma 2.2, we have that

$$K_1 \frac{\nu}{\gamma}\left[E^\gamma(\rho_0) - E^\gamma(\rho_1)\right] \ge \rho_0^\nu - \rho_1^\nu.$$

Hence once

$$\rho_1^\nu = \rho_0^\nu - K_1 \frac{\nu}{\gamma} E^\gamma(\rho_0), \tag{2.27}$$

we get $E(\rho_1) = 0$ and $E(\rho) = 0$ for all $\rho \le \rho_1$. It follows from (2.25) that $b(\rho) = 0$ for $\rho \le \rho_1$, whence $u(x) = 0$ a.e. in $B_\rho(x_0)$ for all $\rho \le \rho_1$. Since ν does not depend on the parameter τ, (2.27) remains true if its right-hand side attains the minimum value as a function of τ. This completes the proof of Theorem 2.1. □

Remark 2.2. The conditions on α and β can be relaxed by imposing the extra assumption $u \in L^\infty_{\mathrm{loc}}(\Omega)$. The arguments used for the proof of the second step show that it suffices to assume that $\alpha \ge 0$, $\beta \le p$, and $C_3 > 0$ is sufficiently small.

We proceed to consider equation (2.1) with a source, i.e., with a prescribed right-hand side $f(x) \ne 0$.

Theorem 2.2. *Let $u(x)$ be a weak solution of equation* (2.1) *and let $f \in L^1_{\mathrm{loc}}(\Omega)$ be such that $f \equiv 0$ in $B_{\rho_1}(x_0)$ with some $0 < \rho_1 < \rho_0$. Assume the fulfillment of the hypotheses of Theorem* 2.1 *and, additionally, let one of the following conditions be true: either*

$$I(\rho_0, f) := \int_{\rho_1-\delta}^{\rho_0} (\tau - \rho_1)_+^{-1/\gamma} \, \|f\|_{L^{(\sigma+1)/\sigma}(B_\tau(x_0))}^{(1+\sigma)(1+\gamma)/\sigma} d\tau < \infty \tag{2.28}$$

with

$$\gamma = \frac{(1-\theta)(p-\sigma-1)}{(p-1)(\sigma+1)}, \qquad \theta = \frac{N(p-\sigma-1)+\sigma+1}{N(p-\sigma-1)+p(\sigma+1)},$$

or

$$\|f\|_{L^{(1+\sigma)/\sigma}(B_\rho(x_0))}^{(1+\sigma)(1-\gamma)/\sigma} \le \varepsilon\,(\rho - \rho_1)_+^{(1-\gamma)/\gamma}, \quad \rho \in (\rho_1 - \delta, \rho_2) \tag{2.29}$$

with some $\delta > 0$ *and* $\rho_2 \in (\rho_1, \rho_0)$. *Then there exist positive constants* I_*, E_*, ε_*, *such that once*

$$I(\rho_0, f) \le I_*, \qquad E(\rho_0, u) \le E_*, \qquad \varepsilon \le \varepsilon_*, \tag{2.30}$$

any weak solution of equation (2.1) *possesses the property*

$$u(x) \equiv 0 \qquad \text{in } B_{\rho_1}(x_0).$$

Theorem 2.2 asserts that if the right-hand side $f(x)$ vanishes fast enough (the admissible rate of vanishing is controlled by the condition of convergence of the integral $I(\rho_0; f)$ or, directly, by (2.30)), then the boundaries of supports of $f(x)$ and the solution $u(x)$ may have common parts or even coincide.

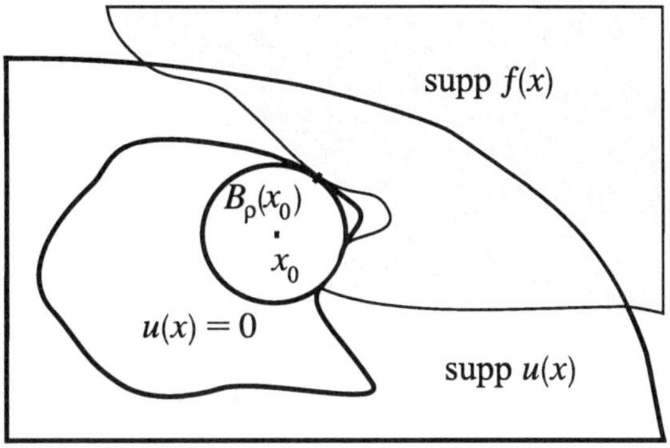

Figure 2.1

Proof. Given a weak solution $u(x)$ of equation (2.1), the inequality

$$\begin{aligned}\int_{B_\rho(x_0)} &\left\{ \mathbf{A}(x, u, \nabla u) \cdot \nabla u + B(x, u, \nabla u) u + C_4 |u|^{\sigma+1} \right\} dx \\ &\le - \int_{S_\rho(x_0)} \mathbf{A}(x, u, \nabla u) u \cdot \mathbf{n} \, dS + \int_{B_\rho(x_0)} f(x) u(x) dx\end{aligned} \tag{2.31}$$

holds. To obtain this inequality we apply the arguments used for proof of Lemma 2.1. Applying the Hölder and Young inequalities, we have that

$$\begin{aligned}\left| \int_{B_\rho} f u \, dx \right| &\le \|f\|_{L^{(\sigma+1)/\sigma}(B_\rho)} b^{1/(\sigma+1)}(\rho) \\ &\le \varepsilon b(\rho) + \frac{\sigma}{\sigma+1} (\varepsilon(\sigma+1))^{-1/\sigma} \, \|f\|^{(\sigma+1)/\sigma}_{L^{(\sigma+1)/\sigma}(B_\rho)}.\end{aligned} \tag{2.32}$$

To estimate the resting terms of (2.31) we make use of (2.19), (2.20), (2.23), (2.25). An appropriate choice of ε in (2.32) leads to the following generalization of inequality (2.25):

$$E + b \leq \left(K_1 \rho^{1-\nu} \frac{dE}{d\rho}\right)^{(p-1)/p} \cdot (E+b)^{1-(p-1)(1-\gamma)/p} + K \|f\|^{(\sigma+1)/\sigma}_{L^{(\sigma+1)/\sigma}(B_\rho)} := I_1 + I_2, \tag{2.33}$$

where only the constant K was changed. We estimate the term I_1 by the Young inequality

$$I_1 \leq \frac{1}{2}(E+b) + \left(K_1 \rho^{1-\nu} \frac{dE}{d\rho}\right)^{1/(1-\gamma)} \tag{2.34}$$

changing, if needed, the constant K in the definition of K_1. Gathering (2.33), (2.34) and raising both sides of the obtained inequality to the power $1-\gamma$, we finally get

$$E^{1-\gamma}(\rho) \leq (E+b)^{1-\gamma} \leq K_1 \rho^{1-\nu} \frac{dE}{d\rho} + K_2 \|f\|^{(\sigma+1)(1+\gamma)/\sigma}_{L^{(\sigma+1)/\sigma}(B_\rho)},$$

whence

$$E^{1-\gamma} \leq \Lambda \frac{dE}{d\rho} + F(\rho) \quad \text{for } \rho \in (\rho_1, \rho_0) \tag{2.35}$$

with

$$\Lambda = \rho_0^{1-\nu}\left(2K \max\left(1, b^{\eta}(\rho_0)\right) \max\left(1, \rho_0^{-\delta\theta}\right)\right)^{p/(p-1)},$$

$$F(\rho) := K_2 \|f\|^{(\sigma+1)(1+\gamma)/\sigma}_{L^{(\sigma+1)/\sigma}(B_\rho)}, \qquad K_2 = K_2(K, \gamma).$$

□

Let us first assume that (2.28) holds. The conclusion will be obtained by using the following generalization of Lemma 2.2.

Lemma 2.3. *Let φ be as in Lemma 2.2. Let $E(\rho) \in W^{1,1}_{\mathrm{loc}}(0, R)$ with $E \geq 0$ and $E' \geq 0$. Assume that the inequality*

$$E'(\rho) + F((\rho - R_0)_+) \geq \varphi(E(\rho)) \quad \text{a.e. in } (R_1, R)$$

holds with some $R_1 \in (0, R)$ and $R_0 \in (R_1, R)$. Given $\mu > 0$, we define the functions

$$\theta_\mu(s) = \int_0^s \frac{d\tau}{\mu\varphi(\tau)}, \qquad \eta_\mu(s) = \theta_\mu^{-1}(s).$$

Assume that there exists $\overline{\mu} \in (0, 1)$ such that

$$(R_0 - R) \geq \theta_{\overline{\mu}}(E(R)) \quad \text{and} \quad F(s) \leq (1 - \overline{\mu})\varphi(\eta_{\overline{\mu}}(s)) \quad \text{if } s \in (R_0, R). \tag{2.36}$$

Then

$$E(\rho) \equiv 0 \quad \text{for any } \rho \in [R_1, R_0].$$

Proof. By construction, the function

$$\overline{E}(\rho) = \eta_{\overline{\mu}}((\rho - R_0)_+)$$

satisfies the equation

$$\overline{E}'(\rho) = \varphi(\overline{E}(\rho)) - (1 - \overline{\mu})\varphi(\eta_{\overline{\mu}}(\rho - R_0)_+). \tag{2.37}$$

At the same time,

$$E(R) \leq \overline{E}(R) \tag{2.38}$$

by virtue of (2.36). It follows then from the comparison principle for ordinary differential equations that

$$E(\rho) \leq \overline{E}(\rho) \quad \text{as } \rho \in [R_1, R].$$

Moreover, since $\overline{E}(R_0) = \eta_{\overline{\mu}}(0) = 0$, we conclude that $E(R_0) = 0$, and since $E'(\rho) \geq 0$ and $E \geq 0$, we get the desired conclusion. □

Remark 2.3. Conditions (2.36) can be illustrated by the following example of homogeneous nonlinearities:

$$\varphi(s) = \delta s^m, \quad (0 < m < 1), \quad \text{and} \quad F(\tau) = \varepsilon \tau_+^{m/(1-m)}.$$

Then

$$\theta_\mu(\tau) = \frac{\tau^{1-m}}{\mu(1-m)}, \qquad \eta_\mu(s) = (s\mu(1-m))^{1/(1-m)},$$

and (2.36) takes the form

$$\varepsilon \leq (1 - \overline{\mu}) d(\overline{\mu}(1-m))^{m/(1-m)} := \delta h(\overline{\mu}). \tag{2.39}$$

Set $\max_{0 \leq \tau \leq 1} h(\tau) := h(m)$,

$$R - R_0 \geq \frac{1}{\overline{\mu}(1-m)}(E(R_0))^{1-m}. \tag{2.40}$$

It is easy to see that condition (2.39) holds if

$$\varepsilon < \delta h(m). \tag{2.41}$$

This conclusion can be interpreted in the following way: the intensity ε of the source F must be small in comparison with the dissipation δ. Condition (2.40) is safely fulfilled if $E(R)$ is small in comparison with $R - R_0$.

Remark 2.4. Under the same conditions on the functions $\phi(\cdot)$ and $F(\cdot)$, the assertion analogous to Lemma 2.3 holds for the nonnegative nonincreasing function $E(t) \in W^{1,1}(T_f - \delta, T_f + \delta)$, $\delta > 0$ satisfying the differential inequality

$$E'(t) + \phi(E(t)) \le F((T_f - t)_+) \quad \text{in } (T_f - \delta, T_f + \delta).$$

For the proof one has to take $\rho = T_f + R_0 - t$ for the new independent variable and apply Lemma 2.2 to the function $E(\rho) \equiv E(T_f + R_0 - t)$. We have: $E(t) \equiv 0$ for all $t \ge T_f$.

In order to deal with the case (2.29), we need the following variant of Lemma 2.3.

Lemma 2.4. *Let $E \in W^{1,1}_{\mathrm{loc}}(0, R)$ with $E \ge 0$ and $E' \ge 0$. Assume that the inequality*

$$\Lambda E'(\rho) + F(\rho) \ge E(\rho)^{1-\mu} \quad \textit{for a.e. } \rho \in (R_0, R) \tag{2.42}$$

holds with some $R_0 \in (0, R)$, where $\mu \in (0, 1)$, $\Lambda = \mathrm{const} > 0$, $F(t) \ge 0$. Assume that the integral

$$I(R) = \int_{R_0}^{R} (\tau - R_0)^{-\mu} F(\tau) d\tau$$

is convergent. Then the function $E(\rho)$ admits the estimate

$$E(\rho) \le G(\rho) \equiv E(\rho) - (\rho - R_0)^{1/\mu} \left(\left(\frac{\mu}{\Lambda}\right)^{1/\mu} - \frac{I(\rho)}{\Lambda} \right) \tag{2.43}$$

for every $\rho \in (R_0, R)$, and $E(\rho_*) = 0$ if there exists $\rho_* \in (R_0, R)$ such that $G(\rho_*) = 0$.

Proof. The function

$$\overline{E}(t) := (\mu/\Lambda)^{1/\mu} (t - R_0)^{1/\mu}$$

satisfies the conditions

$$\overline{E}^{1-\mu} = \Lambda \overline{E}', \qquad \overline{E}(R_0) = 0. \tag{2.44}$$

Subtracting equation (2.44) from inequality (2.42), we get the inequality

$$E^{1-\mu} - \overline{E}^{1-\mu} \le \Lambda (E - \overline{E})' + F(t). \tag{2.45}$$

There holds the identity

$$E^{1-\mu} - \overline{E}^{1-\mu} = (1 - \mu) \left\{ \int \left(\theta E(t) + (1 - \mu \overline{E}(t))\right)^{-\mu} d\theta \right\} \left(E(t) - \overline{E}(t)\right). \tag{2.46}$$

Introduce the function

$$\phi(t) = \exp\left(\alpha \int_{R_0}^{t} \left(\int_0^1 (\theta E(\tau) + (1-\theta)\overline{E}(\tau))^{-\mu} d\theta\right) d\tau\right) \tag{2.47}$$

with $\alpha = -(1-\mu)/\Lambda)$. Making use of (2.46), (2.47), we can rewrite (2.45) in the following equivalent form:

$$\frac{d}{dt}\left(E - \overline{E}\phi(t)\right) \geq -\frac{1}{\Lambda}\phi(t)F(t). \tag{2.48}$$

Integrating now (2.48) over the interval (R_0, t), we arrive at the inequality

$$E(t) \geq \overline{E}(t) + \frac{E(R_0)}{\phi(t)} - \frac{1}{\Lambda\phi(t)} \int_{R_0}^{t} \phi(\tau)F(\tau)d\tau. \tag{2.49}$$

Let us relax inequality (2.49) rewriting it in the form

$$\begin{aligned} E(R) \geq \overline{E}(t) + E(R_0)\frac{1}{\Lambda}\int_{R_0}^{t} F(t) \\ \exp\left(\frac{1}{\Lambda}\int_{\tau}^{t}\left(\int_0^1 (\theta E(\tau) + (1-\theta)\overline{E}(\tau))^{-\mu} d\theta\right) d\tau\right). \end{aligned} \tag{2.50}$$

Next, the following chain of relations is true:

$$\begin{aligned} &\exp\left(\frac{1-\mu}{\Lambda}\int_{\tau}^{t}\left(\int_0^1 (\theta E(s) + (1-\theta)\overline{E}(s))^{-\mu} d\theta\right) ds\right) \\ &\leq \exp\left(\frac{1-\mu}{\Lambda}\left(\int_0^1 (1-\theta)^{-\mu} d\theta\right)\int_{\tau}^{t} ds\right) \\ &= \exp\left(\int_{\tau}^{t} \frac{d\overline{E}}{\overline{E}}\right) = \exp\left(\ln\frac{\overline{E}(t)}{\overline{E}(\tau)}\right)\frac{\overline{E}(t)}{\overline{E}(\tau)}, \qquad \tau < t. \end{aligned} \tag{2.51}$$

Applying (2.44), we infer from (2.50), (2.51) that

$$0 \leq E(R_0) \leq E(R) - \overline{E}(t)\left(1 - \frac{1}{\Lambda}\int_{R_0}^{t} \frac{E(\tau)}{\overline{E}(\tau)d\tau}\right) \equiv G(t).$$

Thus if the equation $G(t) = 0$ has a solution in (R_0, R), then $E(R_0) = 0$. □

Remark 2.5. It is easy to see that solvability of the equation $G(t) = 0$ is guaranteed if, for instance, the condition

$$\Lambda E(R) + \Lambda^{1/\mu-1}(R - R_0)^{1/\mu} I(R) < \mu^{1/\mu}(R - R_0)^{1/\mu}$$

holds.

Completion of proof of Theorem 2.2. Lemma 2.4 yields the estimate

$$E(R) \leq E(_0) - (\rho - \rho_1)^{1/\gamma} \left[\left(\frac{\gamma}{\Lambda} \right) - K_2 \frac{I(\rho)}{\Lambda} \right] := \Theta(\rho),$$

provided that (2.28) is true, e.g., if $\Theta(\rho_*) = 0$ for some $\rho_* \in (\rho_1, \rho_0)$. By virtue of (2.30),

$$\Theta(\rho) \leq E_* - (\rho - \rho_1)^{1/\gamma} \left[\left(\frac{\gamma}{\Lambda} \right) - K_2 \frac{I_*}{\Lambda} \right] := F_*(\rho).$$

Hence, if $I_* < \gamma / K_2$, then there exist $\rho_* \in (\rho_1, \rho_0)$ and E_*,

$$E_* < (\rho_0 - \rho_1)^{1/\gamma} \left[\left(\frac{\gamma}{\Lambda} \right) - K_2 \frac{C_*}{\Lambda} \right],$$

such that $F_*(\rho_*) = 0$. Define the function

$$z(\rho) = E(R_0)\, (R - R_0)^{-1/\gamma}\, (\rho - R_0)_+^{1/\gamma}\, .$$

This function satisfies the equation

$$z^{1-\gamma} = \Lambda z' + \varepsilon\, (\rho - R_0)_+^{(1-\gamma)/\gamma}\, ,$$

provided that

$$\varepsilon\, (\rho_0 - \rho_1)^{(1-\gamma)/\gamma} \leq E^{1-\gamma}(\rho_0) \left(1 - \Lambda \frac{1}{\gamma} E^{\gamma}(\rho_0)(\rho - \rho_1)^{-1} \right). \tag{2.52}$$

On the other hand, $z(\rho)$ is a majorant for $E(\rho)$ because

$$0 \leq E(\rho) \leq E(\rho_0)\, (\rho_0 - \rho_1)^{-1/\gamma}\, (\rho - \rho_1)_+^{1/\gamma}\, .$$

At last, it is not difficult to see that there exist constants $E(\rho_0)$, ε, such that (2.52) is fulfilled, and the proof is thus completed. □

Remark 2.6. The assertion of Theorem 2.2 can be interpreted as follows: if the source $f(x)$ vanishes in a ball B_{ρ_1}, and either (2.28) or (2.29) are fulfilled, then every weak solution $u(x)$ of equation (2.1) in a ball B_{ρ_0} vanishes in the ball B_{ρ_1}, provided that the energy $E(\rho_0)$ is sufficiently small.

Remark 2.7. Condition (2.4) can be generalized in the following way:

$$|B(x, r, q)| \leq \sum_{k=1}^{m} C_{k3} |u|^{\alpha_i} |Du|^{\beta_i}, \quad 0 \leq \beta_i, \quad \alpha_i = \sigma - \frac{\beta_i(1+\sigma)}{p}.$$

The assertions of Theorems 2.1 and 2.2 remain true if we assume the existence of $\varepsilon_i \geq 0$ such that

$$C_5 = \min \left(C_4 - \sum_{i=1}^{m} \varepsilon_i C_{i3} \frac{p - \beta_i}{p}, 1 - \sum_{i=1}^{m} \frac{\beta_i C_{i3}}{p C_2} \varepsilon_i^{-(p-\beta_i)} / \beta_i \right) > 0.$$

This condition holds if, say, C_{i3} are sufficiently small.

3 The weighted diffusion/absorption balance

The results of the previous subsection can be extended to the case when the diffusion or/and absorption terms are strongly nonhomogeneous and their dependence on the space variable is given by some weight function. To be precise, let us consider the equation (for the sake of presentation we assume that $B \equiv 0$)

$$-\operatorname{div} \mathbf{A}(x, u, \nabla u) + C(x, u) = f(x), \tag{3.1}$$

where condition (2.3) on $\mathbf{A}$ is replaced by the weaker condition

$$0 \leq A_0(x)|\mathbf{q}|^p \leq \mathbf{A}(x, r, \mathbf{q}) \cdot \mathbf{q}, \tag{3.2}$$

and the "absorption" term is assumed to satisfy the condition

$$0 \leq Q(x)|r|^{\sigma+1} \leq C(x, r)r. \tag{3.3}$$

The weight functions $A_0(x)$ and $Q(x)$ are allowed to vanish on a subset of Ω which means that $\mathbf{A}$ and C can vanish too. We assume the fulfillment of the degeneracy condition

$$\left(\left\| \frac{1}{Q} \right\|_{L^{n/(n-1)}(B_{\rho_0})}, \left\| \frac{1}{A_0} \right\|_{L^{m/(m-1)}(B_{\rho_0})} \right) \leq C_0^{-1}, \tag{3.4}$$

for some constant $C_0 > 0$, and with $n, m \geq 1$ such that

$$\frac{n-1}{n} < \sigma, \qquad \max\left(\frac{1}{p}, \frac{p-1}{p}, \frac{(N-1)p+1}{Np} \right) < m \leq 1. \tag{3.5}$$

We also assume that the structural condition (2.2) holds so that the concept of weak solution of (3.1) can be easily obtained from Definition 2.1 by setting $B \equiv 0$.

The following result generalizes Theorem 2.2 (similar arguments lead to the generalization of Theorem 2.1).

Theorem 3.1. *Let $B_{\rho_0}(x_0) \subset \Omega$ and let $u(x)$ be a weak solution of equation* (3.1) *with $f \in L^{n(1+\sigma)/(n(1+\sigma)-1)}_{\mathrm{loc}}(\Omega)$ such that $f(x) = 0$ in $B_{\rho_1(x_0)}$, $0 < \rho_1 < \rho_0$. Assume that* (3.2), (3.3), (3.4) *are fulfilled, as well as the key assumption $\sigma < p-1$. Let one of the following conditions be true: either*

$$I(\rho, f) := \int_0^{\rho} (\tau - \rho_1)_+^{-1/\gamma} \, \|f\|^{(1+\sigma)(1-\gamma)m/\sigma}_{L^{n(1+\sigma)/(n(1+\sigma)-1)}(B_\tau)} d\tau < \infty$$

with

$$\gamma = \frac{(1-\theta)(p-1-\sigma)}{(p-1)(1+\sigma)},$$

$$\theta = \frac{N(pm - n(1+\sigma)(pm-p+1)) + n(1+\sigma)(pm-p+1)}{N(pm - n(1+\sigma)) + mnp(1+\sigma)},$$

or

$$\|f\|^{m(1+\sigma)(1-\gamma)/\sigma}_{L^{n(1+\sigma)/(n(1+\sigma)-1)}(B_\rho)} \le \varepsilon\,(\rho-\rho_1)_+^{(1-\gamma)/\gamma} \qquad \textit{for } \rho\in(0,\rho_0).$$

Then there exist positive constants I_*, E_*, C_* such that if

$$I(\rho_0, f)\le I_*, \qquad E(\rho_0,u)\le E_*, \qquad \varepsilon\le\varepsilon_*,$$

then $u(x)\equiv 0$ in $B_{\rho_1}(x_0)$.

Proof. First of all, a simple adaptation of the proof of Lemma 2.1 gives the inequality

$$\int_{B_\rho}\Big\{\mathbf{A}(x,u,\nabla u)\cdot\nabla u + Q(x)|u|^{1+\sigma} - fu\Big\}\,dx \le -\int_{S_\rho}\mathbf{A}(x,u,\nabla u)u\cdot\mathbf{n}dx \quad \text{for } \rho\in(0,\rho_0).$$

Let us introduce the new energy functions

$$E(\rho)=\int_{B_\rho(x_0)}|\nabla u|^{pm}dx = \|\nabla u\|^{pm}_{L^{pm}(B_\rho)},$$
$$b(\rho)=\int_{B_\rho(x_0)}|u|^{n(1+\sigma)}dx = \|u\|^{n(1+\sigma)}_{L^{n(1+\sigma)}(B_\rho)}.$$

Applying the Hölder and Young inequalities and making use of (3.2)–(3.3) we come to the following relations:

$$C_0E^{1/m}\le\int_{B_\rho}A_0(x)|\nabla u|^p dx\le\int_{B_\rho}\mathbf{A}(x,u,\nabla u)\cdot\nabla u dx, \tag{3.6}$$

$$C_0b^{1/n}\le b^{1/n}\left\|\frac{1}{Q}\right\|^{-1}_{L^{n/(1-n)}(B_\rho)}\le\int_{B_\rho}Q(x)|u|^{1+\sigma}dx, \tag{3.7}$$

$$\left|\int_{S_\rho}\mathbf{A}u\cdot\mathbf{n}dS\right| \le C_2\left(\int_{S_\rho}|\nabla u|^{mp}dS\right)^{(p-1)/mp}\|u\|_{L^\lambda(S_\rho)} = C_2\left(\frac{dE}{d\rho}\right)^{(p-1)/pm}\|u\|_{L^\lambda(S_\rho)}, \tag{3.8}$$

$$\left|\int_{B_\rho}fudx\right|\le\varepsilon b^{1/n}+C(\varepsilon,n,\sigma)\|f\|^{(1+\sigma)/\sigma}_{L^{n(1+\sigma)/(n(1+\sigma)-1)}(B_\rho)}, \tag{3.9}$$

where $\lambda=pm/(pm-p+1)$, and $\varepsilon>0$. It follows from the *interpolation-trace inequality* that

$$\|u\|_{L^\lambda(S_\rho)}\le C\left(E^{1/pm}+\rho^\delta b^{1/n(1+\sigma)}\right)^\theta b^{(1-\theta)/n(1+\sigma)}, \tag{3.10}$$

where

$$\theta = \frac{Npm + (1-N)n(1+\sigma)(pm-p+1)}{k},$$

$$\delta = -\frac{k}{nmp(1+\sigma)}, \qquad k = N(pm - n(1+\sigma) + pnm(1+\sigma)).$$

Letting $\varepsilon = C_0/2$ in (3.9) and then gathering (3.6)–(3.10), we have

$$E^{1/m} + b^{1/n} \le C\left(\frac{dE}{d\rho}\right)^{(p-1)/pm}\left(E^{1/pm} + \rho^{\delta} b^{1/n(1+\sigma)}\right)^{\theta} b^{(1-\theta)/n(1+\sigma)} + C\|f\|^{(1+\sigma)/\sigma}_{L^{n(1+\sigma)/(n(1+\sigma)-1)}(B_\rho)} \tag{3.11}$$

with a new constant $C = C(C_0, C_2, N, p, n, m, \sigma)$. Proceeding by analogy with the proofs of Theorems 2.1 and 2.2, we come to the relations

$$\begin{aligned}\left(\frac{dE}{d\rho}\right)^{(p-1)/pm}&\left(E^{1/m} + \rho^{\delta} b^{1/n(1+\sigma)}\right)^{\theta} b^{(1-\theta)/n(1+\sigma)}\\ &\le K_1\left(\frac{dE}{d\rho}\right)^{(p-1)/pm}\left(E^{1/m} + b^{1/n}\right)^{\mu}\\ &\le \frac{1}{2C}\left(E^{1/m} + b^{1/n}\right) + \left(K_1\left(\frac{dE}{d\rho}\right)^{(p-1)/pm}\right)^{1/(1-\mu)},\end{aligned} \tag{3.12}$$

where

$$\mu = \frac{\theta}{p} + \frac{1-\theta}{1+\sigma},$$

$$K_1 = \frac{1}{1-\mu}(2C\mu)^{\mu/(1-\mu)} \max\left(1, \rho_1^{\delta\theta}\right) \cdot \left(\max\left(1, b(\rho_0)\right)\right)^{(p-1-\sigma)\theta/n(1+\sigma)}.$$

Coupling (3.11), (3.12), and then raising both sides of the resulting inequality to the power $mp(1-\mu)/(p-1)$, we arrive at the main inequality

$$E^{1-\gamma}(\rho) \le \Lambda\frac{dE}{d\rho} + \varphi(\rho), \tag{3.13}$$

where

$$\varphi(\rho) = K\|f\|^{(1+\sigma)(1-\gamma)m/\sigma}_{L^{n(1+\sigma)/(n(1+\sigma)-1)}(B_\rho)},$$

$$1-\gamma = \frac{(1-\mu)p}{p-1}, \qquad \Lambda = \Lambda(K_1, m, p, \mu).$$

The conclusion now follows in the same way as in the proof of Theorem 2.2. □

3.1 Global applications: Localized solutions of the associated boundary-value problems: The cylinder domains. The main goal of this section is to present some global applications of the local vanishing results given in Subsections 2.1 and 2.2. We extend them to the case where the solution is not merely a local weak solution to a partial differential equation but also satisfies some prescribed boundary conditions. Such a situation was already described in the introduction to this chapter (see Corollary 1.1) but in a simpler case of a one-dimensional problem.

To fix ideas, we begin by considering the boundary-value problem

$$-\operatorname{div}\mathbf{A}(x,u,\nabla u)+B(x,u,\nabla u)+C(x,u)=f(x)\quad \text{in } \Omega, \tag{3.14}$$

$$u=g\quad \text{on } \Gamma_D, \tag{3.15}$$

$$\mathbf{A}(x,u,\nabla u)\cdot\mathbf{n}=h,\quad \text{on } \Gamma_N, \tag{3.16}$$

where Ω is a regular open subset of $\mathbb{R}^N$, $\partial\Omega=\Gamma_D\cup\Gamma_N$, and $\mathbf{n}$ denotes the unit outer normal vector to $\partial\Omega$.

The solvability of problem (3.14)–(3.16) can be proved under various assumptions on the data f, g, h (see, for instance, the general exposition in Díaz [128]). For our present purposes, it is sufficient to claim that

$$f\in L^{p'}(\Omega),\quad g\in W^{1,p}(\Omega),\quad \text{and}\quad h\in C^0(\overline{\Omega}). \tag{3.17}$$

If we assume the fulfillment of condition (3.17), of the structural assumptions on $\mathbf{A}$, B, and C given in Section 2, and add some additional coercivity and ellipticity/monotonicity conditions on $\mathbf{A}$, the existence of weak solutions to problem (3.14)–(3.16) can be found in the literature (see, e.g., Lions [243], Ladyzenskaya and Ural'tseva [235], and Ferone and Murat [158]). As we said in the introduction to this chapter, global estimates on the location of the vanishing (or null) sets of localized solutions can be obtained provided that the *total energy* is estimated. As usual, the total energy is defined by

$$T(\Omega):=\int_\Omega\left(\mathbf{A}(x,u,\nabla u)\cdot\nabla u+|u|^{1+\sigma}\right)dx. \tag{3.18}$$

To state our next result we introduce the following notation: given a scalar function w, defined on a subset $\mathcal{M}$ of Ω, we call the *null set* and the *support of* w the sets

$$\mathcal{N}(w)=\{x\in\mathcal{M}: w=0\}\quad\text{and}\quad S(w)=\{x\in\mathcal{M}: w\neq 0\},$$

respectively.

Theorem 3.2. *Let $u(x)$ be a weak solution of* (3.14), (3.15), (3.16). *Assume that $\mathbf{A}$, B, and C satisfy the structural conditions* (2.2)–(2.4) *with $\sigma<p-1$ and that B meets the conditions of Theorem* 2.1. *Then the null set $\mathcal{N}(u)$ of $u(x)$ contains, at least, the set of points*

$$x\in\mathcal{N}(f)\cup\mathcal{N}(g|_{\Gamma_0})\cup\mathcal{N}(h|_{\Gamma_N})$$

such that

$$\text{dist}\left(x, S(f) \cup S(g|_{\Gamma_0}) \cup S(h|_{\Gamma_N})\right) \geq \mathbf{M} \tag{3.19}$$

for some **M** *depending on the total energy* $T(\Omega)$. *In particular, if* Ω *is unbounded and the set*

$$S(f) \cup S(g|_{\Gamma_0}) \cup S(h|_{\Gamma_N}) \cup \mathcal{N}(f)$$

is compact, then the support of $u(x)$ *is also compact.*

Remark 3.1. If the set $\mathcal{N}(f) \cup \mathcal{N}(g|_{\Gamma_0}) \cup \mathcal{N}(h|_{\Gamma_N})$ is bounded, for the existence of a nonempty null set $\mathcal{N}(u)$ a suitable balance is required between the "size" of the set $\mathcal{N}(f) \cup \mathcal{N}(g|_{\Gamma_0}) \cup \mathcal{N}(h)|_{\Gamma_N}$ and the total energy $T(\Omega)$: the set of points satisfying (3.19) must be nonempty. An illustration of the global estimate of $\mathcal{N}(u)$ is presented in Figure 3.1.

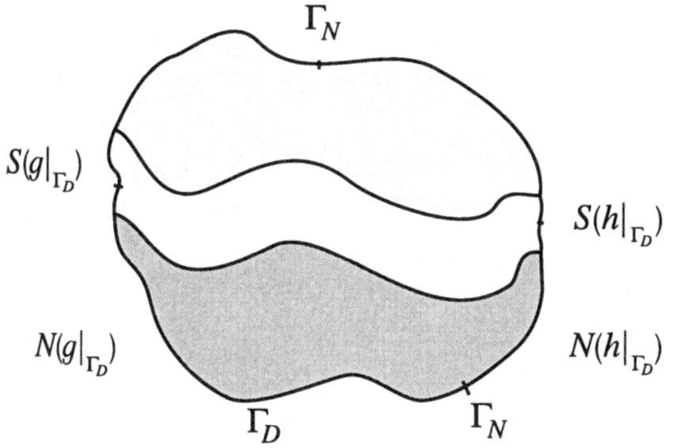

Figure 3.1: The estimate on $\mathcal{N}(u)$.

The above sets are nonempty if $\mathcal{N}(f)$ is unbounded (if we relax condition (2.22)).

Remark 3.2. If $f(x)$ is sufficiently flat near $\partial S(f)$, we can improve estimate (3.19) by applying the local result given in Theorem 2.2 (which we leave to the reader). In this case we obtain that $\mathcal{N}(u)$ contains the set $x \in \mathcal{N}(f) \cup \mathcal{N}(g|_{\Gamma_0}) \cup$, $\mathcal{N}(h|_{\Gamma_N})$ such that

$$\text{dist}(x, S(g|_{\Gamma_0}) \cup S(h|_{\Gamma_N})) \geq M.$$

Proof of Theorem 3.2. Let $x_0 \in \mathcal{N}(f)$ and $\rho_0 = \text{dist}\,(x, S(f) \cup \Gamma_D \cup \Gamma_N)$. Then $B_{\rho_0}(x_0) \subset \mathcal{N}(f)$, and Theorem 2.1 implies that $u \equiv 0$ a.e. in $B_{\rho_1}(x_0)$ with

$$\rho_1^\nu = \left(\rho_0^\nu - C \min(E^\gamma(\rho_0) G(\rho_0))\right)_+ \geq \left(\rho_0^\nu - C \widehat{T}^\gamma(\Omega)\right)_+$$

for some $\widehat{T}(\Omega)$. Thus if $\rho_0 > (C\widehat{T}(\Omega))^{1/\nu}$, we conclude that $u(x_0) = 0$ (in the sense that $u \equiv 0$ a.e. on $B_{\rho_1}(x_0)$) for some $\rho_1 > 0$). To complete the proof we have to extend this property to those points $x_0 \in \mathcal{N}(f)$ which are closer to $\mathcal{N}(g|_{\Gamma_D}) \cup \mathcal{N}(h|_{\Gamma_N})$ than to $S(f) \cup S(g|_{\Gamma_D}) \cup S(h|_{\Gamma_N})$. In that case, if we define $\rho_0 = \operatorname{dist}(x_0, S(f) \cup \Gamma_D \cup \Gamma_N)$, we have that $B_{\rho_0}(x_0) \cup (\mathbb{R}^N \setminus \Omega) \neq \emptyset$ and Theorem 2.1 is not directly applicable. Let us show that the conclusion of Theorem 2.1 remains true for the points $x_0 \in \mathcal{N}(f)$ such that $(B_{\rho_0}(x_0) \cap \Omega) \subset \mathcal{N}(f)$ and $u = 0$ on $\Gamma_D \cap B_{\rho_0}(x_0)$ in the sense of traces of functions from $W^{1,p}(\Omega)$. Let

$$\Omega_\rho(x_0) = B_\rho(x_0) \cap \Omega,$$
$$\Gamma_{D,\rho}(x_0) = (\partial\Omega_\rho(x_0)) \cap \Gamma_D,$$
$$\Gamma_{N,\rho}(x_0) = (\partial\Omega_\rho(x_0)) \cap \Gamma_N,$$

and

$$\widehat{S}_\rho(x_0) = \partial\Omega_\rho(x_0) \setminus \left(\Gamma_{D,\rho}(x_0) \cap \Gamma_{N,\rho}(x_0)\right).$$

A careful revision of the proof of Lemma 2.1 allows us to see that inequality (2.11) must be replaced by

$$\begin{aligned} &\int_{\Omega_\rho(x_0)} [\mathbf{A}(x,u,\nabla u)\cdot\nabla u + B(x,u,\nabla u)u]\,dx + C_4 \int_{\Omega_\rho(x_0)} |u|^{\sigma+1}dx \\ &\quad \le -\int_{\widehat{S}_\rho(x_0)} \mathbf{A}(x,u,\nabla u)u\cdot\nu\,dS \quad \text{for a.e. } \rho \in (0,\rho_1). \end{aligned} \tag{3.20}$$

Indeed, it suffices to observe that the function $\varphi_{n,k}(x)$ is still a test function, i.e., it vanishes on $\Omega_\rho(x_0) \setminus \Gamma_{N,\rho}$, and that $\mathbf{A}(x,u,\nabla)\cdot\nu = 0$ on $\Gamma_{N,\rho}$. Let

$$\tilde{u}(x) = \begin{cases} u(x) & \text{on } \Omega_\rho(x_0), \\ 0 & \text{on } B_\rho(x_0) \setminus \Omega. \end{cases}$$

It follows from (3.20) that inequality (2.11) holds with u replaced by $\tilde{u}$. Consequently, the proof of Theorem 2.1 can now be applied to $\tilde{u}$, which gives the needed conclusion. □

The dependence on the prescribed functions f and g in (3.19) can be given in a more explicit way. It amounts to obtaining a priori estimates on the total energy $T(\Omega)$ in terms of norms of the functions f and g. This estimating renders an easy task if we assume that

$$g \equiv h \equiv 0, \quad f \in L^{(1+\sigma)/\sigma}(\Omega) \quad \text{and } S(f) \text{ is compact.} \tag{3.21}$$

Theorem 3.3. *Under the conditions of Theorem* 3.2 *there exists a positive constant* K *such that for every weak solution of problem* (3.14)–(3.16) *with* $u \in W^{1,p}(\Omega) \cap L^{1+\sigma}(\Omega)$, *we have*

$$T(\Omega) \le K \|f\|^{(1+\sigma)/\sigma}_{L^{(1+\sigma)/\sigma}(\Omega)}. \tag{3.22}$$

Moreover, if $p < N$ *and* $f \in L^{p/(p-1)}(\Omega)$, *then every weak solution with* $u \in L^{1+\sigma}(\Omega)$ *and* $\nabla u \in L^p(\Omega)$ *satisfies the estimate*

$$T(\Omega) \leq \widehat{K} \| f \|^{p/(p-1)}_{L^{p/(p-1)}(\Omega)} \tag{3.23}$$

with a positive constant $\widehat{K}$ *independent of* u.

Proof. We argue as in the the proof of Lemma 2.1 with $x_0 = 0$, but replace the auxiliarly function $\psi(r)$ by $\xi_n(r)$ with

$$\xi_n = \begin{cases} 1 & \text{if } r < n, \\ -r + (n+1) & \text{if } n \leq r < n+1, \\ 0 & \text{if } n+1 \leq r. \end{cases}$$

This leads to the inequality

$$\int_\Omega (\mathbf{A}(x, u, \nabla u) \cdot \nabla u + B(x, u\nabla u)u)\, dx + C_4 \int_\Omega |u|^{1+\sigma} dx \leq \int_\Omega fu\, dx. \tag{3.24}$$

As at Step 2 of the proof of Theorem 2.1, the assumptions on B yield the existence of a constant $C_5 > 0$ such that

$$C_5 \int_\Omega \left(\mathbf{A}(x, u, \nabla u) \cdot \nabla u + |u|^{1+\sigma}\right) dx \leq \int_\Omega fudx. \tag{3.25}$$

Applying Young's inequality we obtain (3.22). If $p > N$, from Sobolev's inequality we have that $u \in L^{pN/(N-p)}(\Omega)$. It follows from the interpolation inequalities (see the appendix) that $u \in L^{1+\sigma}(\Omega) \cap L^p(\Omega)$, which allows us to pass to the limit and to obtain (3.24). To get (3.23) we apply the duality relation $\left[L^{p/(p-1)}, L^p\right]$ and Young's inequality. □

Remark 3.3. Theorem 3.2 can be generalized in many ways: it can be extended to the cases where g and/or h are not identically zero; where f belongs to other functional spaces, etc.

Theorem 3.2 can also be applied to solutions of problems posed on unbounded domains in order to show that the support of the solution is compact, provided that $S(f) \cap S(g|_{\Gamma_D}) \cap S(h|_{\Gamma_N})$ is compact. Nonetheless, in certain cases we can rely on *global energy methods* to obtain estimates of another type. Throughout the rest of this subsection we shall study unbounded *cylinder domains*

$$\Omega = \left\{(y, t) : y \in G \subset \mathbb{R}^N,\ t \in \mathbb{R}^+\right\}, \quad \text{meas}\, G < \infty,$$

or *layer domains* of the form

$$L = \left\{(y, t) : y \in \mathbb{R}^N,\ 0 < t < 1\right\}.$$

From now on, we assume that $G \subset \mathbb{R}^N$ is a bounded domain with the smooth boundary ∂G.

Let us consider a weak solution $u(x)$ of the problem

$$-\operatorname{div} \mathbf{A}(x, u, \nabla u) + B(x, u, \nabla u) + C(x, u) = f(x) \text{ in } \Omega, \tag{3.26}$$

$$u(y, t) = 0, \quad y \in \Gamma_D; \quad \mathbf{A} \cdot \mathbf{n} = 0 \quad \text{for } y \in \Gamma_N, \quad t \in \mathbb{R}^+. \tag{3.27}$$

We use the notation

$$\operatorname{div} \mathbf{A} = \sum_{i=1}^{N+1} \frac{dA_i}{dx_i}, \qquad x_{N+1} = t,$$

$$\nabla u = \left(\frac{\partial u}{\partial x_1}, \ldots, \frac{\partial u}{\partial x_{N+1}} \right),$$

$$\Gamma = \Gamma_D \cup \Gamma_N = \partial G \quad (\operatorname{meas} \Gamma_D > 0),$$

and $\mathbf{n} = (n_1, \ldots, n_N)$ stands for the unit outer normal vector to Γ. We shall assume the fulfillment of the structural conditions similar to those of Section 2:

$$\forall (x, r, \mathbf{q}) \in \Omega \times \mathbb{R} \times \mathbb{R}^N,$$

$$C_2 |\mathbf{q}|^p \le \mathbf{A}(x, , r, \mathbf{q}) \cdot \mathbf{q} \le C_1 |\mathbf{q}|^{p-1}, \tag{3.28a}$$

$$0 \le C_4 Q(x) |r|^{1+\sigma} \le C(x, r) r, \tag{3.28b}$$

$$|B(x, r, \mathbf{q})| \le C_3 Q^{(p-\beta)/p}(x) \sum_{i=1}^{m} |r|^{\alpha_i} |\mathbf{q}|^{\beta_i}. \tag{3.28c}$$

We are interested in the solutions of equation (3.26) which satisfy the boundary condition (3.27), while no boundary condition is posed on $G \times \{0\}$. By a weak solution of this problem we mean, as usual, a function $u \in W^{1,p}(G \times \mathbb{R}_+)$ satisfying the condition $C(\cdot, u) \in L^1(\Omega)$, vanishing on $\Gamma_D \times \mathbb{R}_+$, and satisfying equality (2.7) for any $\varphi \in C^\infty(G \times \mathbb{R}_+) \cap W^{1,p}(G \times \mathbb{R}_+)$. The function φ vanishes on $(\Gamma_D \times \mathbb{R}_+) \cup (G \times \{0\})$ and is compactly supported in $G \times \mathbb{R}_+$. The energy functions are defined as in Section 2 but the energy set is changed:

$$E(\rho) = \int_\rho^\infty \int_G \mathbf{A} \cdot \nabla u \, dy dt, \qquad b(\rho) = \int_\rho^\infty \int_G Q(x) |u|^{\sigma+1} \, dy dt.$$

It is to be noted that under our structural assumptions the *total energy*

$$T(\Omega) := E(0) + b(0)$$

of every weak solution is finite despite the fact that the boundary values on $G \times \{0\}$ are not prescribed. We start by considering the easier case $f \equiv 0$.

Theorem 3.4. *Let $u(x)$ be a weak solution of equation* (3.26) *satisfying the boundary conditions* (3.27). *Assume that $f(x) \equiv 0$, that*

$$\left\| Q^{-1} \right\|_{L^{n/(n-1)}(\Omega)} \le C_0^{-1}, \quad 0 \le \frac{n-1}{n} < \sigma, \tag{3.29}$$

and, additionally, that one of the following conditions is fulfilled:

(i) *Relations* (3.28a)–(3.28c) *hold, with constants satisfying the conditions of Theorem* 2.1;

(ii) *Relations* (3.28a)–(3.28c) *hold with* $0 \le \sigma < p-1$, $\alpha_k + \beta_k + 1 = p$, $0 \le \beta_k$, $0 < p - \beta_k$, *and*

$$\|B_k(\cdot, t)\|_{L^{\lambda_k}(G)} \le \delta(t), \quad \lambda_k = \frac{N}{p - \beta_k}, \tag{3.30}$$

where $\delta(t)$ *is a decreasing function such that* $\delta(\infty) = 0$.

Then there exists a finite $\rho_0 > 0$, *depending only on the constants in the above conditions, such that*

$$u(y, \rho) \equiv 0 \quad \text{for a.e. } y \in G \text{ and } \rho_0 \le \rho.$$

Proof. We proceed as in the proofs of Theorems 2.1 and 2.2. Let us take the function

$$\varphi_{n,m,k} \equiv \psi_n(t)\xi_m(t) \min(k, |u|) sign\, u$$

for the test function in the integral identity of the definition of weak solution. The function ψ_m is defined in Lemma 2.1 and ξ_m was chosen in the proof of Theorem 2.2. First letting $k \to \infty$, then $n \to \infty$, and finally $m \to \infty$, and then arguing as in the proofs of Lemma 2.1 and Theorem 2.2, we obtain the inequality

$$\begin{aligned} E(\rho) + C_4 b(\rho) + \int_\rho^\infty \int_G B(x, u, \nabla u) u \, dy dt \\ \le \int_G A_{N+1}(y, \rho, u, \nabla u) u(y, \rho)\, dy := I(\rho). \end{aligned} \tag{3.31}$$

Let us consider case (i). Similarly to (2.18) we have

$$\begin{aligned} \left| \int_\rho^\infty \int_G Bu\, dydt \right| &\le C_3 \int_\rho^\infty \int_G \left(\varepsilon \frac{p-\beta}{p} Q|u|^{\sigma+1} + \frac{\beta}{pC_2} \varepsilon^{(p-\beta)/\beta} \mathbf{A} \cdot \nabla u \right) dyd\rho \\ &= C_3 \left(\varepsilon \frac{p-\beta}{p} b(\rho) + \frac{\beta}{pC_2} \varepsilon^{-(p-\beta)/p} E(\rho) \right). \end{aligned} \tag{3.32}$$

Relations (3.31), (3.32) yield an inequality similar to (2.19),

$$E(\rho) + b(\rho) \le |I(\rho)|, \tag{3.33}$$

provided that ε is chosen in a suitable way. In case (ii), we make use of the inequality

$$\|u(\cdot, t)\|_{L^{Np/(N-p)}(G)} \le C_0 \|\nabla u(\cdot, t)\|_{L^p(G)}, \quad p < N \tag{3.34}$$

(see Lemma 3.2). Then gathering (3.30), (3.34), we get

$$\left|\int_\rho^\infty \int_G Bu\,dydt\right| \le \int_\rho^\infty \sum_{k=1}^m \|B_k(\cdot,t)\|_{L^k(G)} \|u(\cdot,t)\|^{1+\alpha_k}_{L^{Np/(N-p)}(G)} \cdot \|\nabla u\|^{\beta_k}_{L^p(G)} dt$$
$$\le C\delta(\rho) \int_\rho^G \|\nabla u(\cdot,t)\|^p_{L^P(G)} dt \le C\delta(\rho) E(\rho). \tag{3.35}$$

Choosing ρ large enough to provide the inequality $C\delta(\rho) < 1$, we arrive once again at inequality (3.33). We now have to control the right-hand side of (3.33). Firstly, proceeding in the routine way (see (2.20)), we get the inequality

$$|I| \le C_2 \cdot C_1^{(1-p)/p} \left(-E_\rho\right)^{(p-1)/p} \|u(\cdot,\rho)\|_{L^p(G)}. \tag{3.36}$$

Second, applying Hölder's inequality and (3.34), we find out that

$$\begin{aligned} \|u\|_{L^p(G)} &\le \left\| Q^{-1} \right\|^{1/p\alpha}_{L^{n/(1-n)}(G)} \left\| Q|u|^{1+\sigma} \right\|^{1/\alpha p}_{L^1(G)} \|u\|^{1-(1+\sigma)/p\alpha}_{L^{Np/(N-p)}(\Omega)} \\ &\le C\left((b(\rho) - E(\rho))\right)^{1-\sigma/\alpha p} \end{aligned} \tag{3.37}$$

with

$$\alpha = \frac{N}{p}\left(\frac{1}{n} - \frac{(N-p)(1+\sigma)}{Np}\right).$$

The constant C depends only on C_0 and the constant in (3.34). Gathering (3.36)–(3.37), we have

$$|I| \le C\left(-(b_\rho + E_\rho)\right)^{1/(1-\gamma)}, \tag{3.38}$$

where

$$\frac{1}{1-\gamma} = \left(\frac{N(p-n(1+\sigma)+np)}{N(p-n(1+\sigma))+np(1+\sigma)} + \frac{p+1}{p}\right) > 1.$$

Inequalities (3.33) and (3.38) imply the inequality

$$C\Lambda^{1-\gamma} + \Lambda_\rho \le 0 \quad \text{with} \quad \Lambda(\rho) := E(\rho) + b(\rho)$$

and, correspondingly, the estimate

$$\Lambda^\gamma(\rho) \le \Lambda^\gamma(0) - \gamma C\rho.$$

Hence

$$\Lambda(\rho) = E(\rho) + b(\rho) = 0 \quad \text{if } \rho \ge \Lambda^\gamma(0)/\gamma C,$$

whence the desired assertion. □

Remark 3.4. Consider a nonhomogeneous equation of the type (3.26) posed on $G \times \mathbb{R}$ with the right-hand side f compactly supported in $G \times (0, \rho_0)$, $\rho_0 < \infty$. The total energy is easy to estimate under the assumption that

$$f \in L^{p/(p-1)}(\mathbb{R}^+; L^\lambda(G)) \quad \text{or} \quad f Q^{-1/(1+\sigma)} \in L^{(1+\sigma)/\sigma}(G \times \mathbb{R}^+),$$

where

$$\lambda = \frac{pN}{N(p-1)+p}. \tag{3.39}$$

Indeed, taking $\varphi_{m,k} = \xi_m(t)\min(k, |u|)\operatorname{sign} u$ for the test function in the definition of the weak solution and following the proofs of Theorems 2.2 and 3.1, we arrive at the inequality

$$\int_{\mathbb{R}}\int_G \left(|\nabla u|^p + Q|u|^{1+\sigma}\right) dydt \le C\left|\int_{\mathbb{R}}\int_G f(x)u\, dydt\right| := CI. \tag{3.40}$$

The right-hand side of (3.40) can be estimated in two different ways: either

$$\begin{aligned} I &\le \int_{\mathbb{R}^+} \|u\|_{L^{Np/(N-p)}(G)} \|f\|_{L^\lambda(G)} dt \le C \int_{\mathbb{R}^+} \|\nabla u\|_{L^p(G)} \|f\|_{L^\lambda(G)} dt \\ &\le \varepsilon \int_{\mathbb{R}^+}\int_G |\nabla u|^p dydt + C(\varepsilon)\int_{\mathbb{R}^+} \|f\|^{p/(p-1)}_{L^\lambda(G)} dt \end{aligned} \tag{3.41}$$

or

$$\begin{aligned} I &\le \int_{\mathbb{R}^+}\left(\int_G Q|u|^{1+\sigma} dy\right)^{1/(1+\sigma)} \int_G \left(|f|^{(1+\sigma)/\sigma} Q^{-1/\sigma} dy\right)^{\sigma/(1+\sigma)} dt \\ &\le \varepsilon \int_{\mathbb{R}^+}\int_G Q|u|^{1+\sigma} dydt + \int_{\mathbb{R}^+}\int_G |f|^{(\sigma)/\sigma} Q^{-1/\sigma} dydt. \end{aligned} \tag{3.42}$$

Hence

$$T(\Omega) := \int_{\mathbb{R}^+}\int_G \left(\mathbf{A}\cdot\nabla u + Q|u|^{1+\sigma}\right) dydt \le C\int_{\mathbb{R}^+} F(f)dt, \tag{3.43}$$

where

$$F(f) = \|f(\cdot,t)\|^{p/(p-1)}_{L^\lambda(G)} \quad \text{or} \quad \|fQ^{-1/(1+\sigma)}\|^{(\sigma+1)/\sigma}_{L^{(1+\sigma)/\sigma}(G)}. \tag{3.44}$$

Let us consider now a solution of equation (3.26) in the domain $G \times \mathbb{R}^+$ and show that $\operatorname{supp} u(x) \subset G \times (0, \rho_0)$, provided that $\operatorname{supp} f(x) \subset G \times (0, \rho_0)$ and the energy $E(\rho)$ and the norm of f satisfy certain suitable conditions, which we specify below.

Theorem 3.5. *Let $u(x)$ be a weak solution of equation* (3.26) *in $G\times\mathbb{R}^+$, satisfying* (3.27). *Assume that* $\operatorname{supp} f(x) \subset G \times (0, \rho_0)$ *and, additionally,*

$$\int_\rho^\infty F(f)d\tau \le \varepsilon\left(1 - \frac{\rho}{\rho_0}\right)_+^{1/\gamma} \tag{3.45}$$

for any $\rho \in (\rho_0 - \delta, +\infty)$ *and with some positive constants* ε *and* δ. *Here* $u_+ = \max(0, u)$, *the function* $F(f)$ *is defined by formula* (3.44), *and* γ *is given by* (3.38). *Then there exist positive constants* ε_*, Λ_* *such that if*

$$\varepsilon \le \varepsilon_*, \quad T(\Omega) := \int_0^\infty \int_G \left(\mathbf{A} \cdot \nabla u + Q|u|^{1+\sigma}\right) dydt \le \Lambda_*,$$

then

$$\operatorname{supp} u(x) \subset G \times (0, \rho_0). \tag{3.46}$$

Proof. We proceed as in the proof of Theorem 3.2. The inequality

$$E(\rho) + C_4 b(\rho) + \int_\rho^\infty \int_G Bu\, dydt \le \int_G \mathbf{A}_{N+1} u(y, \rho) dy + \int_\rho^\infty \int_G fu\, dydt \tag{3.47}$$

holds (see (3.31)). By (3.32), (3.33), and (3.38), from (3.47) we obtain the inequality

$$\forall \rho < \rho_0, \quad \Lambda \le C\left(\left(-\frac{d\Lambda}{d\rho}\right)^{1/(1-\gamma)} + \int_\rho^\infty F(f)dt\right),$$

or, correspondingly, the inequality

$$C\Lambda^{1-\gamma} + \Lambda' \le \left(\int_\rho^\infty F(f)dt\right)^{1-\gamma} \le \varepsilon^{1-\gamma}\left(1 - \frac{\rho}{\rho_0}\right)^{(1-\gamma)/\gamma}. \tag{3.48}$$

It is easy to check that the function

$$z(\rho) = \Lambda_0 \left(1 - \frac{\rho}{\rho_0}\right)_+^{1/\gamma}, \quad \Lambda_0 = \Lambda(0),$$

is a solution of the equation

$$Cz^{1-\gamma} + z' = \varepsilon^{1-\gamma}\left(1 - \frac{\rho}{\rho_0}\right)_+^{1/\gamma},$$

provided that

$$\Lambda_0^{1-\gamma}\left(C - \frac{\Lambda_0^\gamma}{\gamma\rho_0}\right) = \varepsilon^{1-\gamma}. \tag{3.49}$$

By the comparison principle for ordinary differential equations, we then get

$$0 \le \Lambda(\rho) \le z(\rho). \tag{3.50}$$

To check the validity of (3.49) we choose Λ_* such that

$$C - \frac{\Lambda_*^\gamma}{\gamma\rho_0} > 0$$

and then define

$$\varepsilon_*^{1-\gamma} = \Lambda_*^{1-\gamma}\left(C - \frac{\Lambda_*^\gamma}{\gamma}\rho_0\right). \qquad \square$$

Remark 3.5. Inequality (3.45) holds under the simpler condition

$$F(f) \le \frac{\gamma}{\rho_0}\left(1 - \frac{\rho}{\rho_0}\right)_+^{(1-\gamma)/\gamma}.$$

Remark 3.6. Analogous results can also be obtained for the layers $L = \{(y,t) : y \in \mathbb{R}^N,\ t \in (0,1)\}$, if the solution of equation (3.26) satisfies one of the boundary conditions

$$u(y,t) = 0, \quad y \in \mathbb{R}^N, \quad t = 0,$$
$$\mathbf{A}(y,t,u,\nabla u)\cdot \mathbf{n} = 0, \quad t = 1.$$

We complete this section by consideration of problems with nonhomogeneous conditions posed on the lateral boundaries. To be precise, we are concerned with the problem

$$-\operatorname{div}\mathbf{A}(x,u,\nabla u) + C(x,u) = f(x) \quad \text{in } \Omega = G \times \mathbb{R}^+, \tag{3.51}$$
$$u(y,t) = g(y,t) \quad y \in \Gamma = \partial G, \quad t > 0. \tag{3.52}$$

As usual, we assume that the structural conditions (3.28a)–(3.28c) are fulfilled. For the sake of presentation we replace (3.28b) by the condition

$$C_4|r|^{1+\sigma} \le C(x,r)r \le \frac{1}{C_4}|r|^{1+\sigma}. \tag{3.53}$$

Definition 3.1. Given functions $g \in W^{1,p}(\Omega)$ and $f \in L^{(1+\sigma)/\sigma}(\Omega)$, a measurable function

$$u \in W^{1,p}(\Omega), \qquad u - g \in W^{1,p}(\Omega) \cap L^p_{\mathrm{loc}}(\mathbb{R}^+;\ W_0^{1,p}(G)),$$

is said to be a weak solution of problem (3.51)–(3.52) if

$$C(x,u) \in L^1(\Omega),$$

and for every test function φ such that $(\varphi - g) \in C^1(\Omega)$, $\varphi = g$ on $\Gamma \times \mathbb{R}^+$, φ vanishes in a neighborhood of the set $G \times \{0\}$ and on G for all t large enough, the equality

$$\int_\Omega \{\mathbf{A}(x,u,\nabla u)\cdot\nabla(\varphi - g) + C(x,u)(\varphi - g) - f(x)(\varphi - g)\}\,dx = 0 \tag{3.54}$$

holds.

Let us assume that the functions f and g satisfy the condition

$$F(\rho; g, f) = \int_\rho^\infty \left\{ \|fg\|_{L^1(G)} + \|f\|_{L^{1+\sigma}(G)}^{(1+\sigma)/\sigma} \right\} d\tau + Q \leq \varepsilon \left(1 - \frac{\rho}{\rho_0}\right)_+^{1/\gamma} \tag{3.55}$$

for any $\rho \in (\rho_0 - \delta, +\infty)$ and with positive constants ρ_0 and δ. In this condition

$$Q(\rho; g) = \int_G \left\{ |\nabla g(y, \rho)|^p + |g(y, \rho)|^{1+\sigma} \right\} dy,$$

$$\frac{1}{1-\gamma} = \left(\frac{N(p-1-\sigma)+p}{N(p-1-\sigma)+p(1+\sigma)} + \frac{p-1}{p} \right).$$

The following theorem holds.

Theorem 3.6. *Assume that* $\sigma < p - 1$ *and that conditions* (3.31), (3.53) *and* (3.55) *hold. Let* $u(x)$ *be a weak solution of problem* (3.51), (3.52). *Then there exist constants* ε_*, Λ_* *such that* $\operatorname{supp} u(x) \subset G \times (0, \rho_0)$ *provided that*

$$\varepsilon \leq \varepsilon_*, \qquad T(\Omega) := \int_0^\infty \int_\Omega \left(\mathbf{A} \cdot \nabla u + C_4 |u|^{1+\sigma} \right) dx \leq \Lambda_*.$$

Proof. Defining the *total energy* $\Lambda(\rho)$ as in the proof of Theorem 3.4 (i.e., $\Lambda(\rho) := E(\rho) + b(\rho)$) and using Definition 4.1, we arrive at the inequality

$$\begin{aligned}\Lambda(\rho) \leq &- \int_G \mathbf{A}(y, \rho, u, \nabla u)(u - g)dy \\ &+ \int_\rho^\infty \int_G (\mathbf{A}(x, u, \nabla u)\nabla g + Cg - fg + fu)\, dx := I_1 + I_2.\end{aligned} \tag{3.56}$$

The terms I_1 and I_2 can be estimated in a routine way by virtue of (3.28a), (3.53), (3.55). We then have

$$\begin{aligned}|I_1| &\leq C_1^{(p-1)/p} \left(\|\nabla u\|_{L^p(G)}^{p-1} \|u - g\|_{L^p(G)} \right) \\ &\leq C \|\nabla u\|_{L^p(G)}^{p-1} \|\nabla(u-g)\|_{L^p(G)}^\theta \times \|u - g\|_{L^{1+\sigma}(G)} \\ &\leq C \left(-\Lambda_\rho + |\nabla g|^p + |g|^{1+\sigma} \right)^{1/(1-\gamma)},\end{aligned} \tag{3.57}$$

where

$$\theta = \frac{N(p-1-\sigma)}{N(p-1-\sigma)+1+\sigma}.$$

In a similar way, we deduce that

$$|I_2| \leq \frac{1}{2}\Lambda(\rho) + CF(\rho : g, f). \tag{3.58}$$

Collecting (3.57)–(3.58), we arrive at an inequality of the type (3.48). More precisely,

$$C\Lambda^{1-\gamma} + \Lambda' \le \varepsilon^{1-\gamma}\left(1 - \frac{\rho}{\rho_0}\right)_+^{(1-\gamma)/\gamma}.$$

The assertion of Theorem 3.6 now follows as in the end of the proof of Theorem 3.5. □

4 Anisotropic equations: Diffusion/absorption balance

4.1 Localization via diffusion/absorption balance. The properties of the diffusion operator may vary with the space direction $x_i, i = 1, \dots, N$. In this subsection, we study the localization property of solutions of such anisotropic equations under the assumption that the diffusion/absorption balance is subject to a suitable condition.

Our aim is to establish a result similar to Theorem 2.1 but for weak solutions of the equation

$$-\operatorname{div} \mathbf{A}(x, u, \nabla u) + B(x, u, \nabla u) + C(x, u) = f(x) \tag{4.1}$$

under the following structure assumptions, which generalize the assumptions made in Section 2: there exist $p_i \in (0, \infty), i = 1, \dots, N$, such that

$$\forall\,(r, \mathbf{q}) \in \mathbb{R} \times \mathbb{R}^N \quad \text{and a.e. } x \in \Omega, \quad 1 \le i \le N,$$

$$C_2|q_i|^{p_i} \le A_i(x, r, \mathbf{q})q_i \le C_1|q_i|^{p_i}, \tag{4.2a}$$

$$|B(x, r, \mathbf{q})r| \le C_3\left(|r|^{1+\sigma} + \sum_{i=1}^N |q_i|^{p_i}\right), \tag{4.2b}$$

$$C_4|r|^{1+\sigma} \le C(x, r)r \le C_5|r|^{1+\sigma} \tag{4.2c}$$

An example of an equation satisfying conditions (4.2a)–(4.2c) is furnished by the model equation

$$-\sum_{i=1}^N \frac{\partial}{\partial x_i}\left(\left|\frac{\partial u}{\partial x_i}\right|^{p_i-2}\frac{\partial u}{\partial x_i}\right) + \sum_{i=1}^N B_i(x)|u|^{\alpha_i-1}u\left|\frac{\partial u}{\partial x_i}\right|^{\beta_i-1}\frac{\partial u}{\partial x_i} + u\,|u|^{\sigma-1}$$
$$= f(x). \tag{4.3}$$

Note that (4.2b) holds if, for instance,

$$\alpha_i = \sigma - \frac{\beta_i}{p_i}(1+\sigma). \tag{4.4}$$

We start by introducing the concept of weak solution corresponding to a given source function $f \in L^{(1+\sigma)/\sigma}_{\mathrm{loc}}(\Omega)$.

Definition 4.1. A function $u(x)$, locally integrable in Ω, is called a weak solution of equation (4.1) if

(i) $u \in L^{1+\sigma}_{\text{loc}}(\Omega)$, $u_{x_i} \in L^{p_i}_{\text{loc}}(\Omega)$ for every $1 \le i \le N$;

(ii) for every test function $\varphi \in C_0^\infty(\Omega)$ the equality

$$\int_\Omega \{\mathbf{A}(x,u,\nabla u)\cdot\nabla\varphi + B(x,u,\nabla u)\varphi + C(x,u)\varphi\}dx = \int_\Omega f\varphi\, dx \tag{4.5}$$

is fulfilled.

The regularity assumptions imposed in (i) make consistent all integrals in (4.5). Indeed, applying the Hölder inequality and using assumptions (4.2a)–(4.2c), we have that

$$\begin{aligned}\left|\int_\Omega \mathbf{A}(x,u,\nabla u)\cdot\nabla\varphi dx\right| &\le C_1\sum_{i=1}^N \|u_{x_i}\|^{(p-1)/p}_{L^{p_i}(\Omega)}\|\varphi_{x_i}\|_{L^{p_i}(\Omega)},\\ \left|\int_\Omega B(x,u,\nabla u)\varphi dx\right| &\le C_3\sum_{i=1}^N \|u_{x_i}\|^{\beta_i}_{L^{p_i}(\Omega)}\|u\|^{\alpha_i}_{L^{1+\sigma}(\Omega)}\|\varphi\|_{L^{1+\sigma}(\Omega)},\\ \left|\int_\Omega C(x,u)\varphi dx\right| &\le C_5\|u\|_{L^{1+\sigma}(\Omega)}\|\varphi\|_{L^{1+\sigma}(\Omega)}.\end{aligned} \tag{4.6}$$

The energy functions are now defined as

$$E(\rho) = \int_{B_\rho}\mathbf{A}(x,u,\nabla u)\cdot\nabla u\, dx = \sum_{i=1}^N E_i(\rho),$$

where

$$E_i(\rho) = \int_{B_\rho} A_i(x,u,\nabla u)\frac{\partial u}{\partial x_i}dx, \qquad b(\rho) := \int_{B_\rho}|u|^{1+\sigma}dx = \|u\|^{1+\sigma}_{L^{1+\sigma}(B_\rho)}. \tag{4.7}$$

Observe that now

$$\begin{cases} C_2\|u_{x_i}\|^{p_i}_{L^{p_i}(B_\rho)} \le E_i(\rho) \le C_1\|u_{x_i}\|^{p_i}_{L^{p_i}(B_\rho)},\\ C_2\|u_{x_i}\|^{p_i}_{L^{p_i}(S_\rho)} \le \dfrac{dE_i}{d\rho} \le C_1\|u_{x_i}\|^{p_i}_{L^{p_i}(S_\rho)}, \end{cases} \tag{4.8}$$

where $S_\rho = \partial B_\rho$. The key assumption on the diffusion/absorption balance is now written as

$$\sigma + 1 < \min_i p_i. \tag{4.9}$$

This relation turns out to be the main hypothesis of the previous subsections once $p_i = p$ for all i. To avoid the technical details we start by considering the case $B \equiv 0$.

Theorem 4.1. *Let $u(x)$ be a weak solution of equation* (4.1) *with $B \equiv 0$, and $f \equiv 0$ on $B_{\rho_0}(x_0)$, $0 < \rho_0 < \operatorname{dist}(x_0, \partial\Omega)$. Assume that conditions* (4.2a) *and* (4.9) *are fulfilled. Then $u(x) \equiv 0$ on $B_{\rho_1}(x_0)$ with*

$$\rho_1^\nu = \rho_0^\nu - E^\gamma(\rho_0) \max\left(1, \rho_0^{\gamma-1}\right) G, \tag{4.10}$$

where exponents ν and γ and a constant G depend on $E(\rho_0)$. The constants ν and γ increase when $E(\rho_0)$ increases. If $C_3 > 0$, the same conclusion is true under a suitable assumption on C_3.

Proof. Take the function $\psi_n(r)$, defined in the proof of Theorem 2.1, for the cutting function in the ball $B_\rho(x_0)$. According to (4.2a)–(4.2c) and (4.6), this function can be substituted into (4.5) as a test function, which leads to the identity

$$\begin{aligned} &\int_{B_\rho} \psi_n \left\{\mathbf{A}(x, u, \nabla u) \cdot \nabla u + (B(x, u, \nabla u) + C(x, u))\, u\right\} dx \\ &\quad = \int_{B_\rho} u\mathbf{A}(x, u, \nabla u) \cdot \nabla \psi_n dx. \end{aligned} \tag{4.11}$$

Passing to the limit $n \to \infty$, we first obtain the energy equality

$$\int_{B_\rho} (\mathbf{A} \cdot \nabla u + (B + C)u)\, dx = \int_{S_\rho} u\mathbf{A}\mathbf{n} dS := I \tag{4.12}$$

and then, using (4.2a)–(4.2c), arrive at the inequality

$$C_5\,(E + b) \le I \tag{4.13}$$

with $C_5 = \min(C_4 - C_3, 1 - C_3/C_2) > 0$. By Hölder's inequality and (4.2a), we get that

$$|I| \le C_1 \sum_{i=1}^N \left(E_i'\right)^{(p_i-1)/p_i} \|u\|_{L^{p_i}(S_\rho)} \le M_1 \sum_{i=1}^N \left(E_i'\right)^{(p_i-1)/p_i} \|u\|_{L^{p_i}(S_\rho)}$$

with

$$M_1 = C_1 \max\left(1, \max_i \rho_0^{(\beta-p_i)(N-1)/\beta p_i}\right), \quad \beta = \max_i p_i.$$

To estimate $\|u\|_{L^\beta(S_\rho)}$ we use the *interpolation-trace inequality*. We have

$$\begin{aligned} |I| &\le CM_1 \sum_{i=1}^N \left(E_i'\right)^{(p_i-1)/p_i} \left(\sum_{i=1}^N E_j^{1/p_j} \rho^{\lambda_j+\delta} + b^{1/(1+\sigma)}\right)^\theta b^{(1-\theta)/(1+\sigma)} \rho^{-\delta\theta} \\ &\le M_2 \rho^{-\delta\theta} \sum_{i=1}^N \left(E_i'\right)^{(p_i-1)/p_i} (E + b)^\mu, \end{aligned} \tag{4.14}$$

where

$$M_2 = C M_1 \left(\max \left(1, \max_j \rho_0^{\lambda_j+\delta} \right) \right)^{\theta} \times \left(\max \left(b^{(\beta-1-\sigma)/(1+\sigma)}(\rho_0), \max_j E_j(\rho_0)^{(\beta-p_j/p_j)} \right) \right),$$

$$\mu = \frac{\theta(\beta)}{\beta} + \frac{1-\theta(\beta)}{1+\sigma},$$

and the constants $\lambda_j, \theta, \delta$ are given in Lemma 3.6 of the appendix. Next, applying Young's inequality to the right-hand side of (4.14), we have that

$$\begin{aligned} |I| &\le M_2 \sum_{i=1}^{N} \left(\frac{dE_i}{d\rho} \right)^{(p_i-1)/p_i(1-\gamma)} \rho^{-(\delta\theta/(1-\gamma))} (E+b)^{\mu/(1-\gamma)} \\ &\le M_3 \left[\varepsilon^{-k} E' \rho^{1-\nu} + \varepsilon^{k'} (E+b)^{1-\gamma} \right]^{1/(1-\gamma)}, \end{aligned} \tag{4.15}$$

where

$$M_3 = C\, M_2 \left[\max \left(1, \max_i \rho_0^{\delta\theta(\alpha/(\alpha-1)-p_i/(p_i-1))} \right) \right]^{1/(1-\gamma)} \times \max_i \left(b(\rho_0) + E(\rho_0) \right)^{\mu p_i/(1+\gamma(p_i-1))},$$

$$k = \frac{\alpha}{(\alpha-1)(1-\gamma)}, \quad k' = \frac{k}{k-1}, \quad 0 < \varepsilon < 1.$$

Letting

$$\varepsilon^{k'} M_3^{1-\gamma} \le \frac{1}{2} C_5^{1-\gamma},$$

in (4.15) and using inequalities (4.13), we come to a counterpart of inequality (2.26),

$$E^{1-\gamma} \le (E+b)^{1-\gamma} \le M_4 \rho^{1-\gamma} E' \tag{4.16}$$

with

$$M_4 = 2^k C_5^{-\alpha/(\alpha-1)} M_3^{\alpha/(\alpha-1)}.$$

Integration of this inequality completes the proof of Theorem 3.1. □

It is worth pointing out that Remark 2.1 remains true in the case of Theorem 3.1 and that the values of ν, γ for $\alpha = \beta = p$ coincide with those of ν and $\gamma(\tau)$ in Theorem 2.1 (for $\tau = 1$).

4.2 One-directional phenomena. In this subsection we show how the anisotropy of an equation can lead to existence of localized solutions even if the equation under study has no absorption terms. This phenomenon is of one-dimensional nature and can be caused by two different reasons: a) the degeneracy of the diffusion operator in some space directions x_i, b) a suitable diffusion/convection balance in the direction x_i. Examples of such equations are furnished by

$$-\sum_{i=1}^{N}\frac{\partial^2 u}{\partial x_i^2}-\frac{\partial}{\partial x_{N+1}}\left(|u|^{\alpha}\left|\frac{\partial u}{\partial x_{N+1}}\right|^{m-2}\frac{\partial u}{\partial x_{N+1}}\right)=f(x)$$

under the assumption that $\alpha+m>2$ and

$$-\sum_{i=1}^{N+1}\frac{\partial^2 u}{\partial x_i^2}+B(x)\frac{\partial}{\partial x_{N+1}}\left(|u|^{\sigma-1}u\right)=f(x)$$

if $0<\sigma<1$.

Because of the one-dimensional character of this phenomenon, its analysis becomes the most transparent in the special case where the domain Ω is a cylinder: $\Omega=G\times\mathbb{R}^+$, where G is an open bounded subset of $\mathbb{R}^N$. For the sake of presentation, in this section we restrict the study to the domains of this class.

First of all, let us see how the one-directional degeneracy of the diffusion operator can cause the occurrence of localized solutions. Consider the problem

$$-\operatorname{div}\mathbf{A}(x,u,\nabla u)+B(x,u,\nabla u)+C(x,u)=f(x)\quad\text{in }\Omega,\tag{4.17}$$

$$u(y,t)=0,\quad y\in\Gamma_D;\quad \mathbf{A}\cdot\mathbf{n}=0,\quad y\in\Gamma_N,\quad t\in\mathbb{R}^+,\tag{4.18}$$

where $\Gamma=\Gamma_D$, and the fulfillment of the following conditions is assumed:

$$\forall\,(r,\mathbf{q})\in\mathbb{R}\times\mathbb{R}^{N+1}\quad\text{and a.e. }x\in\Omega,\quad \widehat{q}^2:=\sum_{i=1}^{N}q_i^2$$

$$C_2\left(|r|^{\alpha}|q_{N+1}|^m+\widehat{q}^p\right)\le\mathbf{A}(x,r,\mathbf{q})\cdot\mathbf{q},\tag{4.19a}$$

$$|\mathbf{A}(x,r,\mathbf{q})|\le C_1\left(|r|^{\alpha}|q_{N+1}|^{m-1}+\widehat{q}^{p-1}\right)\tag{4.19b}$$

$$|A_{N+1}(x,r,\mathbf{q})|\le C_1|r|^{\alpha}|q_{N+1}|^{m-1},\tag{4.19c}$$

for some $p>1$, $m>1$ and $\alpha\in[p-m,\infty)$. The concept of weak solution is similar to that given in Section 2. Nonetheless, we shall require the fulfillment of the extra condition

$$\int_G\left(B(y,\rho,u,\nabla u)+C(y,\rho,u)\right)u|u|^{k-1}\,d\rho\le k\delta(\rho)\int_G|u|^{k-1}\mathbf{A}\cdot\nabla u\,dy\tag{4.20}$$

for $k=\max(1,(N(\alpha+m-p)-p(\alpha+m-1))/p)$. In this condition $\delta(\rho)$ is a monotone decreasing function such that $\delta(\infty)=0$. The parameters α, m, p will be specified later on.

Observe that condition (4.19c) provides the fulfillment of inequalities of the type (3.28a)–(3.28c) for $B + C$ and the validity of conditions (ii) in Theorem 3.4.

Theorem 4.2. *Let* $u(x)$ *be a weak solution of problem* (4.17), (4.18) *with* $f \neq 0$, $\operatorname{supp} f \subset \Omega \times (0, \rho_0)$, *and let*

$$E(\rho) := k \int_\rho^\infty \int_G |u|^{k-1} \mathbf{A}(y, t, u, \nabla u) \cdot \nabla u \, dy dt, \qquad E(0) < \infty, \tag{4.21}$$

where k is given in (4.20). *Assume that conditions* (4.19a)–(4.19c) *hold with*

$$p < \alpha + m. \tag{4.22}$$

Let $f(x)$ *be such that for any* $\rho \in (\rho_0 - \delta, +\infty)$, $\delta > 0$, *and* $\varepsilon > 0$

$$\int_\rho^\infty \|f(\cdot, \tau)\|_{L^\lambda(G)}^{(k-1+p)/(p-1)} d\tau \leq \varepsilon \left(1 - \frac{\rho}{\rho_0}\right)_+^{1/\gamma} \tag{4.23}$$

with the parameters

$$\lambda = \frac{k + \alpha + m - 1}{\alpha + m - 1} = \frac{\kappa}{\theta}, \qquad \gamma = \frac{\alpha + m - p}{m(k - 1 + p) + \alpha + m - p}.$$

Then $\operatorname{supp} u \subset \Omega \times (0, \rho_0)$.

Proof. Applying the routine arguments and using the regularity condition (4.20), we have that for every ρ large enough the inequality

$$\begin{aligned} 0 &\leq (1 - \delta) E(\rho) \leq E(\rho) + \int_\rho^\infty \int_G \left\{ (B + C) u |u|^{k-1} \right\} dy dt \\ &= \int_G A_{N+1}(y, \rho, u, \nabla u) u |u|^{k-1} dy + \int_\rho^\infty \int_G f u |u|^{k-1} dy dt := I_1 + I_2 \end{aligned} \tag{4.24}$$

holds. The terms I_1, I_2 can be estimated as

$$\begin{aligned} |I_1| &\leq C_1 \left(-\frac{dE}{d\rho} \right)^{(m-1)/m} \|u(\cdot, \rho)\|_{L^\kappa(G)}^{\kappa/m} \\ &\leq C \left(-\frac{dE}{d\rho} \right)^{(m-1)/m} \| |u|^{(k-1)/p} |\nabla u| \|_{L^p(G)}^{\kappa/m\mu} \\ &\leq C \left(-\frac{dE}{d\rho} \right)^{1/(1-\gamma)}, \quad \mu = \frac{k - 1 + p}{p}, \end{aligned} \tag{4.25}$$

$$\begin{aligned} |I_2| &\leq C \int_\rho^\infty \|f\|_{L^\lambda(G)} \|u\|_{L^{k+\alpha+m-1}(G)}^k d\rho \\ &\leq C \int_\rho^\infty \|f\|_{L^\lambda(G)} \left(-E_\rho \right)^{\frac{k}{k-1+p}} d\rho \\ &\leq C \tau E(\rho) + C_\tau \int_\rho^\infty |f\|_{L^\lambda(G)}^{(k-1+p)/(p-1)} d\rho \end{aligned} \tag{4.26}$$

with some constant $C = C(k, C_1, p, G)$ and for any $\tau \in (0, 1)$. Here we have used (4.19a), (4.19c), and the inequality

$$\|u\|^{\mu}_{L^{\kappa}(G)} \le C\|\nabla_G(u|u|^{(k-1)/p})\|_{L^p(G)} \tag{4.27}$$

with $C = C(G, \kappa, k, p)$. Gathering (4.24), (4.25), (4.26), and using (4.21), (4.23) we come to the inequality

$$E \le C\left((-E_\rho)^{1/(1-\gamma)} + \varepsilon(1-\rho/\rho_0)_+^{1/(1-\gamma)}\right)$$

and, correspondingly, to an inequality of the type (3.48),

$$CE^{1-\gamma} + E_\rho \le C\varepsilon^{1-\gamma}(1-\rho/\rho_0)_+^{(1-\gamma)/\gamma},$$

which has already been studied. The rest of the proof literally repeats the proof of Theorem 3.4 in the part concerning the analysis of inequality (3.48). □

Now we are going to show that the localization of solutions of elliptic equations of the type (2.1) can also be caused by the terms of the form

$$B(x)\frac{\partial}{\partial x_{N+1}}\left(|u|^{\sigma-1}u\right),$$

which usually appear in modelling of physical processes of convective nature.

Let us consider the equation

$$-\operatorname{div}\mathbf{A}(x, u, \nabla u) + B(x)\frac{\partial}{\partial x_{N+1}}\left(|u|^{\sigma-1}u\right) = 0 \quad \text{in } \Omega = G \times \mathbb{R}^+ \tag{4.28}$$

under the assumption that the following conditions are fulfilled:

$$\forall (x, r, \mathbf{q}) \in \Omega \times \mathbb{R} \times \mathbb{R}^{N+1},$$
$$C_2|\mathbf{q}|^p \le \mathbf{A}(x, r, \mathbf{q}) \cdot \mathbf{q}, \qquad |A_{N+1}(x, r, \mathbf{q})| \le C_1|\mathbf{q}|^p, \tag{4.29a}$$
$$C_4 \le |B(x)| \le C_4^{-1}, \qquad \max\left(B, \frac{\partial B}{\partial x_{N+1}}\right) \le 0 \tag{4.29b}$$

with some $p > 1$ and $\sigma > 0$.

Theorem 4.3. *Assume that conditions* (4.29a)–(4.29b) *are fulfilled. Let $u(x)$ be a weak solution of equation* (4.28) *with $0 < \sigma < p - 1$ which satisfies condition* (4.18) *and has a finite energy, i.e.,*

$$\begin{aligned} E(\rho) + b(\rho) &:= \int_\rho^\infty \int_G (\mathbf{A}(x, u, \nabla u) \cdot \nabla u + |u|^{1+\sigma})dx \\ &\le E(0) + b(0) < \infty. \end{aligned} \tag{4.30}$$

Then there exists $\rho_0 \equiv \rho_0(C_1, C_2, p, \sigma, N, E(0)) \ge 0$ such that

$$\operatorname{supp} u \subset G \times (0, \rho_0).$$

Proof. First, we establish the energy relation

$$\int_\rho \int_G \left(\mathbf{A}(x, u, \nabla u) \cdot \nabla u + \left| \frac{\partial B}{\partial x_{N+1}} \right| \frac{\sigma}{\sigma + 1} |u|^{1+\sigma} \right) dx + \frac{\sigma}{1+\sigma} \int_G |B(y, \rho)||u|^{1+\sigma} dy = - \int_G A_{N+1}(y, \rho, u, \nabla u) u \, dy := I(\rho). \tag{4.31}$$

Second, applying (4.29a), (4.29b), and the inequality

$$\|u(\cdot, \rho)\|_{L^p(G)} \le C \|\nabla u\|^{\theta}_{L^p(G)} \cdot \|u(\cdot, \rho)\|^{1-\theta}_{L^{1+\sigma}(G)},$$
$$\theta = \frac{N(p - 1 - \sigma)}{N(p - 1 - \sigma) + p(1 + \sigma)}$$

we estimate $I(\rho)$ in the following way:

$$|I| \le C_1 \|\nabla u\|^{p-1}_{L^p(G)} \|u\|_{L^p(G)} \le C(-E_\rho)^{\frac{(p-1)}{p} + \frac{\theta}{p}} \left(\int_G |B| \frac{\sigma}{\sigma + 1} |u|^{1+\sigma} dy \right)^{\frac{(1-\theta)}{(1+\sigma)}}$$
$$\le \frac{1}{2} \int_G |B| \frac{\sigma}{\sigma + 1} |u|^{1+\sigma} dy + C(-E_\rho)^{1/(1-\gamma)}, \tag{4.32}$$
$$\frac{1}{1-\gamma} = \left(\frac{p-1}{p} + \frac{\theta}{p} \right) \frac{1+\sigma}{\sigma + \theta} > 1.$$

It now follows from (4.31), (4.32) that the energy function satisfies the already known inequality

$$CE^{1-\gamma} + E' \le 0,$$

whence the desired assertion. □

5 Systems of second-order elliptic equations

In this section we apply energy methods to the study of localization properties of solutions to systems of second-order elliptic equations. We consider systems of the form

$$- \operatorname{div} \mathbf{A}(x, \mathbf{u}, \nabla \mathbf{u}) + \mathbf{B}(x, \mathbf{u}, \nabla \mathbf{u}) + \mathbf{C}(x, \mathbf{u}) = \mathbf{f}(x), \tag{5.1}$$

in an open domain $\Omega \subset \mathbb{R}^N$, $N \ge 1$. Here $\mathbf{u}(x) = (u_1, \ldots, u_m)$, $m > 1$, is a vector-valued function. The coefficients $\mathbf{C}(x, \mathbf{r})$, $\mathbf{B}(x, \mathbf{r}, \mathbf{q})$ are also given vector-valued functions defined on $\Omega \times \mathbb{R}^m$ and $\Omega \times \mathbb{R}^m \times \mathbb{R}^{Nm}$ correspondingly. Throughout the section $\mathbf{A}$ denotes the diffusion matrix with the elements

$$\mathbf{A}(x, \mathbf{r}, \mathbf{q}) = A^k_i(x, \mathbf{r}, \mathbf{q}), \quad k = 1, \ldots, m, \quad i = 1, \ldots, N.$$

The scalar functions A_i^k are defined on $\Omega \times \mathbb{R}^m \times \mathbb{R}^{Nm}$. We adopt the usual notation

$$\nabla \mathbf{u} = \left(\frac{\partial u_k}{\partial x_j}\right), \quad k = 1, \dots, m, \quad j = 1, \dots, N;$$
$$\operatorname{div} \mathbf{A} = \left(\sum_{i=1}^{N} \frac{dA_i^1}{dx_i}, \dots, \sum_{i=1}^{N} \frac{dA_i^m}{dx_i}\right) \tag{5.2}$$

and introduce the following ones:

$$|\mathbf{u}|^2 = \sum_{k=1}^{m} |u_k|^2, \qquad |\nabla \mathbf{u}|^2 = \sum_{k=1}^{m}\sum_{i=1}^{N} \left|\frac{\partial u_k}{\partial x_i}\right|^2,$$
$$\|\mathbf{u}\|_{L^p} = \||\mathbf{u}|\|_{L^p}, \qquad \|\nabla \mathbf{u}\|_{L^p} = \||\nabla \mathbf{u}|\|_{L^p}, \tag{5.3}$$
$$\mathbf{u} \cdot \mathbf{v} = \sum_{k=1}^{m} u_k v_k, \qquad \mathbf{A} : \nabla \mathbf{u} = \sum_{k,i=1}^{m,N} A_i^k \frac{\partial u_k}{\partial x_i}. \tag{5.4}$$

It will be assumed that **A**, **B**, and **C** satisfy the structural conditions

$$\forall (x, \mathbf{r}, \mathbf{q}) \in \Omega \times \mathbb{R}^m \times \mathbb{R}^{Nm}$$
$$C_2 |\mathbf{q}|^p \le \mathbf{A}(x, \mathbf{r}, \mathbf{q}) : \mathbf{q} \le C_1 |\mathbf{q}|^p, \tag{5.5a}$$
$$|\mathbf{B}(x, \mathbf{r}, \mathbf{q})| \le C_3 |\mathbf{r}|^\alpha |\mathbf{q}|^\beta, \tag{5.5b}$$
$$C_4 |\mathbf{r}|^{1+\sigma} \le \mathbf{C}(x, \mathbf{r}) \cdot \mathbf{r}. \tag{5.5c}$$

The following two systems meet all the required conditions:

$$-\sum_{i,j=1}^{N} \frac{d}{dx_i}\left(A_{ij}^k(x, \mathbf{u}, \nabla \mathbf{u}) |\nabla \mathbf{u}|^{p-2} \frac{\partial u_k}{\partial x_j}\right)$$
$$+ \sum_{i,j=1}^{N} B_{ik}(x) |u_i|^{\alpha_{ik}} |u_k|^{\alpha_{ki}} \left|\frac{\partial u_k}{\partial x_i}\right|^{\beta_{ki}} + |\mathbf{u}|^{\sigma-1} u_k = f_k(x), \tag{5.6}$$

$$-\sum_{i,j=1}^{N} \frac{d}{dx_i}\left(A_{ij}^k(x, \mathbf{u}, \nabla \mathbf{u}) |\nabla u_k|^{p-2} \frac{\partial u_k}{\partial x_j}\right)$$
$$+ B_{ik}(x) |u_i|^{\alpha_{ik}} |u_k|^{\alpha_{ki}} \left|\frac{\partial u_k}{\partial x_i}\right|^{\beta_{ki}} + |u|^{\sigma-1} u_k = f_k(x), \tag{5.7}$$

$k = 1, \dots, m$. The matrixes A_{ij}^k must satisfy the *ellipticity condition*: there exist constants $0 < \nu \le \mu < \infty$ such that

$$\forall\, (x, \mathbf{r}, \mathbf{q}, \xi) \in \Omega \times \mathbb{R}^m \times \mathbb{R}^{Nm} \times \mathbb{R}^{mN}$$
$$\nu \sum_{\substack{1 \le k \le m \\ 1 \le i, j \le N}} \xi_{ki}^2 \le \sum_{\substack{1 \le k \le m \\ 1 \le i, j \le N}} A_{ij}^k(x, \mathbf{r}, \mathbf{q}) \xi_{ki} \xi_{kj} \le \mu \sum_{\substack{1 \le k \le m \\ 1 \le i \le N}} \xi_{ki}^2 \tag{5.8}$$

for equation (5.6) and

$$\nu \sum_{i=1}^{N} \xi_i^2 \leq \sum_{\substack{1 \leq k \leq m \\ 1 \leq i,j \leq N}} A_{ij}^k(x, \mathbf{r}, \mathbf{q}) \xi_i \xi_j \leq \mu \sum_{i=1}^{N} \xi_i{}^2 \tag{5.9}$$

for equation (5.7).

Results on the existence of weak solutions for systems of elliptic equations can be found, for instance, in [235]. As in the previous sections, we are concerned here only with qualitative properties of weak solutions, always assuming that these solutions exist.

Let us assume that the structural constants in (5.5a)–(5.5c) are subject to the conditions

$$0 \leq \sigma < p - 1, \quad \alpha = \sigma - \beta \frac{1+\sigma}{p}, \tag{5.10}$$

$$C_3 < \left(C_4 \frac{\beta}{p - \beta} \right)^{(p-\beta)/p} \left(C_2 \frac{p}{\beta} \right)^{\beta/p} \quad \text{if } 0 < \beta < p, \tag{5.11}$$

and

$$C_3 < C_4 \text{ if } \beta = 0 \quad \text{or} \quad C_3 < C_2 \text{ if } \beta = p.$$

Definition 5.1. A locally integrable in Ω vector-function $\mathbf{u}(x)$ is said to be a weak solution of (5.1) if

$$\mathbf{u} \in W_{\text{loc}}^{1,p}(\Omega) \cap L_{\text{loc}}^{1+\sigma}(\Omega), \qquad \mathbf{f} \in L_{\text{loc}}^{(1+\sigma)/\sigma}(\Omega),$$

and for every vector-function $\boldsymbol{\varphi} \in W^{1,p}(\Omega) \cap L^{1+\sigma}(\Omega)$, compactly supported in Ω, the identity

$$\int_{\Omega} (\mathbf{A}(x, \mathbf{u}, \nabla \mathbf{u}) : \nabla \boldsymbol{\varphi} + \mathbf{B}(x, \mathbf{u}, \nabla \mathbf{u}) \cdot \boldsymbol{\varphi} + \mathbf{C}(x, \mathbf{u}) \cdot \boldsymbol{\varphi}) dx = \int_{\Omega} \mathbf{f} \cdot \boldsymbol{\varphi} dx \tag{5.12}$$

holds.

According to (5.2)–(5.3),

$$\mathbf{A} : \nabla \boldsymbol{\varphi} = \sum_{k=1,i=1}^{m,N} A_j^k \frac{\partial \varphi_k}{\partial x_j}; \qquad \mathbf{B} \cdot \boldsymbol{\varphi} = \sum_{k=1}^{m} B_k \varphi_k.$$

It is easy to verify that under assumptions (5.5a)–(5.5c) this definition is consistent (all the terms of (5.12) are defined and bounded). Indeed, applying the Hölder inequality we have the inequalities

$$\begin{cases} \left| \int_{\Omega} \mathbf{A} : \nabla \boldsymbol{\varphi} dx \right| \leq C_1 \|\nabla \mathbf{u}\|_{L^p(\Omega)}^{(p-1)/p} \|\nabla \boldsymbol{\varphi}\|_{L^p(\Omega)}, \\ \left| \int_{\Omega} \mathbf{B} \cdot \boldsymbol{\varphi} dx \right| \leq C_3 \|\mathbf{u}\|_{L^{1+\sigma}(\Omega)}^{\alpha} \|\boldsymbol{\varphi}\|_{L^{1+\sigma}(\Omega)} \|\nabla \mathbf{u}\|_{L^p(\Omega)}^{\beta}, \\ \left| \int_{\Omega} \mathbf{C} \cdot \boldsymbol{\varphi} dx \right| \leq C_5 \|\mathbf{u}\|_{L^{1+\sigma}(\Omega)}^{\sigma} \|\boldsymbol{\varphi}\|_{L^{1+\sigma}(\Omega)}. \end{cases} \tag{5.13}$$

Following the same scheme of arguments that we used in previous sections, let us introduce the energy functions

$$E(\rho) = \int_{B_\rho(x_0)} \mathbf{A}(x, \mathbf{u}, \nabla \mathbf{u}) : \nabla \mathbf{u} dx, \quad b(\rho) = \int_{B_\rho(x_0)} |\mathbf{u}|^{1+\sigma} dx.$$

As in the case of scalar equations, we have that

$$\begin{aligned} C_2 \|\nabla \mathbf{u}\|^p_{L^p(\Omega)} &\leq E(\rho) \leq C_1 \|\nabla \mathbf{u}\|^p_{L^p(\Omega)} = C_1 \int_\Omega |\nabla \mathbf{u}|^p dx, \\ C_2 \int_{S_\rho} |\nabla \mathbf{u}|^p dS &\leq \frac{dE}{d\rho} = \int_{S_\rho} \mathbf{A} : \nabla \mathbf{u} dS \leq C_1 \int_{S_\rho} |\nabla \mathbf{u}|^p dS, \end{aligned} \tag{5.14}$$

(see (2.8)).

The main result of this section is the following.

Theorem 5.1. *Assume that conditions* (5.10)–(5.11) *are fulfilled and let* $\mathbf{u}(x)$ *be a weak solution of system* (5.1) *with* $\mathbf{f}(x) \equiv 0$ *in a ball* $B_{\rho_0}(x_0)$ *with* $0 < \rho_0 < \operatorname{dist}(x_0, \partial\Omega)$. *Then* $\mathbf{u}(x) \equiv 0$ *a.e. in* $B_{\rho_1}(x_0)$, *where*

$$\begin{aligned} \rho_1^\nu &= \rho_0^\nu - C \min_{p(1+\sigma)<\tau\leq 1} \left\{ \frac{E^\gamma(\rho_0)}{\tau p - 1 - \sigma} G(\rho_0) \right\}, \\ G &:= \max\left(1, \rho_0^{\nu-1}\right) \max\left(b^\mu(\rho_0), b^\eta(\rho_0)\right), \\ \nu &= \frac{1}{k}(p-1)(1+\sigma) > 1, \quad \gamma = \frac{1}{k}(\tau p - 1 - \sigma) < 1, \\ \eta &= \left(\frac{p-1-\sigma}{(p-1)(1+\sigma)} + \frac{\tau p - 1 - \sigma}{k} \right) > 0, \\ k &= N(p-1-\sigma) + p(1+\sigma), \qquad \mu = \frac{1}{k} p(1-\tau) > 0, \end{aligned}$$

and $C \equiv C(C_1 - C_4, N, p, \beta, \sigma)$.

Proof. Let $\psi_n(|x - x_0|)$ be the scalar cutoff function for the ball $B_\rho(x_0)$ defined in Theorem 2.1. Using (5.13), it is not difficult to see that the function

$$\boldsymbol{\varphi}(x) := \psi_n(|x - x_0|)\mathbf{u}(x) = \psi_n(|x - x_0|)\,(u_1(x), \ldots, u_m(x)) \tag{5.15}$$

can be substituted into (5.12) as a test function. The standard arguments (truncation and passage to the limit) lead to the equality

$$\int_{B_\rho} \psi_n\,(\mathbf{A} : \nabla \mathbf{u} + \mathbf{B} \cdot \mathbf{u} + \mathbf{C} \cdot \mathbf{u})\, dx = -\int_{B_\rho} \mathbf{A} : \nabla \psi_n \mathbf{u} dx, \tag{5.16}$$

where

$$\mathbf{A} : \nabla \psi_n \mathbf{u} = \sum_{k=1,i=1}^{m,N} A_i^k \frac{\partial \psi_n}{\partial x_i} u_k.$$

Letting in (5.16) $n \to \infty$, analogously to the proof of Theorem 2.1, we conclude that the integral identity

$$\int_{B_\rho} (\mathbf{A} : \nabla \mathbf{u} + \mathbf{B} \cdot \mathbf{u} + \mathbf{C} \cdot \mathbf{u})\, dx = -\int_{S_\rho} \mathbf{A}\mathbf{u} \cdot \mathbf{n} dS := I \tag{5.17}$$

holds, where $\mathbf{A}\mathbf{u} \cdot \mathbf{n} = \sum_{i=1}^{N} A_i^k n_i u_k$. The rest of the argument is analogous to that used in Step 3 of the proof of Theorem 2.1. First, it follows from (5.5a)–(5.5c), (5.10), (5.11), (5.14), (5.17) that

$$C_5\,(E(\rho) + b(\rho)) \leq \int_{B_\rho} (\mathbf{A} : \nabla \mathbf{u} + \mathbf{B} \cdot \mathbf{u} + \mathbf{C} \cdot \mathbf{u})\, dx, \tag{5.18}$$

where

$$C_5 = \min\left(C_4 - \varepsilon C_3 \frac{p-\beta}{p}, 1 - \frac{\beta C_3}{p C_2} \varepsilon^{-(p-\beta)/\beta}\right) > 0.$$

Second, we have the estimate

$$\begin{aligned} |I| &\leq C_1 \left(\int_{S_\rho} |\nabla \mathbf{u}|^p dS\right)^{(p-1)/p} \left(\int_{S_\rho} |\mathbf{u}|^p dS\right)^{1/p} \\ &\leq C_1 C_2^{(1-p)/p} \left(\frac{dE}{d\rho}\right)^{(p-1)/p} \|\mathbf{u}\|_{L^p(\Omega)}. \end{aligned} \tag{5.19}$$

Gathering (5.17)–(5.19) and then applying to the vector-valued function $\mathbf{u}(x)$ an analogue of inequality (2.21), we get the estimate

$$E + b \leq M \left(\frac{dE}{d\rho}\right)^{(p-1)/p} \left(E^{1/p} + \rho^\delta b^{1/(1+\sigma)}\right)^\theta b^{(1-\theta)/(1+\sigma)} \tag{5.20}$$

with the same parameters δ, θ that in the proof of Theorem 2.1. □

Remark 5.1. Theorem 2.2 and Remarks 2.1–2.2 remain true for systems of equations. It is worth stressing here that the methods based on the comparison principles do not generally apply to systems of equations.

The energy method allows for numerous modifications which make it applicable to a much wider class of elliptic systems. For instance, when deriving the principal energy inequality we may take for the test function

$$\varphi(x) = \psi_n(|x - x_0|)\mathbf{F}(\mathbf{u}),$$

with a suitable nonlinear vector-function $\mathbf{F}(\mathbf{u})$.

Not entering into the details, let us illustrate this possibility by the following simple example. Let us consider the system of ordinary differential equations

$$\begin{aligned} -\frac{d}{dx}\left(\frac{du_1}{dx}\right) + u_2|u_2|^{\sigma_1 - 1} &= 0, \\ -\frac{d}{dx}\left(\frac{du_2}{dx}\right) + u_1|u_1|^{\sigma_2 - 1} &= 0 \quad \text{in } \Omega = (-1, 1) \end{aligned} \tag{5.21}$$

with $0 < \sigma_i < 1$. Later on, in Chapters 2 and 3, we shall discuss the questions of genesis of such systems and applicability of other methods to their study (see also [153]. From the formal point of view, system (5.21) does not satisfy condition (5.5c) since

$$\mathbf{C}(x, \mathbf{u}) \cdot \mathbf{u} = u_1 u_2 \left(|u_1|^{\sigma_1 - 1} + |u_1|^{\sigma_2 - 1} \right).$$

For the sake of simplicity, let us confine ourselves to consideration of nonnegative solutions of (5.21). Set

$$E(\rho) = \int_{-\rho}^{\rho} |\mathbf{u}_x|^2 dx, \quad b(\rho) = \int_{-\rho}^{\rho} |\mathbf{u}|^{1+\sigma} dx, \tag{5.22}$$

and assume that

$$0 \le u_i(x) \le M < \infty, \sigma = \max(\sigma_1, \sigma_2) < 1.$$

Multiplying (formally) the first equation of (5.21) by $u_1 + \lambda u_2$ and the second one by $u_2 + \lambda u_1$ with some $\lambda \in (0, 1/2)$, we obtain the energy relation

$$\int_{B_\rho} \left(u_{1x}^2 + 2\lambda u_{1x} u_{2x} + u_{2x}^2 \right) dx + \int_{B_\rho} Q(u_1, u_2) dx := I. \tag{5.23}$$

In this relation,

$$\left\{ \begin{aligned} Q(u_1, u_2) &:= u_2^{\sigma_1} (u_1 + \lambda u_2) u_2 + u_1^{\sigma_2} (u_2 + \lambda u_1), \\ I &:= [u_{1x} (u_1 + \lambda u_2) + u_{2x} (u_2 + \lambda u_1)]|_{-\rho}^{\rho}. \end{aligned} \right. \tag{5.24}$$

Let us assume, for instance, that $\sigma = \sigma_1 = \max(\sigma_1, \sigma_2)$. It follows then that

$$\begin{aligned} Q &\ge \min(1, \lambda)(u_1 + u_2) \left(u_2^{\sigma_1} + u_1^{\sigma_1} u_1^{\sigma_2 - \sigma_1} \right) \\ &\ge \min(1, \lambda) \min(1, M^{\sigma_2 - \sigma_1})(u_1 + u_2) \left(u_1^{\sigma} + u_2^{\sigma} \right) \ge C|\mathbf{u}|^{1+\sigma}. \end{aligned} \tag{5.25}$$

Hence by virtue of (5.22)–(5.25),

$$C(E + b) \le I, \quad C \equiv C(\lambda, M, \sigma). \tag{5.26}$$

To estimate I, we apply the same argument (essentially one-dimensional) that was already used in the introduction:

$$|I| \le C(\lambda) \left(E'(\rho) \right)^{1/2} \cdot \left(b'(\rho) \right)^{1/(1+\sigma)}.$$

Having gathered the last inequality with (5.26), we finally get that

$$E^{1-\gamma} \le (E + b)^{1-\gamma} \le CE', \quad \text{where } 1 - \gamma = \frac{2(1 + \sigma)}{3 + \sigma} < 1 \text{ if } \sigma < 1,$$

whence the localization of $\mathbf{u}$.

6 Higher-order elliptic equations

In the papers [73, 74, 75], F. Bernis offered a modification of the local energy method which made it possible to apply the method to the study of elliptic and parabolic equations of arbitrary order. Roughly speaking, this modification consists in using special weighted energy functions leading to some higher-order ordinary differential inequalities. Following [73], let us explain this modification by considering elliptic equations of the type

$$Lu \equiv Au + |u|^{\sigma-1}u = f(x), \quad x \in \Omega \subset \mathbb{R}^N. \tag{6.1}$$

Here $1 < 1+\sigma < p$, $1 < p < \infty$,

$$\begin{aligned} Au &\equiv (-1)^m \sum_{|\alpha|=m} D^\alpha \left(|D^\alpha u|^{p-2} D^\alpha u \right) \\ &\equiv \sum_{|\alpha|=m} \frac{\partial^{|\alpha|}}{\partial^{\alpha_1} x_1 \cdots \partial^{\alpha_N} x_N} \left(|D^\alpha u|^{p-2} \frac{\partial^{|\alpha|} u}{\partial^{\alpha_1} x_1 \cdots \partial^{\alpha_N} x_N} \right), \end{aligned} \tag{6.2}$$

$$\begin{aligned} |D^\alpha u|^p &= \sum_{|k|=|\alpha|} |D^k u|^p = \sum_{|k|=|\alpha|} \left| \frac{\partial^{|k|} u}{\partial^{k_1} x_1 \cdots \partial^{k_N} x_N} \right|^p, \\ |k| &= k_1 + \cdots + k_N. \end{aligned} \tag{6.3}$$

Clearly, $A = (-1)^m \Delta^m$, for $p = 2$ and if $m = 1$,

$$Au = \Delta_p u = \operatorname{div} \left(|\nabla u|^{p-2} Du \right).$$

In the case $N = 1$, $p = 2$ equation (6.1) with $f(x) \equiv 0$ takes on the form

$$Lu \equiv (-1)^m \frac{d^{2m} u}{dx^{2m}} + |u|^{\sigma-1} u = 0. \tag{6.4}$$

Equation (6.4) admits explicit solutions of the form

$$\begin{aligned} u(x) &= u_0 \left(1 - \frac{x}{x_0} \right)_+^\lambda \quad \text{for } x \in (0, \infty),\ u(0) = u_0, \\ \lambda &= \frac{2-\sigma}{2m}, \quad C = \lambda(\lambda-1)\cdots(\lambda-2m+1), \quad x_0 = u_0^\lambda C^{1/2m}, \end{aligned} \tag{6.5}$$

or

$$u(x) = u_1 \left(\frac{x - x_0}{1 - x_0} \right)_+^\lambda, \quad \text{if } x \in (0, 1), \tag{6.6}$$

with $1 - x_0 = (u_0^\lambda C^{1/2m}) < 1$.

The following equations of the type (6.4) with nonzero source terms,

$$Lu = \varepsilon \left(1 - \frac{x}{x_0}\right)^{\lambda(r-1)}, \quad x \in (0, \infty), \tag{6.7}$$

$$Lu = \varepsilon \left(\frac{x - x_0}{1 - x_0}\right)^{\lambda(r-1)}, \quad x \in (0, 1), \tag{6.8}$$

also admit solutions of the form (6.5), (6.6) provided that the constants u_0, u_1, ε satisfy the relations

$$\varepsilon = u_0^{r-1} \left(1 - u_0^{2-r} x_0^{-2m} C\right), \tag{6.9}$$

$$\varepsilon = u_1^{r-1} \left(1 - u_1^{2-r} (1 - x_0)^{-2m} C\right). \tag{6.10}$$

It follows from (6.5), (6.6), (6.8), (6.9) that a solution $u(x)$ of equation (6.7) is localized in a bounded domain if the source intensity ε is small enough and at the same time the boundary condition u_1 is small (implying that the appropriately defined energy is also small).

The existence of a weak solution $u \in W_0^{m,p}(\Omega)$ to equation (6.4) was proven in [73, 243] under the assumptions

$$f \in W^{-m,p'}(\Omega), \qquad p' = \tfrac{p}{p-1},$$
$$\left(\text{either} \quad f \in L^{(1+\sigma)/\sigma}(\Omega), \quad \text{or} \quad f \in L^{p'}(\Omega)\right).$$

Here $W^{-m,p'}(\Omega)$ denotes the dual space of $W_0^{m,p}(\Omega)$. This means that if we write (for the sake of simplicity of notation)

$$\langle f, v \rangle_{W^{-m,p'}(\Omega), W_0^{m,p}(\Omega)} = \left| \int_\Omega f v \, dx \right|,$$

then

$$\left| \int_\Omega f v \, dx \right| \le \|f\|_{W^{-m,p'}(\Omega)} \|v\|_{W_0^{m,p}(\Omega)}.$$

The solution $u(x)$ satisfies the inequality

$$\int_\Omega \left(|D^\alpha u|^p + |u|^{1+\sigma}\right) dx \le C(p, \sigma, \|f\|^*), \tag{6.11}$$

where

$$\|f\|^* = \begin{cases} \text{either} & \|f\|_{W^{-m,p'}(\Omega)}, \\ \text{or} & \|f\|^{1+\sigma}_{L^{(1+\sigma)/\sigma}(\Omega)}. \end{cases}$$

This estimate is derived by multiplying equation (6.1) by $u \in W_0^{m,p}(\Omega)$, integrating by parts and applying the Young inequality.

Definition 6.1. A function $u \in W_0^{m,p}(\Omega)$ is said to be a weak solution of (6.1) if for every test function $\varphi \in W_0^{m,p}(\Omega)$, the identity

$$\int_\Omega \left(\sum_{|\alpha|=m} |D^\alpha u|^{p-2} D^\alpha u \cdot D^\alpha \varphi + |u|^{\sigma-1} u\varphi \right) dx = \int_\Omega f\varphi \, dx \tag{6.12}$$

holds.

Before starting the presentation, let us stress that the energy method in the form used in Sections 1–3 fails in the case $m > 1$. This is because the integrals over the surface of the sphere $S_\rho(x_0)$ appearing after integration by parts depend now on derivatives of $(2m-1)$th order, while the energy function E only comprises derivatives of the orders not exceeding $m-1$. The latter means that estimates of Sections 1–3 are no longer valid.

6.1 Plane energy sets. We introduce the notation

$$x = (x_1, \dots, x_N), \qquad y = (x_1, \dots, x_{N-1}), \qquad t = x_N.$$

Let us assume that

$$\operatorname{supp} f \subset \Omega, \qquad \operatorname{supp} f \cap \{t > 0\} = \emptyset, \qquad \{t > 0\} \cap \Omega \neq \emptyset. \tag{6.13}$$

It is not difficult to see that for every $t_0 \geq 0$ the function

$$w(x) = (t - t_0)_+^m u(x) = \begin{cases} (t-t_0)^m u(x) & \text{if } t > t_0, \\ 0 & \text{if } t \leq t_0 \end{cases} \tag{6.14}$$

belongs to $W_0^{m,p}(\Omega)$ and hence can be taken as a test function in (6.12). Let us introduce the energy functions

$$E_s(t_0) = K_s(t_0) + \Pi_s(t_0), \quad s = 0, 1, \dots, m, \tag{6.15}$$

where

$$K_s(t_0) = \int_{\Omega, t > t_0} (t - t_0)^s |D^m u|^p \, dx,$$

$$\Pi_s(t_0) = \int_{\Omega, t > t_0} (t - t_0)^s |u|^{1+\sigma} \, dx,$$

where the domain of integration is the set $\Omega \cap \{t > y_0\}$. The following relation holds:

$$\begin{aligned} E_s'(t_0) \equiv \frac{dE_s(t_0)}{dt_0} &= -s \int_{\Omega,\, t>t_0} (t-t_0)^{s-1} \left(|D^m u|^p + |u|^{1+\sigma} \right) dx \\ &= -s E_{s-1}(t_0). \end{aligned}$$

Let us start by applying the method to the special equation which corresponds to the value $m = 1$:

$$Lu \equiv -\operatorname{div}\left(|\nabla u|^{p-2}\nabla u\right) + |u|^{\sigma-1}u = f(x).$$

Formally multiplying it by the function $w(x)$ (with $m = 1$), integrating by parts in $\Omega \cap \{t > t_0\}$, and recalling that $\operatorname{supp} f \cap \{t > 0\} = \emptyset$, we get the equality

$$\begin{aligned} E_1(t_0) &= \int_{\Omega, t>t_0} (t - t_0)\left(|\nabla u|^p + |u|^{1+\sigma}\right) dx \\ &= -\int_{\Omega, t>t_0} u|\nabla u|^{p-2}\, u_t\, dx = I. \end{aligned}$$

In contrast to the energy relations that occurred in previous sections, this one does not contain any integral over a set of dimension $N - 1$. The right-hand side of this energy equality can be estimated as

$$\begin{aligned} |I| &\le \left(\int_{\Omega, t>t_0} |\nabla u|^p dx\right)^{(p-1)/p} \left(\int_{\Omega, t>t_0} |u|^p\, dx\right)^{1/p} \\ &\le K_0^{(p-1)/p} C K_0^{\theta/p} \Pi_0^{(1-\theta)/(1+\sigma)} \le C E_0^{1/(1-\gamma)}, \end{aligned}$$

with

$$\frac{1}{1-\gamma} = \frac{p-1}{p} + \frac{\theta}{p} + \frac{1-\theta}{1+\sigma}, \theta = \frac{N(p-1-\sigma)}{N(p-1-\sigma) + p(1+\sigma)}.$$

We used here the interpolation inequality (Lemma 3.2 of the appendix)

$$\forall\, v \in W_0^{1,p}(\Omega) \qquad \|v\|_{L^p(\Omega)} \le \|\nabla v\|^{\theta}_{L^p(\Omega)} \|v\|^{1-\theta}_{L^{1+\sigma}(\Omega)}.$$

Returning now to the energy equality we get the already familiar first-order ordinary differential inequality

$$E_1(t_0) \le C\left(-E_1'(t_0)\right)^{1/(1-\gamma)},$$

i.e.,

$$C E_1^{1-\gamma} + E_1' \le 0.$$

In the case $m > 1$, the corresponding differential inequality is of mth order.

Theorem 6.1. *Assume that* $1 < 1 + \sigma < p$, *let* (6.13) *be fulfilled, and let* $u \in W_0^{m,p}(\Omega)$ *be a weak solution of equation* (6.1). *Then*

$$E_k(t_0) = 0, \quad k = 0, \dots, m; \quad t_0 \ge a_0, \tag{6.16}$$

(*i.e.*, $\operatorname{supp} u(x) \in \Omega \cap \{t < a_0\}$), *where*

$$a_0 = C E_0^{\gamma/(1-\gamma)}(t_0), \qquad \frac{\gamma}{1-\gamma} = \frac{p-1-\sigma}{N(p-1-\sigma) + pm}. \tag{6.17}$$

Proof of Theorem 6.1: *Step* 1: *Derivation of the energy inequality.* Letting in (6.12) $\varphi = w$, we have

$$\begin{aligned}\Lambda \equiv \sum_{|\alpha|=m} \int_{\Omega, t>t_0} & |D^\alpha u|^{p-2} D^\alpha u D^\alpha \left((t-t_0)^m u\right) dx \\ & + \int_{\omega, t>t_0} (t-t_0)^m |u|^{1+\sigma} dx = 0.\end{aligned} \tag{6.18}$$

Let us calculate $D^\alpha((t-t_0)_+^m u)$. Adopt the notation

$$\begin{aligned} D_t &= \frac{\partial}{\partial t}, \; D^\alpha = D_t^j D_y^\beta, \; |\beta| = |\alpha| - j; \\ \sum_{|\alpha|=m} &= \sum_{j=0}^{m} \sum_{|\beta|=m-j} . \end{aligned} \tag{6.19}$$

It is easy to see that

$$D^\alpha \left((t-t_0)^m u\right) = (t-t_0)^m D^\alpha u + \sum_{i=1}^{\alpha} a_{ijm} (t-t_0)^{m-i} D_t^{j-i} D_y^\beta u \tag{6.20}$$

with some constants a_{ijm}. Besides, since $|\beta| + j = m$, then

$$\left| D_t^{j-i} D_y^\beta u(x) \right| \leq \left| D^{m-i} u(x) \right| . \tag{6.21}$$

It follows from (6.18), (6.20), (6.21), and the Hölder inequality that

$$\begin{aligned} E_m(t_0) &= \int_\Omega (t-t_0)_+^m \left(|D^m u|^p + |u|^{1+\sigma} \right) dx \\ &\leq C_m \sum_{i=1}^m \int_\Omega (t-t_0)_+^{m-i} |D^m u|^{p-1} |D^{m-i} u| \, dx \\ &\leq C_m \sum_{i=1}^m \left(\int_\Omega (t-t_0)_+^{m-i} |D^m u|^p dx \right)^{1/p'} \\ &\quad \times \left(\int_\Omega (t-t_0)_+^{m-i} |D^{m-i} u|^p dx \right)^{1/p} \end{aligned} \tag{6.22}$$

with some constants C_m depending only on m.

Let us make use of the following assertions.

Lemma 6.1 ([74, 72]). *Let* $H = \{x \in R^N : t = x_N > 0\}$ *and* l, j, k *be integers,* $l \leq 1, \quad 0 \leq j < l, \quad 0 \leq k$. *Then*

$$\left(\int_H t^k |D^j u|^p dx \right)^{1/p} \leq C \left(\int_H t^k |D^l u|^p dx \right)^{(1-\theta)/p} \tag{6.23}$$

if the integrals on the right-hand side exist. Here $1 < p < \infty$, $1 \le r \le p$,

$$1/p = j/(N+k) + \theta(1/p - l/(N+k)) + (1-\theta)1/r \tag{6.24}$$

and the constant C *depends only on* N, p, l, j, k, θ.

Lemma 6.2 ([74, 72]). *Let* $\Omega \in R^N$ *be a bounded domain,* $u \in W_0^{l,p}(\Omega)$, *and* $\Omega_+ = \{x \in \Omega : t = x_N > 0\}$, $meas\Omega_+ > 0$. *Let* l, j, k *be integers,* $l \le 1$, $0 \le j < l$, $0 \le k$. *Then*

$$\left(\int_{\Omega_+} t^k |D^j u|^p dx\right)^{1/p} \le C\left(\int_{\Omega_+} t^k |D^l u|^p dx\right)^{(1-\theta)/p} \times \left(\int_{\Omega_+} t^k |u|^r dx\right)^{(1-\theta)/r} \tag{6.25}$$

if the integrals on the right-hand side exist. The constants N, p, l, r, j, k *satisfy the conditions of Lemma* 6.1.

Applying Lemmas 6.1 and 6.2 (see also Lemmas 8.4 and 8.5 in Chapter 3), we arrive at the inequality

$$\begin{aligned}\int_\Omega (t-t_0)_+^\tau |D^j u|^p dx &\le C\left(\int_\Omega (t-t_0)_+^\tau |D^m u|^p dx\right)^{\theta/p} \\ &\quad\times \left(\int_\Omega (t-t_0)_+^\tau |u|^{1+\sigma} dx\right)^{(1-\theta)/(1+\sigma)} \\ &= C\,(K_\tau(t_0))^{\theta/p}\,(\Pi_\tau(t_0))^{(1-\theta)/(1+\sigma)},\end{aligned} \tag{6.26}$$

where $m \ge 1$, $0 \le j \le m$, $\tau > 0$, and

$$\theta_\tau = \frac{(N+\tau)(p-1-\sigma) + jp(1+\sigma)}{(N+\tau)(p-1-\sigma) + mp(+1+\sigma)}.$$

The constant C depends only on N, m, j, τ, p, σ. To estimate the terms on the right-hand side of (6.22) we make use of (6.26) with $\tau = j = m - i$:

$$\begin{aligned}E_m(t_0) &\le C\sum_{i=1}^m (K_{m-i}(t_0))^{1/p'+\theta_i/p} \cdot (\Pi_{m-i}(t_0))^{(1-\theta_i/(1+\sigma))} \\ &\le C\sum_{i=1}^m (E_{m-i}(t_0))^{\lambda_i},\end{aligned} \tag{6.27}$$

where $C = C(N, m, p, \sigma)$,

$$\begin{aligned}\lambda_i &= \frac{1}{p'} + \frac{\theta_i}{p} + \frac{1-\theta_i}{p} = \left(1 + \frac{i}{N+m-i+\frac{pm(1+\sigma)}{p-1-\sigma}}\right) > 1, \\ \theta_i &= \frac{(N+m-i)(p-1-\sigma) + p(m-i)(1+\sigma)}{(N+m-i)(p-1-\sigma) + mp(1+\sigma)}.\end{aligned} \tag{6.28}$$

□

Step 2: The ordinary differential inequality. Set $z(t) = E_m(t)$. Evidently,

$$\begin{cases} z'(t) = D_t z = -mE_{m-1}(t) < 0, \\ z^{(k)}(t) = D_t^k = (-1)^k \frac{m!}{(m-k)!} E_{m-k}(t), \quad k = 1, \dots, m. \end{cases}$$

Moreover, due to (6.27)–(6.28), $z(t)$ satisfies the ordinary differential inequality

$$z(t) \le C \sum_{i=1}^{m} |z^{(i)}|^{\lambda}, \quad C = C(N, m, p, \sigma). \tag{6.29}$$

Using (6.28), we can easily check that

$$\frac{1}{\lambda_i} = \frac{i}{m} \frac{1}{\lambda_m} + \frac{m-i}{m}, \quad i = 1, \dots, m.$$

Let us prove now the following useful auxiliary assertion.

Lemma 6.3 ([72]). *Assume that a nonnegative function*

$$z(t) \in C^m(R^+) \cap L^\infty(R^+), \quad m \ge 1, \quad z^{(m)}(t) \le 0$$

satisfies the inequality

$$z(t) \le C \sum_{i=1}^{m} |z^{(m)}(t)|^{\lambda_i}, \tag{6.30}$$

with positive constants C, λ_i,

$$\frac{1}{\lambda_i} = \frac{i}{m} \frac{1}{\lambda_m} + \frac{m-i}{m}, \quad i = 1, \dots, m, \quad \lambda_m > 1. \tag{6.31}$$

Then $\operatorname{supp} z(t) \subset [0, t_0]$, *where*

$$t_0 \le C_0 |z^{(m)}(0)|^{\gamma/(1-\gamma)} \tag{6.32}$$

and

$$\frac{1}{1-\gamma} = \left(1 + \frac{\lambda m - 1}{m}\right) > 1, \quad C_0 = C_0(C, m, \lambda_1, \dots, \lambda_m).$$

Proof. It follows from the conditions of the lemma that $z^{(i)}(t) \to 0$ as $t \to \infty$, $i = 1, \dots, m$. We will make use of the interpolation inequality

$$\|u\|_{W^{k,q}(\Omega)} \le C \|u\|^{\theta}_{W^{l,q}(\Omega)} \|u\|^{1-\theta}_{L^r(\Omega)}, \tag{6.33}$$

which holds for $\Omega \subseteq R^N$, $u \in W^{l,p}(\Omega) \cap L^r(\Omega)$, $1 \le l$, $1 \le r < \infty$, $1 < p < \infty$ with

$$\theta = \frac{q(kr + N)Nr}{p(N + rl) - Nr} \frac{p}{q}, \quad \frac{k}{l} \le \theta \le 1$$

and the constant C depending only on $l, p, r, k, q, \theta, \Omega$. The constant C may be unbounded if $\Omega = R^N$. Applying (6.33) with $q = p = r = \infty$, $k = i$, $l = m$, $r = (m-i)/m$, $\Omega = R^+$ to the functions

$$|z^{(i)}(t)| = \|z^{(i)}(t)\|_{L^\infty(t,\infty)},$$

we get

$$|z^{(i)}(t)| \leq C|z^{(m)}(t)|^{i/m}|z(t)|^{(m-i)/m}. \tag{6.34}$$

With the use of (6.34), inequality (6.30) can be transformed as follows:

$$|z(t)| \leq C\sum_{i=1}^{m-1}|z^{(m)}(t)|^{\lambda_m i\lambda_i/(\lambda_m m)}|z(t)|^{\lambda_m(m-i)/m} + |z^{(m)}(t)|^{\lambda_m}. \tag{6.35}$$

On the other hand, it follows from (6.31) that

$$\frac{i\lambda_i}{m\lambda_m} + \frac{\lambda_i(m-i)}{m} = 1,$$

and by the Young inequality

$$\begin{aligned} |z(t)| &\leq C|z^{(m)}(t)|^{i\lambda_i/m}|z(t)|^{\lambda_m(m-i)/m} \\ &\leq C(\varepsilon, i)|z^{(m)}(t)|^{\lambda_m} + \varepsilon|z(t)|, \quad \varepsilon > 0. \end{aligned} \tag{6.36}$$

Gathering (6.35) and (6.36), we now have that

$$|z(t)| \leq C|z^{(m)}(t)|^{\lambda_m}. \tag{6.37}$$

Substituting (6.37) into (6.34) with $i = m-1$, we obtain the inequality

$$|z^{(m-1)}(t)| \leq C|z^{(m)}(t)|^{1/(1-\gamma)},\ 1-\gamma(1+\frac{\lambda_m - 1}{m}) > 1, \tag{6.38}$$

or

$$y' + C^{\gamma-1}y^{1-\gamma} \leq 0,\ y(t) = |z^{(m-1)}(t)|. \tag{6.39}$$

Then

$$|z^{(m-1)}(t)|^\gamma \leq |z^{(m-1)}(0)|^\gamma - \gamma C^{\gamma-1}t.$$

But (6.34) with $i = m-1$ and (6.37) imply the inequality

$$|z^{(m-1)}(0)| \leq C|z^{(m)}(0)|^{1/(1-\gamma)}.$$

Hence $z^{(m-1)}(t) = 0$ provided that

$$t \geq a_0 = C|z^{(m)}(0)|^{\gamma/(1-\gamma)}.$$ □

Lemma 6.4. *Let $z(t)$ be a nonnegative function such that $z \in C^m(R^+) \cap L^\infty(R^+)$ with $m \geq 1$ and $z^{(m)}(t) \leq 0$. Assume that $z(t)$ satisfies the inequality*

$$z(t) \leq C \left(\sum_{i=1}^{m} |z^{(i)}(t)|^{\lambda_i} \right) + \varepsilon(\rho_0 - t)_+^{\mu} \tag{6.40}$$

where C, λ_i, ε, μ, ρ_0 are positive constants satisfying the conditions

$$\frac{1}{\lambda_i} = \frac{i}{m} \frac{1}{\lambda_m} + \frac{m-i}{m}, \qquad \mu = \frac{m\lambda_m}{\lambda_m - 1} > 0. \tag{6.41}$$

Then there exist $E_* > 0$, $\varepsilon > 0$ such that

$$\operatorname{supp} z(t) \subset [0, \rho_0] \quad \textit{if } z^{(m)}(0) \leq E_*, \ \varepsilon \leq \varepsilon_*.$$

Proof. Repeating estimates (6.34)–(6.37), we arrive at the inequality (cf. (6.38))

$$Cy^{1-\gamma} + y' \leq \varepsilon^{1/m} (\rho_0 - t)_+^{(1-\gamma)/\gamma},$$

where

$$y(t) = z^{(m-1)}(t), \quad y' = z^{(m)} \geq 0,$$

and

$$y(0) = z^{(m-1)}(0) \leq C|z^{(m)}(0)|^{1/(1-\gamma)}.$$

By Lemma 3.2 and Remark 2.5 of the appendix,

$$\operatorname{supp} z(t) \subset [0, a_0], \quad a_0 \leq C|z^{(m)}(0)|^{\gamma/(1-\gamma)} = C|m! E_0(0)|^{\gamma/(1-\gamma)}$$

with $C = C(m, \lambda_1, \ldots, \lambda_m) \equiv C(N, m, p, \sigma)$. □

Remark 6.1. Localization of the solution $u(x)$ obviously follows from the assertion of Theorem 6.1 if

$$d_0 = \max_{\overline{\Omega}} x_N > a_0.$$

The last condition is surely fulfilled if, for instance, the energy $E_0(0)$ is sufficiently small.

Remark 6.2. Let $u \in W_0^{m,p}(\Omega)$ be a weak solution of equation (6.1) in the domain Ω and

$$\operatorname{supp} f(x) \subset K_{\rho_0} = \{x \in \Omega : |x_i| < \rho_0, \ i = 1, \ldots, N\},$$
$$K_{\rho_0} \subset \Omega, \quad \operatorname{dist}\left(\partial K_{\rho_0}, \partial\Omega\right) > 0.$$

Then there exists $E_* > 0$ such that once

$$E = \int_\Omega \left(|D^m u|^p + |u^{1+\sigma}| \right) dx \le E_*,$$

we have that

$$\operatorname{supp} \subset K_{\rho_1} \subset \Omega$$

with some $\rho_1(E_*) > \rho_0$. To prove this assertion we sequentially apply the arguments used to prove Theorem 6.1 for $t = x_1, \dots, t = x_N$, and with $t_0 \ge \rho_0$. The value of E_* has to be taken so that conditions (6.16) and (6.17) and the conditions of Remark 6.1 are fulfilled.

Remark 6.3. The constants in (6.11), as well as the constants which appeared in the proof of the last theorem, do not depend on Ω. Due to this observation the method is applicable to those solutions of equation (6.1) which have finite energy on the whole of $\mathbb{R}^N$. So, if we assume that

$$\operatorname{supp} f \subset B_{\rho_0}(0), \quad 0 < \rho_0 < \infty,$$

then

$$\operatorname{supp} u \subset B_\rho(0), \quad \rho < \rho_1(E) < \infty.$$

Remark 6.4. In the case $\Omega = \mathbb{R}^N$, inequality (6.28) with

$$\lambda_m = 1 + \frac{m(p-1-\sigma)}{N(p-1-\sigma) + pm(1+\sigma)} \le 1, \quad p \le 1+\sigma,$$

implies that $z^{(i)}(t)$ grows as $t \to \infty$ exponentially or like a power. The complete study of these questions is presented in [72].

Remark 6.5. The assertion of the last theorem applies to equations of the form

$$Lu = \sum_{|\alpha| \le m} (-1)^{|\alpha|} D^\alpha \left(a_\alpha(x,u) |D^\alpha u|^{p-2} D^\alpha u \right) + B(x,u) = f(x),$$

where

$$0 \le a_\alpha(x,r) \le C \quad \forall \alpha \qquad \text{and} \qquad \frac{1}{C} \le a_\alpha \le C \quad \text{for } |\alpha| = m,$$

$$\frac{1}{C}|r|^{1+\sigma} \le B(x,r)r \le C|r|^{1+\sigma} \quad \forall r \in \mathbb{R} \quad \text{and a.e. } x \in \Omega.$$

Let us assume now that

$$0 < \rho_0 < d_0 = \max_{\overline{\Omega}} x_N, \qquad \operatorname{supp} f \subset \Omega \cap \{t < \rho_0\}$$

and

$$\int_\Omega (t-t_0)_+^m |f(x)|^{(1+\sigma)/\sigma}\,dx \le \varepsilon(\rho_0-t_0)_+^\mu, \quad t_0 \ge 0, \tag{6.42}$$
$$\gamma = \frac{p-1-\sigma}{(N+1)(p-1-\sigma)+pm}.$$

Theorem 6.2. *Assume that* (6.2), (6.42) *are fulfilled and* $u \in W_0^{m,p}(\Omega)$ *is a weak solution of equation* (6.1). *Then there exist two positive constants* E_* *and* ε_* *such that*

$$\operatorname{supp} u \subset \Omega \cap \{t < \rho_0\} \quad \textit{if } E_0 \le E_*,\ \varepsilon \le \varepsilon_*.$$

Proof. Following the proof of Theorem 6.1, let us set in (6.6) $\varphi = w$. Then we get a generalization of (6.18),

$$\Lambda = \int_{\Omega,\, t>t_0} (t-t_0)_+^m f u\,dx, \quad t_0 \ge 0. \tag{6.43}$$

By (6.42) and by virtue of Young's inequality, for every $\delta > 0$,

$$\begin{aligned} \left|\int_\Omega (t-t_0)_+^m f u\,dx\right| &\le C(\delta)\int_\Omega (t-t_0)_+^m |f|^{(1+\sigma)/\sigma}|f|^{(1+\sigma)/\sigma}\,dx \\ &\quad + \delta\int_\Omega (t-t_0)_+^m |u|^{1+\sigma}\,dx \\ &\le C(\delta)\varepsilon(\rho_0-t_0)_+^\mu + \delta\int_\Omega (t-t_0)_+^m|u|^{1+\sigma}\,dx. \end{aligned} \tag{6.44}$$

Arguing as in the proof of the previous theorem, we estimate Λ from below and come to the inequalities

$$|z(t_0)| \le C\left(\sum_{i=1}^m |z^{(i)}(t_0)|^{\lambda_i} + \varepsilon(\rho_0-t_0)_+^\mu\right), \tag{6.45}$$

$$|z^{(m-1)}(t)| \le C|z^{(m)}|^{1/(1-\gamma)} + \varepsilon^{1/(1+\gamma(m-1))}(\rho_0-t_0)_+^{1/\gamma}, \tag{6.46}$$

(with $z^m(t) = (-1)^m m! E_0(t)$), which yield the relations

$$Cy^{1-\gamma} + y' \le \varepsilon^\lambda(\rho_0-t_0)_+^{(1-\gamma)/\gamma}, \quad \lambda = \frac{1-\gamma}{1+\gamma(m-1)},$$
$$y(t_0) = |z^{(m-1)}(t_0)|, \quad y' \le 0, \quad t_0 \ge 0. \tag{6.47}$$

As we have seen, the homogeneous equation corresponding to (6.47) has the solution

$$Y(t_0) = y(0)\rho_0^{-1/\gamma}(\rho_0-t_0)_+^{1/\gamma}, \quad y(t_0) \le Y(t_0),$$

provided that

$$\rho^{-(1-\gamma)/\gamma} C y_0^{1-\gamma} \left(1 - \frac{y_0^\gamma}{C\rho_0}\right) = \varepsilon^\lambda.$$

The fulfillment of this condition is guaranteed by an appropriate choice of y_0 and ε. □

Remark 6.6. The results of this section are simply examples of the possibilities of the modification of the method which was proposed and developed in [73, 74] for the higher-order equations. In fact, this modification can be applied to study more complicated equations or to consider problems posed in domains of sophisticated shape (cf. the remarks in Section 2).

6.2 Radial energy sets. We proceed to sketch the proof of the localization property for solutions of equation (6.1) using the energy method with the radial energy set. The details of the proof can be found in [80]. Now the solution u may have nonhomogeneous boundary values.

Let B_R be a ball of radius R and with a fixed center, $0 < R < R_0$. We assume that

$$B_{R_0} \subset \Omega \setminus \operatorname{supp} f.$$

Throughout this section the domain of integration is always the ball B_R.

We introduce the notation (recall that $q = a + 1$)

$$J_s(R) \equiv \int (R - |x|)^s |D^2 u|^2 \, dx + \int (R - |x|)^s |u|^q \, dx.$$

We need the following *Caccioppoli-type inequality*, whose proof can be found in [78]. If $s \geq 2m$, then

$$\begin{aligned} \int (R - |x|)^s u \, (-\Delta)^m u \, dx \geq (1 - \varepsilon) \int (R - |x|)^s |D^m u|^2 \, dx \\ - C \int (R - |x|)^{s-2m} u^2 \, dx, \end{aligned} \tag{6.48}$$

where the constant C is independent of R.

Applying (6.48) to equation (6.1) with $m = 2$ and $s = 4$, we obtain

$$J_4(R) \leq C \int u^2 \, dx. \tag{6.49}$$

Notice that $J_4'(R) = 4J_3(R)$.

Next we need the Gagliardo–Nirenberg inequality (see [263]) for the ball B_R,

$$\begin{aligned} C \left(\int u^2 \, dx \right)^{1/2} \leq \left(\int |D^2 u|^2 \, dx \right)^{\theta/2} \left(\int |u|^q \, dx \right)^{(1-\theta)/q} \\ + R^{-\beta} \left(\int |u|^q \, dx \right)^{1/q}, \end{aligned} \tag{6.50}$$

where C is a positive constant independent of R, while θ and β are defined by

$$\frac{1}{2} = \theta\left(\frac{1}{2} - \frac{2}{n}\right) + \frac{1-\theta}{q}, \qquad \beta = \frac{n}{q} - \frac{n}{2} > 0$$

and θ satisfies $0 \leq \theta \leq 1$.

Taking squares in (6.50), using the numerical inequality $A^a B^b \leq C(A+B)^{a+b}$ and observing that $1/q > \theta/2 + (1-\theta)/q$, it follows that

$$C \int u^2 \leq R^{-2\beta} K(R)(J_0(R))^{\theta+2(1-\theta)/q}, \tag{6.51}$$

where

$$K(R) \equiv R^{2\beta} + \left(\int |u|^q\right)^{\theta(2-q)/q}. \tag{6.52}$$

Notice that the exponents in (6.52) are positive.

From (6.49) and (6.51), we obtain

$$J_4(R) \leq C R^{-2\beta} K(R_0)(J_0(R))^{\theta+2(1-\theta)/q}. \tag{6.53}$$

Since

$$J_0(R) = \frac{1}{4!}\frac{d^4 J_4(R)}{dR^4},$$

(6.53) is a fourth-order differential inequality. Since the integrands of J_s are non-negative, it follows from Hölder's inequality that

$$J_1 \leq (J_4)^{1/4}(J_0)^{3/4}.$$

This and (6.53) imply that

$$J_1(R) \leq C R^{-\beta/2} K(R_0)^{1/4} J_0(R)^{\mu}, \tag{6.54}$$

where

$$\mu = \frac{3}{4} + \frac{1}{4}\left(\theta + \frac{2}{q}(1-\theta)\right) > 1.$$

Relation (6.54) is a first-order differential inequality in the variable R, since $J_0(R) = J_1'(R)$.

The study of the ordinary differential inequality is based on the next lemma, which is proved by explicit integration.

Lemma 6.5. *Let* $0 < \lambda < 1$, $s > 0$, $A > 0$, $y \in C^1$, $y(x) \geq 0$, *and*

$$x^s y(x)^\lambda \leq A y'(x) \quad \text{if } 0 \leq x \leq x_0. \tag{6.55}$$

If $x_1 > 0$, *then* $y(x) = 0$ *for* $0 \leq x \leq x_1$, *where* x_1 *is defined by*

$$x_1^{s+1} = x_0^{s+1} - \frac{s+1}{1-\lambda} A \left(\frac{A y'(x_0)}{x_0^s}\right)^{(1-\lambda)/\lambda}. \tag{6.56}$$

Remark 6.7. The function $y'(x)$ is going to correspond to the usual energy $J_0(R)$, while $y(x)$ corresponds to the weighted energy $J_1(R)$. That is why we present formula (6.56), which involves y' rather than y.

We apply (6.55), (6.56) to (6.54) with

$$\lambda = 1/\mu \quad s = \beta\lambda/2 \quad A = CK(R_0)^{\lambda/4}.$$

In this way, we obtain that

$$u \equiv 0 \quad \text{in } B_R \quad \text{if } R < R_1,$$

where R_1 is defined by

$$R_1^{s+1} = R_0^{s+1} - CK(R_0)^{\lambda/4} \left(\frac{K(R_0)^{\lambda/4} J_0(R_0)}{R_0^s} \right)^{(1-\lambda)/\lambda} . \tag{6.57}$$

We can assure that the solution is localized (i.e., that $R_1 > 0$) in the following two situations.

1. Small energy. $R_1 > 0$ if $J_0(R_0)$ is small enough.

2. Large domains. If the energy is kept bounded, then

$$R_1^{s+1} \quad \text{behaves as} \quad R_0^{s+1} - CR_0^s$$

and hence $R_1 > 0$ for R_0 large enough.

7 Bibliographical notes and open problems

The idea of the local energy method in the form we mostly follow in this book was presented in 1979–1981 in the papers [11, 12, 13, 14, 15] by Antontsev.

The principal Theorem 2.1 is taken from Díaz and Veron [149]. The optimality of the main assumption $\sigma < p - 1$ in this theorem was analyzed by Vazquez in [305] (see also Díaz, Saa, and Thiel [145] and Pucci, Serrin, and Zou [273] for the study of more general diffusion operators). The property of *nondiffusion of the support* given in Theorem 2.2 follows Antontsev and Díaz [26]. This property was first found in Díaz [128] via *local super and subsolutions technique*. Arguments of this sort, gathered with an implicit discretization technique, allow one to analyze the *waiting time effects* for solutions of parabolic problems. See Alvarez and Díaz [5]; see also Alvarez [4] for the study of optimality of the assumptions.

The results of Section 3 are the elliptic version of results of Antontsev [21]. Problems leading to weighted diffusion or/and absorption terms often appear in applications: see Díaz [129], Galdi and Rionero [179], Ivanov [196], Kufner [231], and the references therein.

The global result on the location of the null set $\mathcal{N}(u)$ given in Theorem 3.2 is similar to that obtained via the method of local super and subsolutions (when available); see Díaz [128, Theorem 1.16] and its references. One of the main differences between the methods is that in the energy method the estimates are independent of the L^∞ norm of the solutions and the method is thus applicable to unbounded solutions with a finite total energy. When the domain Ω is unbounded but the data f, g and h are compactly supported, Theorem 3.2 yields compactness of the support of any solution with finite total energy. In fact, the study of solutions with compact support was developed prior to the study of locally vanishing solutions. Haim Brezis initiated the development of the study of solutions with compact support on unbounded domains in his pioneering paper [96] (see also [95]). His motivation came from the study of some concrete obstacle problems arising from the subsonic flows (see Remark 3.3 in Chapter 4). The result of Brezis was then extended to semilinear and quasilinear equations (see Benilan, Brezis, and Crandall [66], Díaz [128], Díaz and Herrero [137, 138], Martinson and Pavlov [250], Redheffer [275], and the references therein on the higher-order equations).

A curious behavior can be observed in solutions of equations with inhomogeneous nonlinearities satisfying the conditions of Theorem 3.2 with r and $\mathbf{q}$ small enough, and under the assumption that the diffusion-absorption balance for large values of r and $\mathbf{q}$ is the contrary (i.e., $\sigma > p - 1$). In that case it can be shown (see Bernis [78] and Díaz and Oleinik [142]) that the total energy is uniformly bounded, independently of the data, and the same happens to the size of the solution support. (A similar result was stated in Díaz [128], and G. Díaz and Letelier [123] for suitable special cases of equation (2.1) with the help of the comparison principle). It is important to remark that local vanishing of solutions of equations like (2.1) is also relevant to the study of singularities in solutions of certain elliptic equations. So, Brezis and Nirenberg [101] use the transformation $u = e^{-v}$ to study the singularity of solutions to the equation

$$-\Delta v + |\nabla v|^2 = h^2(v)$$

for suitable functions $h^2(v)$.

The results of Section 3 on cylinder-like domains $\Omega = G \times (0, \infty)$ and layer-like domains $\Omega = R^N \times (0, 1)$ generalize previous results of Antontsev [16]. We point out that there is a very large literature devoted to the study of the decay of solutions to linear and nonlinear elliptic equations in cylinder-like domains. See, for example, the references to the Saint-Venant principle in the books by Flavin and Rionero [163] and Oleinik and Yosifian [265], as well as in Diaz and Quintanilla [143] and Levine and Quintanilla [241]; see also Berestycki, Caffarelli, and Nirenberg [69] for a free-boundary problem arising in combustion theory.

The existence and uniqueness of weak solutions of nonlinear equations with anisotropic nonlinearities have been treated in the literature by different authors (see, e.g., Lions [243], Attouch and Damlamian [58], and many others). Some results on the existence and location of the free boundary obtained via the comparison principle are due to Díaz and Herrero [138, 137] and Díaz [128] (see also the energy method used in Rykov [281] for the treatment of some anisotropic parabolic problems: his method could be applied to the associated stationary equations).

The one-directional phenomena discussed in Subsection 4.2 are new in the literature. Phenomena of this type are typical for first-order hyperbolic equations (see Díaz and Veron [148]). A previous application of the energy method to some diffusion-convection equations is due to Antontsev [49, 15, 17]. For the free boundaries occurring in solutions of a diffusion-convection equation of Hamilton–Jacobi type see Barles, G. Díaz, and J. I. Díaz [65].

The results on the localization of solutions to systems of nonlinear equations presented in Section 4 are published for the first time. The arguments of this kind are very flexible and could be applied to other systems such as, for example, the systems occurring in the study of the minimum action solutions of some vector field equations (see Brezis and Lieb [100]) and in the study of microstructure of ordered solids (Kinderlehrer and Pedregal [222]), among others. A detailed treatment of these systems will be presented by the authors elsewhere. We also mention that in some systems the free boundary can be generated by a sole unknown function from a set of unknowns. This is what happens in certain systems arising in combustion theory, and what can be analyzed by analogy with the study of the x-dependent nonlinearities in Section 4 (see also Díaz and Hernandez [134] and Pozio and Tesei [271] for an approach based on the comparison argument).

The exposition of Section 6 mostly follows Bernis [73, 80]. Some pioneering results on the compactness of the support of the solution of some one-dimensional fourth-order problem are due to Berkovitz and Pollard [70], Redheffer [274], Hestenes and Redheffer [194], and Bernis [71] (for the associated radial problem, see Bidaut-Veron [84, 85, 86]).

We point out that a totally different energy method allowing the consideration of higher-order equations was introduced by A. Shishkov and collaborators in the series of papers [220, 288, 289]. The main idea of this energy method consists in getting some (nondifferential) inequalities which link different norms of the solution and then deduce some estimates on the null-set of u from a nonlinear implicit inequality of the type $h(s+\Lambda h^{\alpha}(s)) \geq \omega h(s)$ on a suitable energy function $h(s)$, for some $\alpha > 0$ and ω, $\Lambda < 1$ (see also [209]).

As we said in the introduction to this book, there are several alternatives to the general energy method presented in this Chapter (see the comments on the higher-order equations in Section 5). Another general remark concerns the application of those energy methods to the study of free boundaries. For example, if the diffusion operator is degenerate as in (2.1) with $p > 2$, under appropriate conditions on the terms $C(x, u)$, one can study the location of the boundaries of the sets $\{x \in \Omega : u(x) = k\}$ for some constants $k \neq 0$; see Barles, G. Díaz, and J. I. Díaz [122], Díaz [128], Díaz and Kichenassamy (an unpublished manuscript), Kichenassamy and Smoller [221], Guedda and Veron [187], Kamin and Veron [210], Lumer, Redheffer, and Walter [246], García-Melián and Sabina de Lis [181], and Arcoya and Callahorrano [55].

There are many different open problems related to this first chapter. Some of them have been suggested in this section. Some others are in order.

1. Is it possible to introduce an energy method for stationary fully nonlinear equations (i.e., of nondivergent form), as, for instance, the Monge–Ampere equations, in order to study free-boundary properties? The existence of a free boundary can be proved by using comparison arguments: see Díaz [128, Sub-

section 2.4b]; see the energy approach followed in Flavin and Rionero [163, Section 9.2] for the study of the decay of the solution on rectangular domains.

2. It would be interesting to extend the energy methods of this chapter to solutions of the type *very weak solutions* (i.e., when they are not in the natural energy space but in a larger functional space). This situation occurs when the data are not in the dual of the energy but in other spaces as L^1, the space of bounded measures, etc. A study of the free-boundary properties by means of local and global super and subsolutions is available in the literature (see Benilan, Brezis, and Crandall [66] and Díaz [128] and its references). Notice that the problem is trivial once we know that on the null set of the data the solution locally belongs to the energy space.

3. To apply an energy method to the study of the free-boundary properties in a discrete elliptic problems. The interest in the numerical analysis of the problems is obvious. For an approach via the discrete maximum principle, see Garroni and Vivaldi [182].

4. How could the energy method be applied to the study of elliptic equations with nonpower nonisotropic nonlinearities in the presence of nonpower absorption terms (see Chapter 2, Section 3)?

As the last remark, let us mention that various results of this chapter can be extended to associated variational inequalities (which, for instance, formally correspond to the assumption that in Section 2 the exponent $\sigma = 0$), see also the above references on the one-dimensional fourth-order problem. The study of exponents $\sigma < 0$ is sometimes possible (see Bernis [73], Díaz [128, Section 2.3], [59], Levine [240], Deng [120], and their references).

2

Stabilization in a Finite Time to a Stationary State

1 Introduction

In this chapter the way of using the energy method is different from that of Chapter 1. Our aim is to study the property of finite-time stabilization to a stationary profile for solutions to nonlinear evolution problems. To be precise, let $\Omega \subset \mathbb{R}^N$, $N \geq 1$, be an open set (which need be neither bounded nor connected). Denote $Q_\infty = \Omega \times \mathbb{R}_+$, $\Sigma_\infty = \partial\Omega \times \mathbb{R}_+$. To fix ideas, let us consider the general initial and boundary-value problem

$$\begin{cases} u_t + A(u) = f(x,t) & \text{in } Q_\infty, \\ B(u) = g(x,t) & \text{on } \Sigma_\infty, \\ u(x,0) = u_0(x) & \text{on } \Omega \end{cases} \tag{1.1}$$

where $A(u)$ is a differential operator on u in the space variables x, $B(u)$ is the boundary operator, and f, g, u_0 are given functions. Our approach is applicable to the vector-valued solutions $\mathbf{u}$ as well.

The question of possible stabilization as $t \to \infty$ of a solution of this problem to a time-independent state is of significant interest. It is usually assumed that

$$f(x,t) \to f_\infty(x) \quad \text{and} \quad g(x,t) \to g_\infty(x) \quad \text{as } t \to \infty$$

in some sense and a natural aspiration is that

$$u(x,t) \to u_\infty(x) \quad \text{as } t \to \infty$$

in the norm of a suitable function space with the limit function u_∞ being a solution of the stationary problem

$$\begin{cases} A_\infty(u_\infty) = f_\infty(x) & \text{in } \Omega, \\ B_\infty(u_\infty) = g_\infty(x) & \text{on } \partial\Omega. \end{cases} \tag{1.2}$$

The operators A_∞, B_∞ stand for the limits of the operators A and B correspondingly (see, e.g., [131] and the references therein).

We will be interested in the stronger property due to the nonlinear nature of equations under study. To begin with, let us assume that

$$\begin{cases} f(x,t) = f_\infty(x) & \text{for } t \geq T_f, \\ g(x,t) = g_\infty(x) & \text{for } t \geq T_g. \end{cases} \tag{1.3}$$

Definition 1.1. Let $u(x,t)$ be a solution of the initial and boundary-value problem (1.1). We say that $u(x,t)$ *stabilizes in a finite time* to a stationary state $u_\infty(x)$ if there exists $t^* \in (0,\infty)$ such that

$$\forall t \geq t^*, \quad u(x,t) \equiv u_\infty(x) \quad \text{on } \Omega.$$

Introducing the new unknown function $v(x,t) \equiv u(x,t) - u_\infty(x)$ we can write problem (1.1) as

$$\begin{cases} v_t + \widetilde{A}(v) = \widetilde{f}(x,t) & \text{in } Q_\infty, \\ \widetilde{B}(v) = \widetilde{g}(x,t) & \text{on } \Sigma_\infty, \\ v(x,0) = u_0(x) - u_\infty(x) & \text{on } \Omega \end{cases}$$

with

$$\widetilde{A}(v) = A(v + u_\infty) - A_\infty(u_\infty), \qquad \widetilde{B}(v) = B(v + u_\infty) - B_\infty(u_\infty)$$

and

$$\widetilde{f}(x,t) = f(x,t) - f_\infty, \qquad \widetilde{g}(x,t) = g(x,t) - g_\infty.$$

We therefore arrive at a problem similar to (1.1) with

$$f_\infty(x) \equiv 0, \qquad g_\infty(x) = 0, \qquad v_\infty \equiv 0. \tag{1.4}$$

In many physically reasonable cases conditions (1.4) are fulfilled.

Definition 1.2. Let $u(x,t)$ be a solution of problem (1.1) and let conditions (1.4) be fulfilled. We say that $u(x,t)$ has the property of *extinction in a finite time* if there exists $t^* \in (0,\infty)$ such that

$$u(x,t) \equiv 0 \quad \text{for } t \geq t^*.$$

Most of the material collected in this chapter is devoted to the study of the property of finite-time extinction. Subsection 7.4 contains results on the situation where the stationary state is not identically zero. The applications to problems arising from fluid mechanics will be given in Chapter 4.

1.1 Illustrative examples. Let us present the simplest example: an ordinary differential equation whose solution vanishes in a finite time. Let us consider the Cauchy problem

$$\begin{cases} \dfrac{du(t)}{dt} = -\lambda u|u|^{\sigma-1} & \text{for } t > 0, \\ u(0) = u_0 \end{cases}$$

with the parameters $\lambda > 0, \sigma \in (0, 1)$. Introduce the function $E(t) = \frac{1}{2}u^2(t) \geq 0$, which plays the role of the *energy function* and satisfies the Cauchy problem

$$\begin{cases} \dfrac{dE(t)}{dt} = -2^{(\sigma+1)/2}\lambda E^{(1+\sigma)/2} & \text{for } t > 0, \\ E(0) = E_0 \equiv \dfrac{1}{2}u_0^2. \end{cases}$$

The direct integration of the equation for $E(t)$ shows that the function $E(t)$ satisfies the inequalities

$$0 \leq E^{(1-\sigma)/2}(t) \leq \max\left\{0;\ E_0^{(1-\sigma)/2} - \lambda(1-\sigma)2^{(\sigma-1)/2}t\right\}, \quad \text{for all } t \geq 0$$

and, correspondingly, $E(t) = 0$ and $u(t) = 0$ for all $t \geq t^* \equiv \frac{u_0^{(1-\sigma)}}{\lambda(1-\sigma)}$.

The same argument shows that if a function $u(t)$ is a solution of the Cauchy problem

$$\begin{cases} \dfrac{du(t)}{dt} = -\phi(u(t)), \\ u(0) = u_0. \end{cases}$$

with a function ϕ satisfying the conditions $\phi(\tau) > 0, \phi'(\tau) \geq 0$ for all $\tau > 0$ and, moreover, $\phi(0) = 0$, then the condition

$$\int_{0+} \frac{d\tau}{\phi(\tau)} < \infty,$$

is necessary and sufficient for $u(t)$ to vanish at the instant t_* defined by

$$t_* = \int_0^{u_0} \frac{d\tau}{\phi(\tau)}.$$

Let us consider now the Cauchy problem for the nonhomogeneous ordinary differential equation

$$\begin{cases} \dfrac{du(t)}{dt} = -\lambda u|u|^{\sigma-1} + \varepsilon\left(1 - \dfrac{t}{T^*}\right)_+^{\sigma/(1-\sigma)} & \text{for } t > 0, \\ u(0) = u_0 > 0, \end{cases} \tag{1.5}$$

with some $T^* \geq t^*$, $\varepsilon > 0$, and under the standard notation $v_+ = \max(0, v)$. If the data satisfy the relation

$$\lambda u_0^\sigma = \varepsilon + \frac{u_0}{T^*(1-\sigma)},$$

then problem (1.5) admits the explicit solution

$$u(t) = u_0 \left(1 - \frac{t}{T^*}\right)_+^{1/(1-\sigma)}.$$

A simple analysis of this solution shows that if $(\lambda u_0^\sigma - \varepsilon) > 0$, then the solution of problem (1.5) vanishes at the instant

$$T^* = \frac{u_0}{(1-\sigma)(\lambda u_0^\sigma - \varepsilon)}.$$

(Notice that $T^* \downarrow t^*$ if $\varepsilon \downarrow 0$.)

Let us show, still formally and without any rigorous justification, the main idea of the method used throughout this chapter. To start with, we consider the model initial-boundary value problem for a nonlinear degenerate parabolic equation with a single space variable. Denote $Q_T = \Omega \times (0, T)$, $\Omega = (-L, L)$, $T \in \mathbb{R}_+$. Let $u(x, t)$ be a solution of the problem

$$\left(u|u|^{\gamma-1}\right)_t - \left(|u_x|^{p-2}u_x\right)_x + \lambda u|u|^{\sigma-1} = f(x,t) \quad \text{in } Q_T, \tag{1.6}$$

$$u(\pm L, t) = 0 \quad \text{for } t \in (0, T), \quad u(x, 0) = u_0(x) \quad \text{in } \Omega. \tag{1.7}$$

Notice that the equation

$$v_t - \left(\gamma^{1-p}|v|^{m-1}|v_x|^{p-2}v_x\right)_x + \lambda v|v|^{q-1} = f(x,t) \tag{1.8}$$

with the parameters

$$m = 1 + \frac{(1-\gamma)(p-1)}{\gamma}, \quad q = \frac{\sigma}{\gamma}$$

transforms into (1.6) after the change of the unknown function $v = u|u|^{\gamma-1}$.

Equation (1.8) is usually referred to as the nonlinear heat equation with absorption. If $v(x, t)$ is interpreted as the temperature in some fluid, then the first and the second terms on the right-hand side of (1.8) represent the diffusion and the volume absorption of heat. The term $f(x, t)$ models an external source or sink of heat. We will assume that the structural constants defining equation (1.6) satisfy the conditions

$$\gamma > 0, \quad \sigma > 0, \quad 1 \leq p < \infty, \quad \lambda > 0. \tag{1.9}$$

Let us also assume that the solution $u(x,t)$ of problem (1.6), (1.7) is a weak solution from a suitable function space, $V(Q_T)$, such that for almost all $t \in (0, T)$ the energy equality

$$\frac{\gamma}{1+\gamma}\frac{d}{dt}\int_{\Omega}|u|^{1+\gamma}dx + \int_{\Omega}\left(|u_x|^p + \lambda|u|^{1+\sigma} - fu\right)dx = 0 \tag{1.10}$$

holds. Equality (1.10) formally appears as the result of integration by parts of equation (1.6) multiplied by $u(x,t)$ and the use of the boundary conditions. It is natural to term the weak solutions from $V(Q_T)$ "the energy solutions."

We will show that in fact relation (1.10) is true for any energy solution of problem (1.6)–(1.7) from the space

$$V(Q_T) \equiv L^{\infty}\left(0, T; L^{1+\gamma}(\Omega)\right) \cap L^p\left(0, T; W_0^{1,p}(\Omega)\right) \cap L^{1+\sigma}(Q_T).$$

The definitions of the spaces $L^q(0, T; L^p(\Omega))$ are given in the appendix.

Let us introduce *the energy functions*

$$\begin{aligned} y(t) &= \int_{\Omega}|u(x,t)|^{1+\gamma}dx \equiv \|u(\cdot,t)\|^{1+\gamma}_{L^{1+\gamma}(\Omega)}, \\ D(t) &= \int_{\Omega}|u_x(x,t)|^{p}dx \equiv \|u_x(\cdot,t)\|^{p}_{L^{p}(\Omega)} \\ A(t) &= \int_{\Omega}|u(x,t)|^{1+\sigma}dx \equiv \|u(\cdot,t)\|^{1+\sigma}_{L^{1+\sigma}(\Omega)}. \end{aligned} \tag{1.11}$$

Note that given any function $u \in V(Q_T)$, the functions y, D, A are defined for almost all $t \in (0, T)$ and, moreover, $A, D \in L^1(0, T)$.

With this notation the energy equality (1.10) takes on the form

$$\frac{\gamma}{1+\gamma}\frac{dy}{dt} + D(t) + \lambda A(t) = \int_{\Omega} fu\, dx. \tag{1.12}$$

Let us begin with the simple case when in (1.6)

$$\lambda = 0, \quad f(x,t) \equiv 0.$$

According to the Newton–Leibnitz formula

$$|u|^m u = (m+1)\int_{-1}^{x}|u|^m u_x\, dx. \tag{1.13}$$

Letting $m = 0$ and applying Hölder's inequality we get from (1.13) that

$$\frac{\gamma C}{1+\gamma}\, y^{\nu}(t) \leq D(t) \tag{1.14}$$

with the constants

$$C = (1+\gamma)\,(\text{meas}\,\Omega)^{-(p-1+\nu)}, \quad \nu = \frac{p}{1+\gamma}.$$

Gathering (1.12) and (1.14) we obtain the ordinary differential inequality

$$y' + C\, y^{\nu} \le 0. \tag{1.15}$$

Integrating (1.15) and assuming that

$$\nu = \frac{p}{\gamma + 1} < 1, \tag{1.16}$$

we arrive at the following estimate for the energy function $y(t)$: for all $t \ge 0$

$$0 \le y^{1-\nu}(t) \le \Big[0;\ y^{1-\nu}(0) - (1-\nu)Ct\Big]_+ . \tag{1.17}$$

Hence,

$$y(t) = 0 \qquad \Longrightarrow \qquad u(x,t) = 0 \text{ for all } x \in \Omega,\ t \ge t^* \tag{1.18}$$

with

$$t^* = \frac{y^{1-\nu}(0)}{(1-\nu)C}, \qquad y(0) = \|u(\cdot,0)\|^{1+\gamma}_{L^{1+\gamma}(\Omega)}. \tag{1.19}$$

Let us briefly analyze condition (1.16). It is easy to see that if $p = 2$, which corresponds to the linear diffusion, condition (1.16) is equivalent to the inequality

$$\gamma > 1. \tag{1.20}$$

If $\gamma = 1$ (the evolution term in equation (1.6) is linear) condition (1.16) reads as

$$p < 2. \tag{1.21}$$

Let us assume now that condition (1.16) may fail but under the influence of a suitable absorption term

$$\frac{p}{1+\gamma} \ge 1 \quad \text{and} \quad \lambda > 0.$$

With the use of Hölder's and Young's inequalities from (1.11) and (1.13) with $m = \frac{(p-1)}{p}(1+\sigma)$, we obtain that

$$\begin{aligned} |u|^{m+1} &\le (m+1)\lambda^{-(p-1)/p} \left(\int_\Omega |u_x|^p d\Omega\right)^{1/p} \left(\int_\Omega \lambda |u|^{1+\sigma} d\Omega\right)^{(p-1)/p} \\ &\le (m+1)\lambda^{-(p-1)/p}\,(D + \lambda\, A) \end{aligned}$$

or, equivalently,

$$|u| \le (m+1)^{1/(m+1)} \lambda^{-(p-1)/p(m+1)}\,(D + \lambda\, A)^{1/(m+1)}. \tag{1.22}$$

It follows then that

$$
\begin{aligned}
y(t) &= \int_\Omega |u(x,t)|^{1+\gamma}\,dx = \int_\Omega |u(x,t)|^{1+\sigma}\,|u(x,t)|^{\gamma-\sigma}\,dx \\
&\le \lambda^{-1} \max |u(t)|^{\gamma-\sigma} A(t) \\
&\le \left(\frac{(m+1)^{1/(m+1)}}{\lambda^{(p-1)/p(m+1)}}\right)^{\gamma-\sigma} \lambda^{-1}\,(D(t)+\lambda\,A(t))^{1+(\gamma-\sigma)/(m+1)} \\
&\equiv \left(\frac{1+\gamma}{\gamma C}\right)^{1/\nu} (D(t)+\lambda\,A(t))^{1/\nu}.
\end{aligned}
\tag{1.23}
$$

In these inequalities,

$$
C = \frac{(1+\gamma)}{\gamma}\lambda^{-1}\left((m+1)^{1/(m+1)}\lambda^{-(p-1)/p(m+1)}\right)^{-\nu(\gamma-\sigma)}, \qquad \nu = \frac{1}{1+\frac{\gamma-\sigma}{m+1}}.
$$

Now let

$$
\nu < 1, \tag{1.24}
$$

which is equivalent to the condition $\sigma < \gamma$. As in the case $\lambda = 0$, gathering (1.12) with (1.23) leads to the ordinary nonlinear differential inequality (1.15) and, respectively, to estimate (1.17). Hence, for $u(x,t)$ (1.18) holds and the solution vanishes in a finite time.

Let us consider now the limit combinations of the parameters γ, p, σ which provide the fulfillment of condition (1.24).

1. $p = 2$, $\gamma = 1$; (the principal part of equation (1.6) is linear). Then (1.24) holds if $\sigma < 1$.

2. $\sigma = 1$, $p = 2$; (the diffusion and absorption are linear). Then (1.24) is provided by the condition $\gamma > 1$.

3. $\sigma = 1$, $\gamma = 1$; (only the diffusion is not linear). Then (1.24) is guaranteed by the inequality $p < 2$.

It is worth observing that the above-presented conditions on the structural constants (and the conditions which we derive below for more complicated equations) appear as the technical byproduct of the method. Notwithstanding that, in the particular situations where other methods are applicable the conditions given via the energy method coincide with those already known in the literature (see Kalashnikov [204]).

Now let the absorption be absent ($\lambda = 0$) but let 0the equation have a prescribed source term $f(x,t) \neq 0$ subject to the condition

$$
\|f(\cdot,t)\|_{L^1(\Omega)}^{(p-1)/p} = \left(\int_\Omega |f(x,t)|\,dx\right)^{(p-1)/p} \le \varepsilon\left(1-\frac{t}{t_f}\right)_+^{\nu/(1-\nu)} \tag{1.25}
$$

with $\varepsilon = \text{const} > 0$, $t_f > t^*$, and ν, t^* defined in (1.16), (1.19). It follows from (1.12) that

$$\frac{\gamma y'(t)}{1+\gamma} + D(t) \leq \left| \int_\Omega f\,u\,dx \right| \equiv I(t). \tag{1.26}$$

Using (1.13), (1.24) and the inequalities of Hölder and Young, one readily obtains the estimate

$$\begin{aligned} |I(t)| &\leq \|f\|_{L^1(\Omega)} \cdot (\text{meas}\,\Omega)^{(p-1)/p} \cdot D^{1/p}(t) \\ &\leq \delta\, D(t) + \frac{C_1 \gamma\, \varepsilon}{1+\gamma} \left(1 - \frac{t}{t_f}\right)_+^{\nu/(1-\nu)}, \end{aligned} \tag{1.27}$$

with the constants

$$C_1 = \frac{1+\gamma}{\gamma} (\text{meas}\,\Omega)\, \frac{p-1}{p}\, (p\,\delta)^{-1/(p-1)} \quad \text{and arbitrary } \delta \in (0,1).$$

Gathering (1.14), (1.15), (1.26), we arrive at the nonhomogeneous ordinary differential inequality

$$y'(t) + C\,(1-\delta)\, y^\nu(t) \leq C_1\, \varepsilon \left(1 - \frac{t}{t_f}\right)_+^{\nu/(1-\nu)} \tag{1.28}$$

with the constant C defined in (1.14).

Notice that if $\varepsilon = 0$, (the source term is absent), one may set in (1.28) $\delta = 0$, $t_f = t^*$ and revert to (1.15).

Introduce the function

$$G(\delta) \equiv \left(t_f - \frac{y^{1-\nu}(0)}{(1-\nu)(1-\delta)C}\right) \frac{C(1-\delta)}{t_f\, C_1(\delta)}\, y^\nu(0) > 0 \quad \text{for } \delta \in (0, \delta_0) \tag{1.29}$$

with

$$\delta_0 = 1 - \frac{y^{1-\nu}(0)}{C t_f (1-\nu)}$$

and assume that

$$\varepsilon \leq \max_{(0,\delta_0)} G(\delta). \tag{1.30}$$

The latter condition means that the source intensity is small. Since $G(0) = 0$, the equation $G(\delta) = \varepsilon$ has at least one solution δ_*. Let in (1.27) $\delta = \delta_*$. It is easy to check then that the function

$$z(t) = y(0) \left(1 - \frac{t}{t_f}\right)_+^{1/(1-\nu)}, \quad z(0) = y(0),$$

satisfies the equation

$$z' + C\,(1-\delta_*)\,z^{\nu} = \varepsilon\,C_1(\delta_*)\left(1-\frac{t}{t_f}\right)_+^{\nu/(1-\nu)}.$$

Let us check that $z(t)$ is a majorant for $y(t)$. Let $w(t) = z(t) - y(t)$. We have

$$w' + C(1-\delta_*)\nu \int_0^1 (\theta z + (1-\theta)y)^{\nu-1}\,d\theta\, w \geq 0 \quad \text{for } t \in (0, t_f), \qquad w(0) = 0.$$

Writing this inequality in the form

$$\frac{d}{dt}\left[w(t)\exp\left(C(1-\delta_*)\nu \int_0^t \int_0^1 (\theta z + (1-\theta)y)^{\nu-1}\,d\theta d\tau\right)\right] \geq 0$$

and integrating in t we obtain that $w \geq 0$ for all $t \in [0, t_f]$. Thus $y(t)$ and $u(x,t)$ both vanish in Ω beginning with the moment when the source $f(x,t)$ vanishes:

$$u(x,t) = 0 \quad \text{for a.e.} \quad x \in \Omega,\ t \geq t_f.$$

Notice that the condition of positivity of $G(\delta)$ claimed in (1.29) connects the three parameters which characterize the problem: the instant T_* of vanishing of the source, the source intensity ε, and the initial value $y(0)$. For this reason, given an arbitrary intensity $0 < \varepsilon < \infty$, the effect of vanishing of the solution can be provided by an appropriate choice of t_f and $y(0)$.

The preceding arguments are summarized as follows.

Proposition 1.1. *Let $u \in V(Q_T)$ be a weak solution of problem* (1.5)–(1.6). *If $f(x,t) \equiv 0$ and either*

$$1 < p < \gamma + 1 \quad \text{and} \quad \lambda = 0 \tag{1.31}$$

or

$$1 < p, \quad \sigma < \gamma \quad \text{and} \quad 0 < \lambda, \tag{1.32}$$

then there exists $t^* < \infty$ defined in (1.19) *such that*

$$u(x,t) \equiv 0 \quad \textit{for a.e.} \quad x \in \Omega,\ t \geq t^*.$$

If $f(x,t) \not\equiv 0$ and, additionally to (1.31), (1.32), *conditions* (1.29), (1.25), (1.30) *hold, then*

$$u(x,t) \equiv 0 \quad \textit{for a.e.} \quad x \in \Omega,\ t \geq t_f.$$

2 Second-order parabolic equations

2.1 Equations in bounded domains. Let us consider the class of quasilinear degenerate parabolic equations of the form

$$\frac{\partial \psi(x,u)}{\partial t} + P(x,\, t,\, u,\, Du,\, D^2u) = f(x,t) + \operatorname{div} \mathbf{g}(x,t), \tag{2.1}$$

where

$$\begin{aligned} P(x,\, t,\, u,\, Du,\, D^2u) &= -\operatorname{div} \mathbf{A}(x,t,u,Du) + B(x,t,u,Du) + C(x,t,u) \\ Du &= \left(\frac{\partial u}{\partial x_1}, \dots, \frac{\partial u}{\partial x_N}\right), \qquad \mathbf{A} = (A_1, \dots, A_N), \\ \operatorname{div} \mathbf{A} &= \sum_{i=1}^{N} \frac{d}{d x_i} A_i(x,t,u,Du) \\ &\equiv \sum_{i=1}^{N} \left(\frac{\partial A_i}{\partial x_i} + \frac{\partial A_i}{\partial u}\frac{\partial u}{\partial x_i} + \frac{\partial A_i}{\partial u_{x_k}} \frac{\partial^2 u}{\partial x_k x_i} \right). \end{aligned}$$

Equation (2.1) will be considered in a cylinder $Q = \Omega \times (0,T)$, $T \in \mathbb{R}_+$ arbitrary, where $\Omega \subset \mathbb{R}^N$ is a bounded and, generally speaking, multi-connected domain with the boundary Γ. We will claim that in the domain Ω the embedding

$$W_0^{1,p}(\Omega) \subset L^q(\Omega) \cap L^\gamma(\Gamma)$$

takes place with some $q, \gamma \geq 1$ (see the appendix, Lemma 3.1). To this end, it is sufficient to assume that the boundary Γ is Lipschitz-continuous [235, Chapter 2]. The case $\Omega = \mathbb{R}^N$ will be treated separately.

It is assumed that the coefficients of equation (2.1) satisfy the following structural conditions:

$$\forall\, (t, r, \mathbf{q}) \in \times\mathbb{R}_+ \times \mathbb{R} \times \mathbb{R}^N \quad \text{and} \quad \text{a.e. } x \in \Omega,$$

$$|\mathbf{A}(x,t,r,\mathbf{q})| \leq C_1\, |\mathbf{q}|^{p-1}, \tag{2.2}$$

$$C_2\, |\mathbf{q}|^p \leq \mathbf{A}(x,t,r,\mathbf{q}) \cdot \mathbf{q}, \tag{2.3}$$

$$|B(x,t,r,\mathbf{q})| \leq C_3\, |r|^\alpha\, |\mathbf{q}|^\beta, \tag{2.4}$$

$$C_4\, |r|^{1+\sigma} \leq C(x,t,r)\, r, \tag{2.5}$$

$$C_6\, |r|^{\gamma+k} \leq G(r,k) \leq C_5\, |r|^{\gamma+k}, \tag{2.6}$$

$$G(x,r,k) = \psi(x,r)\, |r|^k \operatorname{sign} r - k \int_0^r \psi(x,\tau)\, |\tau|^{k-1}\, d\tau \tag{2.7}$$

where $C_1 - C_6$, p, α, β, σ, γ, k are positive constants which will be specified later on.

A special case of (2.1) is the equation

$$\frac{\partial}{\partial t}\left(u\, |u|^{\gamma-1}\right) - \operatorname{div}\left(|\nabla u|^{p-2}\, \nabla u\right) + |u|^{\sigma-1}\, u = f + \operatorname{div} \mathbf{g}. \tag{2.8}$$

The equations

$$\frac{\partial v}{\partial t} - \operatorname{div}\left(\left(\frac{1}{\gamma}\right)^{p-1} |v|^{(p-1)(1-\gamma)/\gamma} |\nabla v|^{p-2}\nabla v\right) + v\,|v|^{(\sigma-\gamma)/\gamma} = f + \operatorname{div}\mathbf{g},$$

$$\frac{\partial v}{\partial t} - \operatorname{div}\left(\left|\nabla\left(v\,|v|^{(1-\gamma)/\gamma}\right)\right|^{p-2} \nabla\left(v|v|^{(1-\gamma)/\gamma}\right)\right) + v\,|v|^{(\sigma-\gamma)/\gamma} = f + \operatorname{div}\mathbf{g}$$

reduce to (2.8) by the change of the sought function $v = u\,|u|^{\gamma-1}$.

The question of existence of weak solutions to equations like (2.1) were considered in many works (see, for example, [151, 204, 233, 243]). In our presentation we will follow [15, 25, 32, 30, 149]. In this chapter we study the properties of weak solutions from the appropriate function classes assuming their existence.

Let us consider the initial-boundary value problem for equation (2.1),

$$u(x,0) = u_0(x), \quad x \in \Omega; \quad u(x,t) = 0 \quad (x,t) \in \Sigma_T = \Gamma \times (0,T). \tag{2.9}$$

We assume that the data of problem (2.1), (2.9) satisfy the conditions

$$\begin{aligned} &f \in L^{(p+k-1)/(p+k-2)}\left(0, T; L^{(\gamma+k)/\gamma}(\Omega)\right), \\ &u_0 \in L^{\gamma+k}(\Omega), \\ &\mathbf{g} \in L^{(p+k-1)/(p-1)}\left(0, T; L^{p(\gamma+k)/(p-1)(\gamma+1)}(\Omega)\right) \end{aligned} \tag{2.10}$$

with the parameter

$$k = \begin{cases} 1 & \text{if } N \le p \text{ or } (\gamma+1) \le \dfrac{Np}{N-p}, \\ \dfrac{N-p}{p}\left(1+\gamma - \dfrac{p(N-1)}{N-p}\right) > 1 & \text{if } 1 < p < N \text{ and } \gamma + 1 > \dfrac{Np}{(N-p)}. \end{cases}$$

We will study the weak solutions $u(x,t)$ of problem (2.1), (2.9) for which

$$v = u\,|u|^{(k-1)/p} \in V(Q),$$

where now

$$V(Q) = L^p(0,T; W_0^{1,p}(\Omega)) \cap L^\infty\left(0, T; L^{p(\gamma+k)/(p+k-1)}(\Omega)\right) \cap L^{(\sigma+k)p/(p+k-1)}(Q).$$

We will mostly be interested in the case $p \le N$. Let us note that for the functions

$$v = u\,|u|^{(k-1)/p} \in V(Q),$$

the function

$$E(t) = C_6 \int_\Omega |u|^{\gamma+k}\,dx + \int_Q \left(C_2\,k\,|u|^{k-1}\,|D\,u|^p + C_4\,|u|^{\sigma+k}\right) dxdt$$

is bounded by E_0.

Definition 2.1. A measurable-in-Q function $u(x,t)$ is said to be a weak solution of problem (2.1), (2.9) if

(i) $v = u\,|u|^{(k-1)/p} \in V(Q)$;

(ii) $\mathbf{A}(\cdot,\cdot,u,Du)$, $B(\cdot,\cdot,u,Du)$, $C(\cdot,\cdot,u) \in L^1(Q)$ and for every test-function $\varphi \in C^\infty([0,T]; C_0^\infty(\Omega))$ the following identity holds:

$$\begin{aligned}\int_Q \{\psi(x,u)\varphi_t - \mathbf{A}\cdot D\varphi - B\varphi - C\varphi\}\,dx\,dt - \int_\Omega \psi(x,u)\,\varphi\,dx\Big|_{t=0}^{t=T} \\ = \int_Q (\mathbf{g}\cdot D\varphi - f\varphi)\,dx\,dt.\end{aligned} \tag{2.11}$$

2.2 The energy relation. We will rely on the following property of the weak solutions of problem (2.1), (2.9): according to the integration-by-parts formula, for a.e. $t \in \mathbb{R}_+$ each of them satisfies the relation

$$\frac{d}{dt}y(t) + (\mathbf{A},\,Dv)_\Omega + (B+C,\,v)_\Omega \le (f,\,v)_\Omega - (\mathbf{g},\,Dv)_\Omega \tag{2.12}$$

where

$$\begin{gathered} y(t) = \int_\Omega G(x,u(x,t),k)\,dx, \\ G(x,u,k) \equiv \psi(x,u)\,|u|^{k-1}u - k\int_0^u \psi(x,\tau)\,|\tau|^{k-1}\,d\tau, \\ v = u\,|u|^{k-1}, \qquad (u,\,v)_\Omega = \int_\Omega uv\,dx. \end{gathered} \tag{2.13}$$

The proof of this inequality is the first step in the study of the vanishing properties of the solutions belonging to $V(Q)$. When formulating results we will use the constants K_1, K_2 from the following interpolation inequalities (see the appendix, Section 3.1):

$$\forall\, v \in W_0^{1,p}(\Omega) \quad \|v\|_{L^q(\Omega)} \le K_1\,\|Dv\|_{L^p(\Omega)}, \tag{2.14}$$

where K_1 depends on q, p, and Ω if $q < \frac{Np}{N-p}$, and $K_1 = \dfrac{p(N-1)}{N-p}$ if $q = \frac{Np}{N-p}$ (in the latter case, K_1 does not depend on Ω),

$$\begin{gathered} v \in W_0^{1,p}(\Omega), \qquad \|v\|_{L^q(\Omega)} \le K_2\,\|Dv\|^{\delta}_{L^p(\Omega)}\,\|v\|^{1-\delta}_{L^r(\Omega)}, \\ \delta = \frac{1/r - 1/q}{1/r - (N-p)/(Np)}, \qquad K_2 = \left(\frac{p(N-1)}{N-p}\right)^{\delta}. \end{gathered} \tag{2.15}$$

To prove (2.12), we assume that

$$p = \alpha + \beta + 1, \quad 1 < 1 + \sigma \le p, \quad \alpha \ge 0, \quad \beta \ge 0; \tag{2.16}$$

$$\mathbf{g} \in L^{p/(p-1)}(Q), \quad f \in L^{p/(p-1)}(0,T;\,L^{Np/(Np-N+p)}(\Omega)). \tag{2.17}$$

We will also assume that

(i) $\psi(x, r)$ is a Caratheodory function (measurable in x for all $r \in \mathbb{R}$ and continuous in r for almost all $x \in \Omega$);

(ii) $\psi(x, r)$ is nondecreasing in r for almost all $x \in \Omega$;

(iii) $\forall\,(r, s)\mathbb{R} \times \mathbb{R}$ and a.e. $x \in \Omega$, $\psi(x, 0) = 0$,

$$C_0|r|^\gamma \le \psi(x, r)\operatorname{sign} r, \qquad |C(x, t, r)| \le C_0|r|^\gamma,$$

$$|\psi(x, r) - \psi(x, s)| \le \begin{cases} C_0|r - s|^\gamma & \text{if } 0 < \gamma \le 1, \\ C_0\dfrac{|r - s|}{(|r| + |s|)^{1-\gamma}} & \text{if } 1 < \gamma < \infty \end{cases} \tag{2.18}$$

with some constant $C_0 > 0$.

Lemma 2.1. *Let $u(x, t)$ be a weak solution of problem* (2.1), (2.9) *in the sense of Definition* 2.1. *Then for almost all $t \in (0, T)$ the energy relation* (2.12) *holds.*

Proof. Using assumptions (2.2)–(2.6), (2.16)–(2.18), (2.14), and Hölder's inequality, it is easy to get the following estimates:

$$|(\mathbf{A}, D\varphi)_Q| \le C_1\|Du\|_{L^p(Q)}^{(p-1)}\|D\varphi\|_{L^p(Q)}, \tag{2.19}$$

$$|(B, \varphi)_Q| \le C_3K_1^{1+\alpha}\|Du\|_{L^p(Q)}^{(\beta+\alpha)}\|D\varphi\|_{L^p(Q)}, \tag{2.20}$$

$$|(C, \varphi)_Q| \le C_4^*K_1^{1+\alpha}T^{(p-1-\sigma)/(p-1)}\|Du\|_{L^p(Q)}^\sigma\|D\varphi\|_{L^p(Q)}, \tag{2.21}$$

$$|(\mathbf{g}, D\varphi)_Q| \le \|D\varphi\|_{L^p(Q)}\|\mathbf{g}\|_{L^{p/(p-1)}(Q)}, \tag{2.22}$$

$$|(f, \varphi)_Q| \le K_1\|D\varphi\|_{L^p(Q)}\|f\|_{L^{p/(p-1)}(0,T;\, L^{Np/(Np-N+p)}(\Omega))}. \tag{2.23}$$

It follows from these inequalities and (2.11) that for any function

$$\varphi \in C^\infty(0, T;\; C_0^\infty(\Omega)), \quad \varphi(x, 0) = \varphi(x, T) = 0,$$
$$|(\psi, \varphi_t)_Q| \le C\,\|\varphi\|_{L^p(0,T;\, W_0^{1,p}(\Omega))} \tag{2.24}$$

with a constant C depending on $\|u\|_{L^p(0,T;\, W_0^{1,p}(\Omega))}$. Since $C^\infty([0, T];\; C_0^\infty(\Omega))$ is dense in $L^p(0, T;\; W_0^{1,p}(\Omega))$, it follows that (2.24) is true for all $\varphi \in L^p(0, T;\; W_0^{1,p}(\Omega))$. Inequality (2.24) implies that the function $\psi(t) \equiv \psi(x, v(x, t))$ is absolutely continuous as a map from $[0, T]$ to $\mathbb{R}$ and has the weak derivative $\psi_t \in L^{p'}(0, T;\; W_0^{-1,p'}(\Omega))$, $p' = p/(p-1)$. Identity (2.11) is then equivalent to

$$\frac{\partial\psi}{\partial t} - P(u) = f + \operatorname{div}\mathbf{g} \in L^{p'}(0, T;\; W_0^{-1,p'}(\Omega)), \tag{2.25}$$

$\psi(0) = \psi(x, u_0)$ for $x \in \Omega$. Under assumptions (2.16)–(2.18), the formula of integration by parts

$$\int_s^t (\psi_t,\; v(\tau))_\Omega\, d\tau = (G(x, v(x, \tau)),\; 1)_\Omega|_{\tau=s}^{\tau=t} \equiv y(t) - y(s)$$

is true for all $(s, t) \in [0, T]$, $s < t$, and, moreover, for almost all $t \in (0, T)$,

$$\frac{dy}{dt} = (\psi_t, v(t))_\Omega.$$

Multiplying (2.25) by $v = u|u|^{k-1} \in L^p(0, T; W_0^{1,p}(\Omega))$, invoking (2.19)–(2.22) and using the formulas of integration by parts in x, we arrive at the identity

$$\frac{dy}{dt} + (\mathbf{A}, Dv)_\Omega + (B + C, v)_\Omega = (f, v)_\Omega - (\mathbf{g}, Dv)_\Omega. \qquad \square$$

Remark 2.1. Lemma 2.1 remains true if the conditions (2.9) are replaced by

$$u(x, 0) = u_0(x) \quad \text{in } \Omega, \tag{2.26}$$

$$u(x, t) = 0 \quad \text{on } \Sigma_{DT} = \Gamma_D \times (0, T), \tag{2.27}$$

$$(\mathbf{A}, \nu) = 0 \quad \text{on } \Sigma_{NT} = \Gamma_N \times (0, T), \tag{2.28}$$

where $\Gamma = \Gamma_D \cup \Gamma_N$ and $\mathbf{n}$ is the unit outer normal vector to Γ.

Remark 2.2. Another proof of the energy relation (2.12) can be found in [3, 60, 76, 106].

2.3 Finite time extinction. Our aim is to prove the following assertions.

Theorem 2.1. *Let conditions* (2.2)–(2.6), (2.9) *be fulfilled and let* $u(x, t)$ *satisfy the energy relation* (2.12) *for almost all* $t \in \mathbb{R}_+$.

(a) *Assume that* $f = \mathbf{g} = 0$, $C_3 = 0$ *and one of the following conditions holds: either*

$$p < 1 + \gamma, \quad (C_4 \geq 0) \tag{2.29}$$

or

$$1 + \sigma < 1 + \gamma \leq p, \qquad (C_4 > 0). \tag{2.30}$$

Then there exist positive constants $\nu = \nu(\gamma, p, \sigma, N) < 1$, $C = C(\gamma, p, \sigma, N, \Omega)$ *such that*

$$y(t) \equiv \left(\int_\Omega G(x, u(x, t), 1)\, dx\right) \leq \left(y(0)^{1-\nu} - C\,(1 - \nu)\, t\right)^{1/(1-\nu)} \tag{2.31}$$

and

$$\|u(\cdot, t)\|_{L^{1+\gamma}(\Omega)} = 0 \quad \textit{for}\ t \geq t^* = \frac{1}{C\,(1 - \nu)}\, \|u(\cdot, 0)\|^{1-\nu}_{L^{1+\gamma}(\Omega)}.$$

(b) *Let* (2.29) *or* (2.30) *be true and* $f \neq 0$, $\mathbf{g} \neq 0$. *Then for every* $t_f > t^*$ *there exists a positive constant* ε *such that if*

$$\begin{aligned} f^*(t) &\equiv \left(\|f(\cdot, t)\|^{(p+k-1)/(p+k-2)}_{L^{(\gamma+k)/\gamma}(\Omega)} + \|g(\cdot, t)\|^{(p+k-1)/(p+k-2)}_{L^{\lambda p/(p-1)}(\Omega)}\right) \\ &\leq \varepsilon \left(1 - \frac{t}{t_f}\right)_+^{\nu/(1-\nu)}, \qquad \lambda = \frac{k + \gamma}{1 + \gamma}, \end{aligned} \tag{2.32}$$

(ν *is defined in* (2.44) *if* (2.29) *holds and in* (2.45) *if* (2.30) *is true*), *the estimate*

$$\|u(\cdot,t)\|^{\gamma+k}_{L^{\gamma+k}(\Omega)} \le \|u(\cdot,0)\|^{\gamma+k}_{L^{\gamma+k}(\Omega)} \left(1-\frac{t}{t_f}\right)_+^{1/(1-\nu)}$$

holds, meaning that $u(x,t)\equiv 0$ *in* Ω *for all* $t\ge t_f$.

(c) *The assertions of items* (a) *and* (b) *remain valid if*

$$0 < C_3 \le C_3^* < C_2\, k\left(K_1\,\frac{p+k-1}{p}\right)^{\beta-p}, \tag{2.33}$$

$$\alpha+\beta = p-1, \quad 0\le\beta\le p \tag{2.34}$$

with the constant K_1 *from* (2.14).

(d) *The assertions of items* (a) *and* (b) *remain valid if the condition* (2.29) *or* (2.30) *is substituted by the condition*

$$p = 1+\gamma = 1+\sigma \tag{2.35}$$

or

$$p > 1+\gamma, \quad 1+\gamma \le 1+\sigma, \tag{2.36}$$

respectively.

Proof. Relying on (2.3) and (2.4), we derive the inequalities

$$C_6\,\|u(\cdot,t)\|^{\gamma+k}_{L^{\gamma+k}(\Omega)} \le y(t) \le C_5\,\|u(\cdot,t)\|^{\gamma+k}_{L^{\gamma+k}(\Omega)}, \tag{2.37}$$

$$k\,C_2\int_\Omega |u|^{k-1}\,|D\,u|^p\,dx \le k\left(\mathbf{A},\,|u|^{k-1}\,D\,u\right)_\Omega, \tag{2.38}$$

$$C_4\,\|u(\cdot,t)\|^{k+\sigma}_{L^{k+\sigma}(\Omega)} \le (C(x,t,u),\,v)_\Omega, \tag{2.39}$$

$$|\,(B,\,v)_\Omega\,| \le C_3\int_\Omega |u|^{k+\alpha}\,|D\,u|^\beta\,dx. \tag{2.40}$$

Relations (2.37)–(2.40) allow one to rewrite the energy relation (2.12) in the form

$$\begin{aligned}\frac{dy}{dt} &+ k\,C_2\int_\Omega |u|^{k-1}\,|D\,u|^p\,dx + C_4\,\|u(\cdot,t)\|^{k+\gamma}_{L^{k+\gamma}(\Omega)}\\ &\le C_3\int_\Omega |u|^{\alpha+k}\,|D\,u|^\beta\,dx + \int_\Omega |f|\,|u^k|\,dx \\ &+ k\int_\Omega |\mathbf{g}|\,|u|^{k-1}\,|Du|\,dx.\end{aligned} \tag{2.41}$$

We start with case (a). Let $1+\gamma \le \frac{Np}{N-p}$. Then $k=1$ and (2.14) with $v=u$, $q=\gamma+1$ together with (2.38) give

$$C\,y^\nu \le C_2\,\|D\,u\|^p_{L^p(\Omega)} \le (\mathbf{A},\,D\,u)_\Omega \tag{2.42}$$

with

$$\nu = \frac{p}{1+\gamma} < 1, \qquad C = C_2 K_1^{-p} C_5^{-p/(1+\gamma)}.$$

Relations (2.41), (2.42) imply the ordinary differential inequality for the energy function $y(t)$

$$y' + C y^{\nu} \le 0 \quad \text{with } 0 < \nu < 1. \tag{2.43}$$

If $1+\gamma > \frac{Np}{N-p}$, we use the formula

$$\left(\frac{p}{p+k-1}\right)^p \int_\Omega \left|D\left(u\,|u|^{(k-1)/p}\right)\right|^p dx = \int_\Omega |u|^{k-1} |D u|^p \, dx$$

and apply (2.38), (2.14) with $v = u|u|^{(k-1)/p}$, $q = p(\gamma+k)/(p+k-1)$. This leads to the estimate

$$C y^{\nu} \le k C_2 \int_\Omega |u|^{k-1} |D u|^p \, dx \le (A,\ D v)_\Omega, \tag{2.44}$$

$$C = k C_2 \left(\frac{p}{p+k-1}\right)^p K_1^{-p} C_5^{-\nu}, \qquad \nu = \frac{k+p-1}{\gamma+k} < 1 \quad \text{if } p < 1+\gamma.$$

Hence we arrive once again at inequality (2.43) but with different constants C and $\nu < 1$. Let us now consider case (a) assuming that (2.30) is true. Since $1+\sigma < 1+\gamma \le p$, we have that $1+\gamma < \frac{Np}{N-p}$, $k = 1$. Using (2.15) with $v = u$, $q = \gamma+1$, $r = \sigma+1$, we obtain the estimate

$$C y^{\nu} \le \int_\Omega \left(C_2 |D u|^p + C_4 |u|^{\sigma+1}\right) dx \tag{2.45}$$

with $C = C(C_2, C_4, C_5, K_2, \delta, p, \sigma, \gamma)$,

$$\delta = \frac{\gamma-\sigma}{\gamma+1} \frac{N p}{Np - (1+\sigma)(N-p)} < 1, \qquad \nu = \frac{1}{1+\gamma} \frac{p(1+\sigma)}{\delta(1+\sigma)+p(1-\delta)} < 1,$$

which once again leads to (2.43). Thus case (a) is reduced to the investigation of the ordinary differential inequality (2.43). Integrating it, we obtain estimate (2.31) which completes the proof. Let us consider case (b). The last two terms on the right-hand side of (2.41) are estimated with the use of (2.14), (2.37), (2.38):

$$\begin{aligned}
\int_\Omega |f|\,|u|\, dx &\le \|f\|_{L^{(\gamma+k)/\gamma}(\Omega)} \|u\|^k_{L^{\gamma+k}(\Omega)} \\
&\le M \|f\|_{L^{(\gamma+k)/\gamma}(\Omega)} \left(\int_\Omega |u|^{k-1} |D u|^p \, dx\right)^{k/(k+p-1)} \\
&\le \varepsilon_1 k C_2 \int_\Omega |u|^{k-1} |D u|^p \, dx + \varepsilon_1^{-k/(p-1)} M \|f\|_{L^{\gamma+k/\gamma}(\Omega)},
\end{aligned} \tag{2.46}$$

$$
\begin{aligned}
&k\int_\Omega |\mathbf{g}|\,|u|^{k-1}\,|D\,u|\,dx \\
&\quad\le k\left(\int_\Omega |u|^{k-1}\,|D\,u|^p\,dx\right)^{1/p}\times\|\mathbf{g}\|_{L^\mu(\Omega)}\left(\int_\Omega |u|^{\gamma+k}\,dx\right)^{(k-1)(p-1)/p(\gamma+k)} \\
&\quad\le M\|\mathbf{g}\|_{L^\mu(\Omega)}\left(\int_\Omega |u|^{k-1}\,|D\,u|^p\,dx\right)^{k/(k+p-1)} \\
&\quad\le \varepsilon_1\,k\,C_2\int_\Omega |u|^{k-1}\,|D\,u|^p\,dx+\varepsilon_1^{-k/(p-1)}\,M\,\|\mathbf{g}\|_{L^\mu(\Omega)}^{(p+k-1)/p-1}
\end{aligned}
\tag{2.47}
$$

with

$$
\mu=\frac{(k+\gamma)p}{(p-1)(\gamma+1)},\qquad \varepsilon_1\in(0,1).
$$

Here and throughout the section, M, $\widetilde{M}$ denote different constants depending on γ, p, σ, N, Ω. Gathering (2.41), (2.44), (2.47) with $\varepsilon_1 < 1/2$, and then using (2.32), (2.44) we come to the ordinary nonhomogeneous differential inequality

$$
y'(t)+C\,(1-2\varepsilon_1)\,y^\nu(t)\le\varepsilon_1^{-k/(p-1)}\,\varepsilon\,\widetilde{M}\left(1-\frac{t}{t_f}\right)_+^{\nu/(1-\nu)},
\tag{2.48}
$$

in which ν is defined by (2.44) or (2.45), and C is the same as in (2.43). Therefore for $\varepsilon = 0$ (the source is absent) we can set in (2.48) $\varepsilon_1 = 0$ and conclude that the extinction time t^* is the same as in (2.43).

Let us introduce the function

$$
H(s)=\left(t_f-\frac{y(0)^{1-\nu}}{C\,(1-\nu)(1-\delta)}\right)\frac{C(1-\delta)y(0)^{1-\nu}}{t_f\widetilde{C}}\left(\frac{\delta}{2}\right)^{p/(k-1)}.
$$

Let

$$
\max H(\delta)=H(\delta_*),\qquad \delta\in\left(0,1-\frac{y(0)^{1-\nu}}{C(1-\nu)t_f}\right).
$$

Then the function

$$
z(t)=y(0)\left(1-\frac{t}{t_f}\right)_+^{1/(1-\nu)},\qquad z(0)=y(0),
$$

is a majorant for $y(t)$ provided that

$$
\varepsilon\le H(\delta_*)
\tag{2.49}
$$

(see the proof of Proposition 2.1). This completes the proof of assertion (b). To prove (c) it suffices to make use of (2.14), (2.33), (2.34), (2.40) and apply the estimate

$$
\begin{aligned}
|(B, v)_\Omega| &\le C_3 \left(\int_\Omega |u|^{k-1} |D u|^p \, dx \right)^{\beta/p} \left(\int_\Omega |u|^{k+p-1} \, dx \right)^{(p-\beta)/p} \\
&\le C_3^* \left(K_1 \frac{p+k-1}{p} \right)^{p-\beta} \int_\Omega |u|^{k-1} |D u|^p \, dx
\end{aligned}
$$

with $C_3^*(K_1(p+k-1)/p)^{p-\beta} < C_2 k$. Let us consider case (d). It follows from the previous arguments that the function

$$
y(t) = \int_\Omega H(u(x,t)) \, dx
$$

obeys either the differential inequality

$$
y' + C\, y \le M\, f^*(t), \tag{2.50}
$$

if (2.35) holds or

$$
y' + C\, y^\nu \le M\, f^*(t), \qquad \nu = \frac{p}{1+\gamma} > 1 \tag{2.51}
$$

if (2.36) is the case. Here the function f^* and the constants C, M are defined in (2.12), (2.32), (2.43)–(2.47) correspondingly. Integration of inequality (2.50) leads to the estimate

$$
y(t) \le \left(y(0) + \int_0^t M \exp(C\tau)\, f^*(\tau)\, d\tau \right) \exp(-C\, t) .
$$

To deal with inequality (2.51), we introduce the functions

$$
z(t) = y_0 \left(1 + (\nu - 1)\, C\, t\right)^{-1/(\nu-1)}
$$

and $\omega(t)$, which solve the problems

$$
\begin{aligned}
&\omega' + C\, \omega = M\, f^*, \qquad \omega(0) = y_0 = y(0), \\
&z' + C\, z^\nu = 0, \qquad\qquad z(0) = y_0.
\end{aligned} \tag{2.52}
$$

It is easy to verify that

$$
y(t) \le \omega(t), \qquad z(t) \le \omega(t) \quad \forall t \ge 0. \tag{2.53}
$$

Integrating (2.52) and using the second of inequalities (2.53), we have that

$$
\begin{aligned}
\frac{1}{\nu-1}\left[-\omega^{1-\nu}(t) + y_0^{1-\nu} \right] &= \int_0^t \frac{f^*(\tau)}{\omega^\nu(\tau)}\, d\tau - C\, t \le \int_0^t \frac{f^*(\tau)}{z^\nu(\tau)}\, d\tau - C\, t \\
&\equiv -g^*(t).
\end{aligned}
$$

Assuming that $g(t) \geq 0$ and employing the first of inequalities (2.53), we derive the estimate

$$y(t) \leq y_0 \left(1 + (\nu - 1)\, g(t)\, y_0^{\nu-1}\right)^{-1/(\nu-1)}. \tag{2.54}$$

Inequality (2.54) gives asymptotic estimates when $t \to \infty$ for the functions satisfying the energy inequality (2.12) and as well for the solutions of equation (2.1) endowed with appropriate boundary and initial data. □

Let us assume that the energy relation (2.12) only holds on a finite interval $[0, T]$.

Theorem 2.2. *Let the function $u \in V(Q)$ satisfy the energy relation* (2.12) *for almost all $t \in (t_0, T)$, $t_0 > 0$. Let us assume that conditions* (2.2)–(2.6), (2.9), (2.29) *or* (2.30), (2.33), *and* (2.34) *are fulfilled. If $t_f \in (t_0, T)$, there exist positive constants ε, ε_0, C_0 such that if* (2.32) *holds for $t \in (t_0, T)$ and*

$$\|u(\cdot, t_0)\|_{L^{\gamma+k}(\Omega)} \leq \varepsilon_0 \left(1 - \frac{t}{t_f}\right)_+^{1/(1-\nu)},$$

then

$$\|u(\cdot, t)\|_{L^{\gamma+k}(\Omega)}^{\gamma+k} \leq C_0 \left(1 - \frac{t}{t_f}\right)_+^{1/(1-\nu)}, \quad \forall t \in (t_0, T), \tag{2.55}$$

whence $u(x, t) = 0$ for $x \in \Omega$, $t \geq t_f$.

The proof is analogous to the proof of Theorem 2.1. We first derive the ordinary differential inequality (2.48). Next, in the definition of the function H we replace $y(0)$ by $\varepsilon_0(1 - t/t_f)^{1/(1-\nu)}$ and then preliminarily choose ε_0 according to the condition $H(\delta) > 0$. The constant ε has to be chosen then from inequality (2.49).

Remark 2.3. The assertions of Theorems 2.1 and 2.2 remain true for unbounded domains Ω, and, in particular, for $\Omega = \mathbb{R}^N$. This is the case if, additionally to condition (2.29), the structural constants p and γ and the dimension N satisfy the relation $\frac{p(N-1)}{N-p} < 1 + \gamma$.

For the proof, instead of (2.42) and (2.44), one has to use inequality (2.14) with $q = \frac{Np}{N-p}$:

$$\left(\int_\Omega |u|^{\gamma+k}\, dx\right)^{1-p/N} \leq K_1^p \int_\Omega |u|^{k-1}\, |D\, u|^p\, dx, \tag{2.56}$$

with

$$1 + \gamma = \frac{pN + p(k-1)}{N - p}, \quad k > 0, \quad K_1 = K_1(N, p).$$

Then all the constants in the differential inequalities (2.43), (2.48) are independent of Ω and all the above arguments are applicable.

Remark 2.4. Theorems 2.1 and 2.2 still hold if the function $u(x,t)$ satisfies conditions (2.26)–(2.28) with meas $\Gamma_D > 0$. It is sufficient to claim the fulfillment of (2.14), (2.15). They hold if $u_0 \in W^{1,p}(\Omega)$ and either

$$u|_{\Gamma_D} = 0, \quad \Gamma_D \subset \Gamma, \quad \text{meas}\,\Gamma_D > 0, \quad \text{or} \quad \int_\Omega u\,dx = 0. \tag{2.57}$$

In these cases the constants K_1, K_2 in (2.14), (2.15) may depend on Ω. The case meas $\Gamma_D = 0$ is more complicated. To be precise, let us consider the problem

$$\begin{cases} u_t = \operatorname{div} \mathbf{A}(x,t,u,\nabla u), \quad (x,t) \in Q = \Omega \times (0,T), \\ \mathbf{A}(x,t,u,\nabla u) = |u|^\alpha |\nabla u|^{p-2} \nabla u, \\ u(x,0) = u_0(x) \ \text{ in } \Omega, \quad (\mathbf{A},\nu) = 0 \ \text{ on } \Sigma_T = \Gamma \times (0,T), \quad \Gamma = \partial\Omega. \end{cases} \tag{2.58}$$

It is easy to verify that

$$\int_\Omega u(x,t)dx = \int_\Omega u_0(x)dx = C_0 = \text{const}, \quad t \in (0,T). \tag{2.59}$$

Let $C_0 = 0$. According to (2.57), in this case inequalities (2.14), (2.15) are applicable and therefore for the solutions of problem (2.58) the assertions of Theorems 2.1, and 2.2 hold (with $\gamma = (p-1)/(p+\alpha-1)$). Moreover,

$$u(x,t) \equiv 0, \ x \in \Omega, \ t \geq t^*.$$

Let $C_0 \neq 0$ and $\alpha = 0$. If this is the case, the assertions of Theorems 2.1, 2.2 (with $\gamma = 1$ and $p < 2$) hold for the function

$$v(x,t) = u(x,t) - \frac{1}{\text{meas}\,\Omega} \int_\Omega u_0(x)dx$$

and

$$v(x,t) \equiv 0, \ x \in \Omega, \ t \geq t^*.$$

A similar result is true for an arbitrary value of α provided

$$\frac{1}{M} \leq |u(x,t)| \leq M < \infty, \ t \geq t^0,$$

for some $t^0 < \infty$.

Remark 2.5. Inequalities of the type (2.49) connect the three parameters of the problem: the norm of the initial data $\|u(\cdot,0)\|_{L^{\gamma+k}(\Omega)}$, the source intensity ε_*, the instant t_f when the source and the solution vanish. If the source is absent, (in (2.32) $\varepsilon = 0$), the initial data define in a unique way the instant of vanishing of the solution t^*. Conversely, given some t^*, one may indicate the value of $y(0)$ providing the effect.

In the presence of sources ($\varepsilon \neq 0$) and under the assumption that the norm of the initial data is arbitrary, one may either first fix the value $t_f > t^*$ and then define ε, or fix ε and then choose $t_f > t^*$.

Note that under the conditions of Theorems 2.1 and 2.2 for the arbitrary function $u \in V(Q)$ satisfying the energy relation (2.12) the estimate

$$\sup_{0\le t\le T} \|u(\cdot,t)\|^{\gamma+k}_{L^{\gamma+k}(\Omega)} + \int_0^t \int_\Omega \left(|u|^{k-1}\,|D\,u|^p + |u|^{\sigma+k}\right) dx\,dt \\ \le K\left(\|u_0\|^{\gamma+k}_{L^{\gamma+k}(\Omega)} + \varepsilon_1 \left(1-\frac{t}{t_f}\right)_+^{\nu/(1-\nu)}\right)$$

holds with a constant K independent of T.

We proceed to study the properties of weak solutions of equation (2.1) under the boundary conditions (2.9), (2.28). It follows from Lemma 2.1 and Remark 2.1 that for such solutions the energy relation (2.12) is true, whence the validity of Theorem 2.1.

Theorem 2.3. *Let $u \in V(Q)$ be a weak solution of problem* (2.1), (2.9). *Let us assume that conditions* (2.2)–(2.6), (2.16)–(2.18) *hold and, additionally, that one of the conditions*

$$\begin{aligned} p < 1+\gamma, &\qquad C_4 \ge 0, \\ 1+\sigma < 1+\gamma \le p, &\qquad C_4 > 0 \end{aligned} \tag{2.60}$$

is fulfilled. Let

$$C_3 \le C_3^* = C_2\,k\left(K_1\,\frac{p+k-1}{p}\right)^{\beta-p}, \qquad \alpha+\beta = p-1, \quad 0 \le \beta \le p,$$

with the constant K_1 from (2.14). *Then for each $t_f \in (0,T)$ there exist positive constants ε, ε_0, C_0 such that if*

$$\|f(\cdot,t)\|^{(p+k-1)/(p+k-2)}_{L^{(\gamma+k)/\gamma}(\Omega)} + \|\mathbf{g}(\cdot,t)\|^{(p+k-1)/(p-1)}_{L^{\lambda p/(p-1)}(\Omega)} \le \varepsilon_1 \left(1-\frac{t}{t_{f\,+}}\right)^{\nu/(1-\nu)},$$

with $\lambda = (k+\gamma)/(1+\sigma)$ and ν defined in (2.44) *for* (2.55) *and in* (2.45) *for* (2.60), *and*

$$\|u(\cdot,t)\|^{\gamma+k}_{L^{\gamma+k}(\Omega)} \le \varepsilon_0 \left(1-\frac{t}{t_f}\right)_+^{1/(1-\nu)},$$

then

$$\|u(\cdot,t)\|^{\gamma+k}_{L^{\gamma+k}(\Omega)} \le C_0 \left(1-\frac{t}{t_f}\right)_+^{1/(1-\nu)} \qquad \forall t \in (t_0,T).$$

In particular, $u(x,t) = 0$ for $x \in \Omega$, $t \ge t_f$.

Remark 2.6. The assertions of Theorems 2.1–2.3 can be spread to more general equations of the form (2.1) with the coefficient $\mathbf{A}(x, t, u, \nabla u)$ replaced by

$$\mathbf{A}(x, t, u, \nabla u) = \mathbf{A}_0(x, t, u, \nabla u) + \mathbf{A}_1(x, t, u, \nabla u). \tag{2.61}$$

Here $\mathbf{A}_0(x, t, u, \nabla u)$ satisfies conditions (2.2), (2.3), while $\mathbf{A}_1(x, t, u, \nabla u)$ is subject to the inequality

$$|\mathbf{A}(x, t, r, \mathbf{q})| \le M(x, t)\,|r|^\alpha |\mathbf{q}|^\beta \quad \forall (t, r, \mathbf{q}) \in \mathbb{R}_+ \times \mathbb{R} \times \mathbb{R}^N \quad \text{and} \quad \text{a.e. } x \in \Omega$$

with $M(x, t) \ge 0$, $\alpha > 0$, $p = \beta + 1 + \alpha$.

For the proof we consider the energy relation (2.12) corresponding to the representation (2.61). Let us limit ourselves to the case (a) of Theorem 2.1. We assume that

$$f = 0, \quad \mathbf{g} = 0, \quad B = 0, \quad p < 1 + \gamma < \frac{Np}{N - p}.$$

Then the energy relation takes the form

$$\frac{dy}{d\tau} + (A_0, \nabla u)_\Omega \le (A_1, \nabla u)_\Omega \equiv I.$$

The additional term I on the right-hand side of this relation can be estimated as

$$\begin{aligned} |I| &\le \left(\int_\Omega |\nabla u|^p\right)^{(\beta+1)/p} \left(\int_\Omega |u|^{\lambda p}\right)^{\alpha/\lambda p} \mu(t) \\ &\le K_2^\alpha \left(\int_\Omega |\nabla u|^p\, dx\right)^{(p-(1-\delta)\alpha)/p} \left(\int_\Omega |u|^{\gamma+1}\right)^{(1-\delta)\alpha/(1+\gamma)} \mu(t) \\ &\le \frac{\varepsilon^{\tau'}}{\tau'} \int_\Omega |\nabla u|^p\, dx + \frac{1}{\tau}\varepsilon^{-\tau} \left(\int_\Omega |u|^{\gamma+1}\, dx\right)^{p/(\gamma+1)} K_2^{\alpha\tau}\, \mu^\tau. \end{aligned} \tag{2.62}$$

Here $\varepsilon > 0$ is arbitrary,

$$\mu(t) = \left(\int_\Omega |M(x, t)|^{\lambda p/\alpha}\, dx\right)^{\alpha/\lambda p}, \quad \frac{N}{p} < \alpha,$$

and the constants K_2, $\delta \in (0, 1)$ are defined in (2.15). Letting in (2.62)

$$\varepsilon^{\tau'} = \tau' \frac{C_2}{2}, \qquad \tau = \frac{p}{\alpha(1 - \sigma)}$$

and using (2.37), (2.38), (2.42), as in the proof of Theorem 2.1, we arrive at the ordinary differential inequality

$$y' + a(t)\, y^\nu \le 0, \quad \nu = \frac{p}{1 + \gamma} \tag{2.63}$$

with

$$a(t) = \frac{C_2}{2} K_1^{-p} C_5^{-p/(1+\gamma)} - \frac{1}{\tau} (\frac{\tau'}{2} C_2)^{1-\tau} K_2^{p/(1-\delta)} C_6^{p/(1+\gamma)} \mu^\tau. \tag{2.64}$$

Integrating (2.63), we come to the estimate

$$y^{1-\nu}(t) \leq \left(y^{1-\nu}(0) - (1-\nu) \int_0^t a(\tau)\, d\tau \right).$$

Hence the energy function $y(t)$ and the function $u(x,t)$ vanish in a finite time if there exists $t^* < \infty$ such that

$$y^{1-\nu}(0) = (1-\nu) \int_0^{t^*} a(\tau)\, d\tau.$$

According to (2.64), the last condition is fulfilled if the rate of growth of the function $\mu^\tau(t)$ is less than $t^{1-\varepsilon}$ for some $\varepsilon > 0$.

3 The weighted diffusion-absorption balance

Let us consider now the situation where the diffusion and absorption terms are nonhomogeneous with respect to the independent variables (in other words, the process involving absorption and diffusion in a nonhomogeneous medium). We will assume that conditions (2.2), (2.3), (2.5) are replaced by the conditions

$$\forall\, (t, r, \mathbf{q}) \in \mathbb{R}_+ \times \mathbb{R} \times \mathbb{R}^N \quad \text{and} \quad \text{a.e. } x \in \Omega,$$

$$|\mathbf{A}(x,t,r,\mathbf{q})| \leq C_1\, a\, |\mathbf{q}|^{p-1}, \tag{3.1}$$

$$C_2\, a\, |\mathbf{q}|^p \leq \mathbf{A}(x,t,r,\mathbf{q}) \cdot \mathbf{q}, \tag{3.2}$$

$$C_4\, Q\, |r|^{1+\sigma} \leq C(x,t,r)\, r. \tag{3.3}$$

In these conditions $a(x,t)$, $Q(x,t)$ are nonnegative measurable functions possessing some additional properties which we describe below. For the sake of convenience, let us derive first some generalizations of inequalities (2.14) and (2.15). Let us start with (2.14) assuming that $p < 1+\gamma$. Set in (2.14) $v = u\,|u|^{(\kappa-1)/p}$ and $p = m$. It follows from Hölder's inequality that

$$\|u\|^\alpha_{L^{q\alpha}(\Omega)} \leq K_1 \alpha \left(\int_\Omega a\, |u|^{\kappa-1}\, |Du|^p\, dx \right)^{1/p} \left(\int_\Omega a^{-m/(p-m)}\, dx \right)^{(p-m)/pm} \tag{3.4}$$

with

$$\alpha = \frac{\kappa - 1 + p}{p}, \qquad q \leq \frac{Nm}{N-m}, \qquad 1 < m \leq p.$$

In (3.4), let

$$q = \frac{(\gamma+\kappa)p}{\kappa+p-1} \leq \frac{Nm}{N-m}. \tag{3.5}$$

This is true for some $\kappa > 0$ if the parameter m satisfies the conditions

$$\max\left(1, \frac{N\,p}{N+p}\right) < m < \frac{pN\gamma}{N(p-1)+\gamma p}. \tag{3.6}$$

Notice that if (3.5) renders the equality, the constant K_1 in (2.14) does not depend on Ω.

Introduce the function

$$\rho(t) = M\left(\int_\Omega (a(x,t))^{-m/(p-m)}\,dx\right)^{-(p-m)/m} \tag{3.7}$$

with the constant

$$M = K\,C_2\left(K_1\,\alpha\,C_5^{1/q}\right)^{-p}$$

and assume that in (3.4), (3.7) all the integrals are well defined. Using (2.37) we may rewrite the inequality (3.4) in the form

$$\rho(t)\,y^\nu(t) \le K\,C_2\int_\Omega a\,|u|^{\kappa-1}\,|D\,u|^p\,dx, \qquad \nu = \frac{\kappa+p-1}{\gamma+\kappa} < 1. \tag{3.8}$$

Notice that for $m = p$, $a \equiv 1$ inequality (3.8) coincides with (2.44).

With the help of Hölder's inequality, it is easy to generalize (2.15) as

$$\begin{aligned} \|u\|_{L_2(\Omega)} \le K_2 &\left(\int_\Omega a\,|Du|^p\,dx\right)^{\delta/p}\left(\int_\Omega Q\,|u|^{\sigma+1}\,dx\right)^{(1-\delta)/(1+\sigma)} \\ &\times\left(\int_\Omega a^{-m/(p-m)}\,dx\right)^{\delta(p-m)/pm}\left(\int_\Omega Q^{-r/(1+\sigma-r)}\,dx\right)^{\frac{(1-\delta)(1+\sigma-r)}{r(1+\sigma)}} \end{aligned} \tag{3.9}$$

where $1 < m \le p$, $1 \le r \le 1+\sigma$, and δ, K_2 are defined in (2.15). Consider the function

$$\begin{aligned} \rho^{1/\kappa}(t) = M &\left(\int_\Omega a^{-m/(p-m)}\,dx\right)^{\delta(p-m)/pm} \\ &\times\left(\int_\Omega Q^{-r/(+1+\sigma-r)}\,dx\right)^{(1-\delta)(1+\sigma-r)/(1+\sigma)r} \end{aligned} \tag{3.10}$$

with

$$\begin{aligned} M &= K_2^{-1}\,C_2^{\delta/p}\,C_4^{(1-\delta)(1+\sigma-r)/(1+\sigma)r}\,C_5^{-1/(\gamma+1)}, \\ \kappa &= \frac{p(1+\sigma)}{\delta(1+\sigma)+p(1-\delta)}, \qquad \delta = \frac{1+\gamma-r}{1+\gamma}\,\frac{Nm}{Nm-r(N-m)}. \end{aligned}$$

Let all the integrals in (3.9), (3.10) be finite. Then (3.9) with $q = 1+\gamma < \frac{Nm}{N-m}$ gives

$$\rho(t)\, y^{\nu}(t) \le \int_{\Omega} \left(C_2\, a\, |Du|^p + C_4\, Q\, |u|^{\sigma+1} \right) dx \tag{3.11}$$

where

$$\nu = \frac{\kappa}{1+\gamma} < 1 \tag{3.12}$$

if

$$1 \le r \le 1+\sigma \le p, \qquad \max\left(1, \frac{N(1+\gamma)}{N+1+\gamma}\right) < m \le p.$$

Moreover, the constant M in (3.10) does not depend on Ω.

We will consider the class of functions with bounded "energy." Denote by $V_*(Q)$ the class of function with the "energy"

$$E = \sup_{t\in(0,T)} \int_{\Omega} |u|^{\gamma+\kappa} dx + \int_{Q} \left(K\, C_2\, |u|^{\kappa-1}\, |Du|^p + C_4\, Q\, |u|^{\sigma+\kappa} \right) dx dt,$$

bounded by a constant E_0. It will be assumed that initial data $u(x,0)$ and the function $\rho(t)$ defined either by (3.7) if $C_4 = 0$, or (3.10) if $C_4 > 0$, satisfy the condition

$$y^{1-\nu}(0) < (1-\nu) \int_0^{\infty} \rho(\tau)\, d\tau. \tag{3.13}$$

Theorem 3.1. *Let* $u \in V_*(Q)$ *satisfy the energy relation* (2.12) *for almost all* $t \in \mathbb{R}_+$, *and assume that* (2.4) *with* $C_3 = 0$, (2.6), (3.3), (3.13) *hold, and that* $f = \mathbf{g} = 0$. *Let one of the following conditions be fulfilled:*

$$\text{(i)} \qquad p < 1+\gamma, \quad (3.6), \quad C_4 = 0, \tag{3.14}$$

$$\text{(ii)} \qquad 1+\gamma \le p, \quad (3.12), \quad C_4 > 0. \tag{3.15}$$

Then $u(x,t) = 0$ *for* $x \in \Omega$, $t \ge t^*$, *where the value of* t^* *is defined by the relation*

$$y^{1-\nu}(0) \equiv \left(\int_{\Omega} G(x, u_0(x), 1)\, dx \right)^{1-\nu} = (1-\nu) \int_0^{t^*} \rho(\tau)\, d\tau.$$

Proof. It follows from (2.12) and (3.8) in case (i) (from (3.11) in case (ii)) that the energy function $y(t)$ satisfies the generalization of the ordinary differential inequality (2.45)

$$y' + \rho(t)\, y^{\nu} \le 0.$$

Integration of the last inequality leads to the estimate

$$y^{1-\nu}(t) \le y^{1-\nu}(0) - (1-\nu) \int_0^t \rho(\tau)\, d\tau$$

and the assertion follows. □

A weak solution of problem (2.1), (2.8) from the class $V_*(Q)$ can be defined by analogy with Definition 2.1. Arguing as in Lemma 2.1, one can show that this solution satisfies the energy relation (2.12). This leads to the following assertion.

Theorem 3.2. *In the conditions of Theorem* 3.1, *let* $u \in V_*(Q)$ *be a weak solution of problem* (2.1), (2.8). *Then for* $u(x,t)$ *the assertion of Theorem* 3.1 *is true.*

Remark 3.1. The case $f \neq 0$, $\mathbf{g} \neq 0$ is considered similarly to Theorems 2.1–2.3 and 3.2.

Remark 3.2. Let us assume that condition (2.6) is replaced by

$$C_6\, a(x)\, |r|^{\widetilde{\gamma}+\kappa} \leq G(x,r,\kappa) \leq C_5\, a(x)\, |r|^{\widetilde{\gamma}+\kappa},$$

where $a(x) \geq 0$ is a measurable function such that

$$0 < \rho = \left(\int_\Omega a(x)^{(\gamma+1)/(\gamma-1)}\, dx \right)^{(\gamma-\widetilde{\gamma})/(1+\gamma)} < \infty, \quad 0 < \widetilde{\gamma} < \gamma.$$

Let γ, $\nu < 1$ be the parameter defined in (2.6), (2.45). Lastly, let us assume that

$$\widetilde{\nu} = \nu\, \frac{1+\gamma}{1+\widetilde{\gamma}} < 1.$$

It is easy to verify that, first,

$$\widetilde{y}(t) \equiv \int_\Omega a\, |u|^{1+\widetilde{\gamma}}\, dx \leq \rho \left(\int_\Omega |u|^{1+\gamma}\, dx \right)^{(1+\widetilde{\gamma})/(1+\gamma)} = \rho\, y^{(1+\widetilde{\gamma}/(1+\gamma))}(t),$$

and, according to (2.45),

$$C\, \rho\, \widetilde{y}^{\widetilde{\nu}}(t) \leq \int_\Omega \left(C_2\, |Du|^p + C_4\, |u|^{1+\sigma} \right) dx, \quad \widetilde{\nu} < 1.$$

Beginning with this step, the proof of a finite-time vanishing of the function $\widetilde{y}(t)$ is a literal repetition of the proof of Theorem 2.1.

4 The Cauchy problem

Let us consider equation (2.1) in unbounded domains. We begin with the Cauchy problem. It is required to find a weak solution $u(x,t)$ of equation (2.1) satisfying the initial condition

$$u(x,0) = u_0(x), \quad x \in \mathbb{R}^N. \tag{4.1}$$

Definition 4.1. A measurable-in-$Q = \mathbb{R}^N \times (0,T)$ function $u(x,t)$ is said to be a weak solution of the Cauchy problem (2.1), (4.1) if

(i) $u \in V(Q)$,

(ii) $(\mathbf{A}(\cdot,\cdot,u,Du),\ B(\cdot,\cdot,u,Du),\ C(\cdot,\cdot,u)) \in L^1(Q)$,

and for each test function $\varphi \in C^\infty\left([0,T];\ C_0^\infty(\mathbb{R}^N)\right)$ the identity

$$\int_Q \{\psi(x,u)\,\varphi_t - \mathbf{A}\cdot\nabla\varphi - B\,\varphi - C\,\varphi\}\,dx\,dt + \int_\Omega \psi(x,u)\,\varphi\,dx\Big|_{t=0}^{t=T} = \int_Q (\mathbf{g}\cdot D\varphi - f\,\varphi)\,dx\,dt \tag{4.2}$$

holds.

Lemma 4.1. *For almost all $t \in (0,T)$ the weak solution of the Cauchy problem* (2.1), (4.1) *satisfies the energy relation* (2.12).

From now on, we will assume that

$$C_3 \le C_3^*\,K\left(K_1\,\frac{p+k-1}{p}\right)^{\beta-p} \tag{4.3}$$

with the constant $K_1(p,N)$ from (2.14) for $q = \frac{Np}{N-p}$. The exponent ν we either take from (2.45) if $C_4 > 0$, or set $\nu = 1 - \frac{p}{N}$ if $C_4 = 0$.

Theorem 4.1. *Let $u \in V(Q)$, $Q = \mathbb{R}^N \times (0,T)$ be a weak solution of the Cauchy problem* (2.1), (4.1). *Let us assume that conditions* (2.2)–(2.6), (4.3) *hold and, additionally, that one of the conditions is true: either*

$$\frac{p(N-1)}{N-p} < 1+\gamma, \quad C_4 \ge 0,$$

or

$$1+\sigma < 1+\gamma \le p, \quad C_4 > 0.$$

Then for every $t_f \in (0,T)$ there exist positive constants ε, ε_0, C_0 such that if

$$\|f(\cdot,t)\|_{L^{(\gamma+k)/\gamma}(\mathbb{R}^N)}^{(p+k-1)/(p+k-2)} + \|\mathbf{g}(\cdot,t)\|_{L^{\lambda p/(p-1)}(\mathbb{R}^N)}^{(p+k-1)/(p-1)} \le \varepsilon\left(1-\frac{t}{t_f}\right)_+^{\nu/(1-\nu)}$$

and

$$\|u(\cdot,t_0)\|_{L^{\gamma+k}(\mathbb{R}^N)}^{\gamma+k} \le \varepsilon_0\left(1-\frac{t_0}{t_f}\right)_+^{\nu/(1-\nu)}, \quad t_0 \in [0,T),$$

then

$$\|u(\cdot,t)\|_{L^{\gamma+k}(\mathbb{R}^N)}^{\gamma+k} \le C_0\left(1-\frac{t}{t_f}\right)_+^{\nu/(1-\nu)}, \quad \forall t \in [t_0,T) \tag{4.4}$$

and $u(x,t) = 0$ for a.e. $x \in \mathbb{R}^N$, $t \ge t_f$.

The proof follows from Remark 2.3 and Lemma 4.1.

Let us illustrate the assertion of Theorem 4.1 considering the prototype of equation (2.1)

$$\frac{\partial}{\partial t}\left(u\,|u|^{\gamma-1}\right) = \operatorname{div}\left(C_2\,|\nabla u|^{p-2}\nabla u\right) - C_4\,u\,|u|^{\sigma-1},$$

or, equivalently,

$$\frac{\partial v}{\partial t} = \operatorname{div}\left(C_2|v|^{(p-1)(1-\gamma)/\gamma}\,|\nabla v|^{p-2}\,\nabla v\right) - C_4\,v\,|v|^{(\sigma-\gamma)/\gamma}$$

if $v = u\,|u|^{\gamma-1}$ is taken for the new unknown function.

Let $C_4 = 0$. According to Theorem 3.1, the solutions of the Cauchy problem (2.1), (4.1) vanish in a finite time if

$$\frac{p(N-1)}{N-p} < 1+\gamma. \tag{4.5}$$

Besides, the extinction moment is given by the formula

$$t^* = \|u(\cdot,0)\|^{p/N}_{L^{k+\gamma}(\mathbb{R}^N)}\,\frac{N}{p}\left(\frac{(N-1)(p+k-1)}{N-p}\right)^p$$

with

$$k = \frac{N-p}{p}\left(1+\gamma-\frac{p(N-1)}{N-p}\right) > 0.$$

For $p = 2$ (3.12) provides the well-known result for the porous medium equation

$$\frac{\partial v}{\partial t} = C_2\,\Delta\,v^m, \qquad m = 1/\gamma, \qquad v \geq 0. \tag{4.6}$$

Under the condition

$$\|v(\cdot,0)\|_{L^{(\gamma+k)/\gamma}(\mathbb{R}^N)} < \infty$$

the nonnegative solution of the Cauchy problem for equation (4.6) vanishes in a finite time if (see [67, 284])

$$0 < m = \frac{1}{\gamma} < \frac{N-2}{N}.$$

If $\gamma = 1$, for the equation

$$\frac{\partial u}{\partial t} = C_2 \operatorname{div}\left(|\nabla u|^{p-2}\,\nabla u\right)$$

we get from (4.5) the following known result [204]: the solution of the Cauchy problem vanishes in a finite time if $u(x,0) \in L^{1+\kappa}(\mathbb{R}^N)$ and $p < \frac{2N}{N+1}$. It is shown in [204] that these conditions are also necessary.

Remark 4.1. By analogy with Section 3, in the case $\Omega = \mathbb{R}^N$ one may assume that the diffusion and absorption are nonhomogeneous with respect to the independent variables. The boundary-value problems for equation (2.1) in layer or cylinder domains are considered by analogy with Chapter 1.

5 Equations with nonpower and isotropic nonlinearities

Let us proceed to apply the energy method to equations with nonlinearity of general form, which need not be a power.

5.1 Isotropic nonlinear terms. We consider the following initial and boundary-value problem

$$u_t = \Delta\,\varphi(u), \quad \text{in } Q = \Omega \times (0, T), \tag{5.1}$$

$$u(x,0) = u_0(x) \geq 0 \quad \text{in } \Omega, \qquad u(x,t) = 0 \quad \text{on } \Sigma_T = \Gamma \times (0,T) \tag{5.2}$$

under the assumption

$$\varphi \in C^0([0,\infty)) \cap C^1(0,\infty), \quad \varphi(0) = 0, \quad \varphi'(r) > 0 \quad \text{for } r > 0. \tag{5.3}$$

Problem (5.1), (5.2) is studied in the works [122, 67]. In particular, in [122] the finite extinction time of nonnegative solutions of problem (5.1), (5.2) was proven on the basis of another method and under the assumption that, additionally to (5.3),

$$\int_0^1 \frac{d\tau}{\varphi(\tau)} < \infty. \tag{5.4}$$

Let us introduce the functions

$$v = \varphi(u), \quad u = \beta(v) = \varphi^{-1}(v), \quad \omega(u) = \int_0^u \varphi(\tau)\,d\tau.$$

Formally multiplying equation (5.1) by $\varphi(u)$ and integrating by parts, we come to the energy equality

$$\frac{d}{dt}\int_\Omega \omega(u)\,dx + \int_\Omega |\nabla\varphi(u)|^2\,dx = 0. \tag{5.5}$$

Next, let the function

$$g(\omega) = \varphi^q\,(u(\omega))$$

be convex if

$$q \in \left[1, \frac{2N}{N-2}\right] \quad \text{if } N > 2,$$

$$q \in [1,\infty) \qquad \text{if } N = 2.$$

With the use of Jensen's inequality (see the appendix, Section 2) and (2.14), we may write the chain of relations

$$\varphi^2\left(\int_\Omega \omega\,dx\right) \leq C\left(\int_\Omega |\varphi(u)|^q\,dx\right)^{2/q} \leq C\,K_1 \int_\Omega |\nabla\varphi(u)|^2 dx \tag{5.6}$$

with $C = (\text{meas}\,\Omega)^{-2/q}$. Gathering (5.5) with (5.6), we obtain the ordinary nonlinear differential inequality

$$y' + C_0\,\varphi^2(y) \le 0, \quad C_0 = C\,K_1 \tag{5.7}$$

for the energy function

$$0 \le y(t) = \int_\Omega \omega(u(x,t))\,dx \le y(0).$$

Integrating (5.7) in t, for $y(t)$, we get the inequality

$$\int_{y(0)}^{y(t)} \frac{d\tau}{\varphi^2(\tau)} + C_0\,t \le 0. \tag{5.8}$$

According to (5.8), $y(t)$ and, correspondingly, $u(x,t)$, extinct in a finite time if there holds the condition (cf. with (5.4))

$$\int_0^{\omega(1)} \frac{d\omega}{\varphi^2(u(\omega))} = \int_0^1 \frac{du}{\varphi(u)} < \infty.$$

The property of extinction in a finite time of solutions to problem (5.1), (5.2) follows.

5.2 Logarithmic nonlinearity. Let us consider the problem

$$u_t = \Delta(\ln u) = \operatorname{div}\left(\frac{1}{u}\nabla u\right), \quad x \in \Omega, \quad t \in (0,T) \quad (u(x,t) \ge 0), \tag{5.9}$$

$$u(x,0) = u_0(x) \ge 0 \quad \text{in } \Omega, \qquad u(x,t) = 0 \quad \text{on } \Gamma_T = \Gamma \times (0,T). \tag{5.10}$$

Equation (5.9) describes the evolution of the density of the electronic bundle which follows Maxwell distribution (if $N = 3$), or the expansion of a superfine liquid film exposed to the Van der Waals forces. Equation (5.9) also has applications in geometry.

Multiplying equation (5.9) by u^{2k-1}, $k \ge 1$, and integrating by parts, we can obtain the energy relation

$$\frac{1}{2k}\frac{d}{dt}\int_\Omega u^{2k}dx + (2k-1)\int_\Omega u^{2k-3}|\nabla u|^2 dx = 0.$$

Letting

$$y(t) = \|u(\cdot,t)\|^{2k}_{L^{2k}(\Omega)} = \int_\Omega u^{2k}dx$$

and using the inequality

$$\left(\int_\Omega u^{2k}dx\right)^{\nu} \le C(k,N,\Omega)\int_\Omega u^{2k-3}|\nabla u|^2 dx \tag{5.11}$$

(see (2.56)) with $\nu = 1 - \frac{1}{2k}$ and $4k \geq N$, we come to the standard ordinary differential inequality

$$y' + Cy^{\nu} \leq 0. \tag{5.12}$$

Therefore, any solution of problem (5.9), (5.10) such that

$$\int_{\Omega} |u|^{2k} dx + \int_0^T \int_{\Omega} |u|^{2k-3} |\nabla u|^2 dx dt < \infty$$

possesses the property of extinction in a finite time.

We point out that the constant C in inequality (5.11) (and, respectively, in (5.12)) does not depend on Ω if $4k = N$. Hence, in this case any solution of the Cauchy problem also possesses the property of vanishing in a finite time provided that $\|u_0\|_{L^{2k}(\mathbb{R}^N)} < \infty$.

5.3 Equations with anisotropic nonlinearity. Let us consider now a more general problem

$$\frac{\partial u}{\partial t} = \sum_{i=1}^{N} \frac{\partial}{\partial x_i} \left[\psi_i \left(\left| \frac{\partial \varphi(u)}{\partial x_i} \right| \right) \operatorname{sign} \left(\frac{\partial \varphi(u)}{\partial x_i} \right) \right] \quad \text{in } Q, \tag{5.13}$$

$$u(x,0) = u_0(x) \geq 0 \quad \text{in } \Omega; \qquad u(x,t) = 0 \quad \text{on } \Gamma_T. \tag{5.14}$$

A thorough study of problem (5.13), (5.14) delivers serious technical difficulties and relies on some properties of Orlicz spaces $L_N^*(\Omega)$ (see [228]). (The subindex N here is not related to the dimension; we simply retain the traditional notation.) Following [281], we present here an application of the energy method that allows one to establish the property of vanishing in a finite time of the weak solutions.

It will be assumed that the data of the problem satisfy the following conditions:

$$\varphi(s),\ \psi_i(s) \text{ are increasing functions of } s \geq 0, \quad \varphi(0) = 0, \quad \psi_i(0) = 0,$$

$$\varphi \in C[0,\infty] \cap C^1(0,\infty), \quad \psi_i \in C[0,\infty), \quad \varphi(u_0(\cdot)) \equiv \varphi_0(\cdot) \in H^1(\Omega).$$

Definition 5.1. By a weak solution of problem (5.13), (5.14), we mean a nonnegative function $u \in C(\overline{Q})$, bounded along with its generalized derivatives

$$\frac{\partial \varphi(u(x,t))}{\partial x_i}, \quad i = 1, \ldots, N,$$

which satisfies conditions (5.14) and the integral identity

$$\int_Q \left\{ u(x,t) \frac{\partial \omega}{\partial t} - \sum_{i=1}^{N} \psi_i \left(\left| \frac{\partial \varphi(u)}{\partial x_i} \right| \right) \operatorname{sign} \left(\frac{\partial \varphi(u)}{\partial x_i} \right) \frac{\partial \omega}{\partial x_i} \right\} dx\, dt = 0$$

for every $\omega \in C(\overline{Q}) \cap C^1(Q)$ that vanishes on Σ_T and for $t = T$.

References to the works where the question of existence of weak solutions to problem (5.13), (5.14) were studied can be found in [204, 281].

Let us consider first the case where the functions φ and ψ_i are powers. Let

$$\varphi(s) = s^{1/\gamma}, \quad \psi_i(s) = s^{p_i - 1}, \quad p_i \geq 1, \quad \gamma > 0.$$

Then equation (5.13) can be written in the form

$$\frac{\partial v^\gamma}{\partial t} = \sum_{i=1}^{N} \frac{\partial}{\partial x_i}\left(\left|\frac{\partial v}{\partial x_i}\right|^{p_i-2} \frac{\partial v}{\partial x_i}\right).$$

Formally multiplying it by v and integrating by parts we obtain the energy relation

$$\frac{\gamma}{1+\gamma}\frac{d}{dt}\int_\Omega v^{1+\gamma}\,dx + \int_\Omega \sum_{i=1}^{N}\left|\frac{\partial v}{\partial x_i}\right|^{p_i} dx = 0. \tag{5.15}$$

Since the derivatives $\frac{\partial \varphi(u)}{\partial x_i}$ are bounded, we may rely on an inequality of the type (2.14) to obtain

$$\left(\int_\Omega v^{1+\gamma}\,dx\right)^{\widetilde{p}/(1+\gamma)} \leq K_1^{\widetilde{p}}\int_\Omega |\nabla v|^{\widetilde{p}}\,dx \leq K_1^{\widetilde{p}}\, C\int_\Omega \sum_{i=1}^{N}\left|\frac{\partial v}{\partial x_i}\right|^{p_i} dx \tag{5.16}$$

with

$$\widetilde{p} = \max\, p_i, \quad C = C\left(p_i;\ \max_{\overline{\Omega}}\, |\nabla v|\right).$$

Gathering (5.15) with (5.16) leads to the standard ordinary differential inequality

$$y' + \widetilde{C}\,y^\nu \leq 0, \quad y(t) = \int_\Omega v^{1+\gamma}\,dx, \quad \nu = \frac{\widetilde{p}}{1+\gamma}.$$

It follows that for $\nu < 1$ the solution $v(x,t)$ vanishes at a finite time.

Let us assume now that φ, ψ_i need not be powers. Accept the notation

$$\overline{\psi}_i(s) = \int_0^s \psi_i(\xi)\,d\xi, \qquad z = \int_0^{B(z)} \frac{h(\xi)}{\xi}\,d\xi, \tag{5.17}$$

$$h(z) = \left(1/z\Pi_{i=1}^{N}\,\overline{\psi}_i^{\,-1}(z)\right)^{1/N}, \quad \mu(u) = \int_0^u F(\varphi(\xi))d\xi, \tag{5.18}$$

$$F(s) = \int_0^{|s|} f(\xi)\,d\xi, \qquad V(t) = \int_\Omega \mu(u(x,t))\,dx, \tag{5.19}$$

where $f \in C[0,\infty)$ is an increasing function defined for $\xi \geq 0$ and such that $f(0) = 0$, $f(\xi) \leq 1$ for $\xi \leq 1$.

Let us establish the following assertion similar to Lemma 2.1.

Lemma 5.1. *Let* $u(x,t)$ *be a weak solution of problem* (5.13), (5.14). *Then for all* $(t_1, t_2) \subset (0, T)$ *the following inequality holds:*

$$V(t_1) - V(t_2) \leq \int_{t_1}^{t_2}\int_\Omega \sum_{i=1}^{N} \frac{\partial F(\varphi(u))}{\partial x_i}\,\psi_i\left(\left|\frac{\partial \varphi}{\partial x_i}\right|\right)\operatorname{sign}\left(\frac{\partial \varphi(u)}{\partial x_i}\right) dxdt. \tag{5.20}$$

Inequality (5.20) is a byproduct of the formula

$$\frac{dV(t)}{dt} \equiv \frac{d}{dt}\int_\Omega \left(\int_0^u F(\varphi(\xi))\,d\xi\right)dx$$
$$= -\int_\Omega \sum_{i=1}^N \frac{\partial F(\varphi(u))}{\partial x_i}\,\psi_i\left(\left|\frac{\partial \varphi}{\partial x_i}\right|\right)\operatorname{sign}\left(\frac{\partial\varphi}{\partial x_i}\right)dx,$$

which follows after multiplication of equation (5.13) by $F(\varphi(u))$ and integration by parts.

From now on we use the notation (see [228] for the details)

$$\|u\|_M \equiv \|u\|_{L_M^*(\Omega)} = \inf\left\{k > 0 : \int_\Omega M\left(\frac{u(x)}{k}\right)dx \le 1\right\} \tag{5.21}$$

where M is some N-function (see [228] for the detailed definitions).

Theorem 5.1. *Let $u(x,t)$ be a weak solution of problem* (5.13), (5.14). *Let us assume that*

(1)
$$u(x,t) \ge 0, \qquad \left|\frac{\partial \varphi(u)}{\partial x_i}\right| \le C_i, \quad i = 1, \dots, N;$$

(2) *on the interval* $[0, C]$, *the function* $\varphi(s)$ *satisfies the inequality*

$$C_0\,X\left(\int_0^s \varphi(\xi)\,d\xi\right) \le \varphi(s), \quad C_0 = \text{const} > 0, \tag{5.22}$$

where $X(s)$, ($X(0) = 0$), is a monotone increasing continuous function;

(3) *for every $i = 1, \dots, N$, there exists a constant $K_i > 0$ such that*

$$\psi_i(a\,|s|) \le a^{k_i}\,\psi_i(|s|), \quad |s| \le p_i, \quad \forall a \ge 1; \tag{5.23}$$

(4) *the function B defined in* (5.17) *satisfies the inequality*

$$C_1\,\frac{B(s)}{B^\lambda(a)} \le B\left(\frac{s}{a}\right) \quad \forall s \ge 0 \tag{5.24}$$

for some $\lambda > 0$ and $a > 0$ sufficiently small;

(5)
$$\Phi(s) = \left[h\left(B(F(C_0\,X(s)))\right)^{1/\lambda}\right]^{1+\widetilde{k}}, \tag{5.25}$$

$\widetilde{k} = \max_{1\le i\le N} k_i$, and the function $F(s)$ is chosen so that $B(F(C_0\,X(s))) = B(s)$ is an N-function;

(6)

$$\int_0^{\varepsilon} \frac{d\sigma}{\Phi(\sigma)} < \infty, \quad \varepsilon > 0.$$

Then there exists $t^* < \infty$ *such that* $V(t) \equiv 0$ *for* $t \geq t^*$ *and* $u(x,t) \equiv 0$ *for* $x \in \Omega$, $t \geq t^*$.

Proof. Introduce the notation

$$G_i(t) = \int_\Omega f(\varphi(u)) \left| \frac{\partial \varphi(u(t))}{\partial x_i} \right| \psi_i \left(\left| \frac{\partial \varphi(u)}{\partial x_i} \right| \right) dx$$

assuming, without loss of generality, that

$$G_i \leq 1, \quad \left\| \frac{\partial F(\varphi(u))}{\partial x_i} \right\|_{\overline{\psi}_i} = \left\| f \frac{\partial \varphi}{\partial x_i} \right\|_{\overline{\psi}_i} \leq 1.$$

Using (5.17), (5.19) and employing (5.23), we obtain

$$\begin{aligned} &\int_\Omega \overline{\psi}_i \left(\left| \frac{\partial F(\varphi(u))}{\partial x_i} \right| G_i^{-1/(k_i+1)} \right) dx \\ &\quad \leq \int_\Omega \left| \frac{\partial F}{\partial x_i} \right| G_i^{-1/(k_i+1)} G_i^{-k_i/(1+k_i)} \psi_i \left(\left| \frac{\partial F(\varphi(u))}{\partial x_i} \right| \right) dx \leq 1. \end{aligned} \tag{5.26}$$

Next, making use of (5.26), properties of ψ_i and f, and the analogue of inequality (1.20) for Orlicz spaces

$$\|F(\varphi(u))\|_B \leq C \sum_{i=1}^N \left\| \frac{\partial F(\varphi(u))}{\partial x_i} \right\|_{\overline{\psi}_i},$$

we have that

$$-\sum_{i=1}^N G_i \leq -C \left(\sum_{i=1}^N \left\| \frac{\partial F(\varphi(u))}{\partial x_i} \right\|_{\overline{\psi}_i} \right)^{k^*+1} \leq -C \, \|F(\varphi(u))\|_B^{1+k^*} \tag{5.27}$$

with some constant $C = C(k_1, \ldots, k_N)$.

It follows from definitions (5.17), (5.18) of the functions B, h, μ and from assumption (5.22) that

$$B\left(\frac{1}{2} h(s)\right) \leq s, \quad F\left(C_0 X(\mu(u))\right) \leq F\left(\varphi(u)\right). \tag{5.28}$$

The last inequality and (5.24) yield

$$\begin{aligned} &\int_\Omega B\left[F\left(C_0 X(\mu(u))\right)/a\right] dx \\ &\quad \geq C_1 \int_\Omega B\left(F\left(> C_0 X(\mu(u))\right)\right) dx / B^\lambda(a) \geq C_1 \end{aligned} \tag{5.29}$$

because

$$a = \frac{1}{2} h \left(\left(\int_\Omega B(F(C_0 X(\mu(u)))) \, dx \right)^{1/\lambda} \right),$$

$$B^\lambda(a) \le \int_\Omega B(F(C_0 X(\mu(u)))) \, dx.$$

Using (5.21) and the properties of B, we infer from (5.25), (5.28), (5.29) that

$$\begin{aligned} \Phi(V) &= h \left(\left(\int_\Omega B(F(C_0 X(\mu(u)))) \, dx \right)^{1/\lambda} \right) \\ &\le C(C_1) \, \|F(C_0 X(\mu(u)))\|_B \le C \, \|F(\varphi(u))\|_B. \end{aligned} \tag{5.30}$$

Gathering (5.20), (5.27), (5.30), we come to the inequality

$$V(t_2) - V(t_1) \le C \int_{t_1}^{t_2} \Phi(V(\tau)) \, d\tau \quad \forall (t_1, t_2),$$

which means that $V(t)$ is a nonincreasing function and hence its derivative is defined almost everywhere and satisfies the inequality

$$\frac{dV}{dt} \le C \, \Phi(V(t)).$$

The assertion now follows in the usual way. □

5.4 Equations that degenerate in a separate direction. The effect of extinction in a finite time may be caused as well by the degeneracy of the equation in only one space direction. For instance, let us consider the problem

$$\frac{\partial u}{\partial t} = \frac{\partial}{\partial x} \left(|u|^\alpha \left| \frac{\partial u}{\partial x} \right|^{(p-2)} \frac{\partial u}{\partial x} \right) + \frac{\partial^2 u}{\partial^2 y} \quad \text{in } \Omega \times \mathbb{R}_+, \Omega = (0,1) \times (0,1), \tag{5.31}$$

$$u|_\Sigma = 0, \quad \Sigma = \partial\Omega \times (0, \infty), \quad u(x, y, 0) = u_0(x, y), \tag{5.32}$$

under the assumption that the structural constants α, p satisfy the conditions

$$1 \le p < \infty, \qquad 0 < \alpha + p < 2.$$

We introduce the energy functions

$$y(t) = \int_\Omega |u|^2 dx, \qquad D(t) = \int_\Omega \left(|u|^\alpha |u_x|^p + |u_y|^2 \right) dx$$

and assume that the energy relation

$$\frac{1}{2} \frac{dy(t)}{dt} + D(t) \le 0 \tag{5.33}$$

takes place.

There hold the inequalities

$$|u(x,y,t)| = \left|\int_0^1 u_y(x,\eta,t)d\eta\right| \le \left(\int_0^1 |u_y(x,\eta,t)|^2 d\eta\right)^{1/2},$$

$$|u| \le C\left(\int_0^1 |u|^\alpha |u_x(\xi,y,t)|^p d\xi\right)^{1/(p+\alpha)}, \quad C = \left(\frac{p+\alpha}{p}\right)^{p/(p+\alpha)},$$

whence

$$y(t) = \int_\Omega |u|^2 dx \le C D^{1/\nu}(t), \quad \nu = \frac{2(p+\alpha)}{\alpha+p+2} < 1. \tag{5.34}$$

Plugging it into (5.33), we arrive at the differential inequality for the energy function $y(t)$,

$$y'(t) + 2C^{-\nu} y^\nu(t) \le 0.$$

It follows from this inequality that the solutions of problem (5.31), (5.32) vanish in a finite time.

6 Systems of equations of combined type

We devote this section to establishing the property of finite time extinction of solutions of general systems of equations of combined type. In such systems different components of the sought vector-solution may satisfy equations of different types and orders. A few examples of these systems are presented below.

Let us consider the initial and boundary-value problem

$$\mathbf{L}(x,t,\mathbf{u},\mathbf{u}_t) = \mathbf{M}(x,t,\ \mathbf{u},\ D^1\mathbf{u},\ldots,D^l\mathbf{u}) + \mathbf{f}(x,t), \tag{6.1}$$

where $l \ge 1$, $(x,t) \in Q = \Omega \times (0,T)$, $T > 0$,

$$\mathbf{N}(x,t,\mathbf{u},\mathbf{u}_t,D^{l-1}\mathbf{u}) = 0 \quad \text{on } \Sigma_T, \qquad \mathbf{P}(x,\mathbf{u}(x,0)) = 0 \quad \text{in } \Omega \tag{6.2}$$

for the vector-function

$$\mathbf{u}(x,t) = \left(\mathbf{u}^1(x,t),\mathbf{u}^2(x,t)\right) \equiv \left(u_1^1,\ldots,u_p^1,u_1^2,\ldots,u_q^2\right).$$

Here $m = p+q$ and $(u_1^1,\ldots,u_p^1)$ are those components of the sought solution which are expected to vanish at a finite time. The components $(u_1^2,\ldots,u_q^2)$ need not possess this property, generally speaking. In our notation,

$$\mathbf{f}(x,t) = \left(f_1^1,\ldots,f_p^1,f_1^2,\ldots,f_q^2\right) \equiv \left(\mathbf{f}^1,\mathbf{f}^2\right)$$

is a given vector-function satisfying the condition

$$\exists\, T_f \in (0,T): \quad \mathbf{f}^1(x,t) \equiv 0 \quad x \in \Omega, \quad \ge T_f.$$

We will assume that problem (6.1), (6.2) admits a weak solution $\mathbf{u}(x,t)$ from some class $W(Q) \in L^\infty(0,T;\ L^p(\Omega))$, $p \geq 1$. It will also be assumed that given an arbitrary vector $\mathbf{u}(x,t) \in W(Q)$, the operators

$$\mathbf{M} = (\mathbf{M}^1, \mathbf{M}^2) = \left(M_1^1, \ldots, M_p^1, M_1^2, \ldots, M_q^2\right), \quad \mathbf{L} = (\mathbf{L}^1, \mathbf{L}^2),$$
$$\mathbf{N} = (\mathbf{N}^1, \mathbf{N}^2), \quad \mathbf{P} = (\mathbf{P}^1, \mathbf{P}^2)$$

satisfy the following conditions:

$$\frac{dE(t)}{dt} \leq \left(\mathbf{L}^1(x,t,\mathbf{u},\mathbf{u}_t), \mathbf{u}^1(x,t)\right)_\Omega \equiv \int_\Omega \mathbf{L}^1 \mathbf{u}^1\, dx, \tag{6.3}$$

$$E(t) = \int_\Omega \Lambda(x,t,\mathbf{u}(x,t))\, dx,$$

$$\|\mathbf{u}^1(\cdot,t)\|^p_{L^p(\Omega)} \leq E(t) \quad 1 \leq p < \infty, \tag{6.4}$$

$$\mathbf{P}^1(x,\mathbf{u}(x,0)) \equiv \mathbf{u}^1(x,0) - \mathbf{u}_0(0) = 0 \quad x \in \Omega; \tag{6.5}$$

$\mathbf{u}_0 \in L^p(\Omega)$ is a given function,

$$\left(\mathbf{L}^1(x,t,\mathbf{u},D^l\mathbf{u}), \mathbf{u}^1\right)_\Omega \leq -\varphi(E(t)) + F((T_f - t)_+), \tag{6.6}$$

where $\varphi(s) \geq 0$ is a nondecreasing function such that

$$\varphi(0) = 0, \quad \varphi^{-1}(\cdot) \in L^1(0,1), \quad \text{i.e.,} \quad \int_{0+} \frac{ds}{\varphi(s)} < \infty. \tag{6.7}$$

Given arbitrary $\mu > 0$, $\tau > 0$, let us define the functions

$$\theta_\mu(\tau) = \int_0^\tau \frac{ds}{\mu\,\varphi(s)} \tag{6.8}$$

and $\nu_\mu(s) = \theta_\mu^{-1}(s)$. We will assume that there exists $\overline{\mu} < 1$ such that

$$F(s) \leq (1 - \overline{\mu})\, \varphi(\nu_{\overline{\mu}}(s)) \tag{6.9}$$

and

$$\theta_{\overline{\mu}}(E(\theta)) \leq T_f, \quad E(0) = (\Lambda(x,0,\mathbf{u}_0(x)), 1)_\Omega. \tag{6.10}$$

Theorem 6.1. *Let conditions* (6.3)–(6.10) *be true and* $\mathbf{u} \in W(Q)$ *be a weak solution of problem* (6.1), (6.2). *Then*

$$E(t) \equiv 0 \quad \textit{for } T_f \leq t$$

and

$$\mathbf{u}^1(x,t) \equiv 0 \quad x \in \Omega, \quad T_f \leq t. \tag{6.11}$$

Proof. Let us multiply the first p components of equation (6.1) by $\mathbf{u}^1(x,t)$ and integrate over Ω. By virtue of (6.3), (6.6) we come to the nonlinear differential inequality for the function $E(t)$,

$$\frac{dE}{dt} + \varphi(E(t)) \le F\left((T_f - t)_+\right)$$

and the assertion follows from Remark 2.4 of Chapter 1. □

Remark 6.1. Beginning with the moment $t = T_f$, one has to study problem (6.1), (6.2) with $\mathbf{u}^1(x,t) \equiv 0$.

Remark 6.2. Conditions (6.6)–(6.10) in the case when the functions ϕ and F are powers are analyzed in Remark 2.3 in Chapter 1. We derive a connection between the "intensity" of F, the "dissipation" φ, and the instant T_f which provide the effect of the finite time extinction of $E(t)$.

Remark 6.3. The systems of equations of the form

$$\mathbf{L}(x, t, \mathbf{v}_t^0, \dots, \mathbf{v}_t^{(k)}) = \mathbf{M}(x, t, \mathbf{v}, D^1\mathbf{v}), \dots, D^l\mathbf{v}) + \mathbf{f}(x,t),$$
$$\mathbf{v}_t^{(k)} = \frac{\partial^k \mathbf{v}}{\partial t^k}, \quad \mathbf{v}_t^{(0)} = \mathbf{v},$$

can be reduced to (6.1) by means of the change of variables $\mathbf{u}(x,t) = (\mathbf{v}_t^0, \dots, \mathbf{v}_t^{k-1})$.

Examples of systems of equations which meet the conditions of Theorem 6.1 are furnished by fluid mechanics where the operator $\mathbf{L}^1$ often has the form

$$\mathbf{L}^1 \equiv \rho \frac{d\mathbf{u}^1}{dt} \equiv \rho \left(\frac{\partial \mathbf{u}^1}{\partial t} + (\mathbf{v} \cdot \nabla)\, \mathbf{u}^1 \right) \tag{6.12}$$

and $\mathbf{u} = (\mathbf{u}^1, \mathbf{u}^2) \equiv (u_1^1, \dots, u_p^1, u_1^2, \dots, u_q^2)$, $\mathbf{u}^2 \equiv (\rho, v_1, \dots, v_N)$, $q = N+1$, is the sought vector-solution. Here $(u_1^1, \dots, u_p^1)$ are those components which are expected to vanish at a finite time. They may represent densities, concentrations, temperatures, velocities of some phases, etc. The components of the vector $\mathbf{u}^2 = (\rho, v_1, \dots, v_N)$ are, correspondingly, the density and the velocity of the medium.

We assume that the fluid is nonhomogeneous and incompressible (Chapter 4, Section 2). Then the density $\rho(x,t)$ and the velocity $\mathbf{v}(x,t)$ must satisfy the following system of equations:

$$\frac{d\rho}{dt} \equiv \frac{\partial \rho}{\partial t} + \mathbf{v} \cdot \nabla \rho = 0, \quad \operatorname{div} \mathbf{v} = 0.$$

These two equations already form a system of combined type and may form a part of the system

$$\mathbf{L}^2(x, t, \mathbf{u}, \mathbf{u}_t) = \mathbf{M}^2(x, t, \mathbf{u}, \dots, D^l\mathbf{u}) + \mathbf{f}^2(x,t).$$

Let the boundary operator **N** contain the dissipative boundary condition

$$\int_\Gamma \rho\, \mathbf{v}\cdot\mathbf{n}\, |\mathbf{u}^1|^2\, d\Gamma \geq 0, \tag{6.13}$$

where $\mathbf{n}$ is the unit outer normal vector to $\Gamma = \partial\Omega$. Applying (6.13) and the formula

$$\int_\Omega \rho \frac{d\mathbf{u}^1}{dt}\cdot \mathbf{u}^1\, dx = \frac{1}{2}\frac{d}{dt}\int_\Omega \rho\, |\mathbf{u}^1|^2\, dx + \frac{1}{2}\int_\Gamma \rho\, |\mathbf{u}^1|^2\, \mathbf{v}\cdot\mathbf{n}\, d\Gamma,$$

we obtain the following counterpart of (6.3):

$$\frac{dE(t)}{dt} \leq \int_\Omega \rho \frac{d\mathbf{u}^1}{dt}\cdot\mathbf{u}^1\, dx \equiv \int_\Omega \mathbf{L}^1\cdot\mathbf{u}^1\, dx,$$
$$E(t) = \frac{1}{2}\int_\Omega \rho\, |\mathbf{u}^1|^2\, dx \equiv \int_\Omega \Lambda(\mathbf{u}^1)\, dx.$$

If the solution is known to possess the property

$$\frac{1}{C} \leq \rho(x,t), \quad C = \text{const},$$

we immediately get the following counterpart of (6.4):

$$\frac{2}{C}\, \|\mathbf{u}^1(\cdot,t)\|^2_{2,\Omega} \leq E(t).$$

Thus, for the considered class of solutions all the conditions of Theorem 6.1 are fulfilled. Besides, it is unimportant what exactly is the type of the equations for the components of vector $\mathbf{u}^2$.

These properties of operators $\mathbf{L}^1$ of the form (6.12) will be repeatedly used in Chapter 4 in the study of systems of equations of fluid mechanics.

Let us consider the system of equations

$$\begin{aligned}\frac{\partial\psi(x,\mathbf{u})}{\partial t} &= \operatorname{div}\mathbf{A}(x,t,\mathbf{u},D\mathbf{u}) + \mathbf{B}(x,t,\mathbf{u},D\mathbf{u})\\ &\quad + \mathbf{C}(x,t,\mathbf{u}) + \mathbf{f}(x,t) + \operatorname{div}\mathbf{g}(x,t)\end{aligned} \tag{6.14}$$

where $\mathbf{u} = (u_1,\dots,u_m)$ is the sought vector-valued function, $\psi(x,\mathbf{u})$, $\mathbf{C}$, $\mathbf{B}$ and $\mathbf{f}$ are given vector-valued functions, $\mathbf{A}$ and $\mathbf{g}(x,t)$ are given matrices, and $(x,t)\in Q = \Omega\times(0,T)$.

Let us point out in advance that the validity of the operations we perform below is justified in the same way as in Section 1. We assume that the operators **A**, **B**, **C** satisfy for almost all $t\in(0,T)$ conditions (5.5a)–(5.5c) of Chapter 1, while the operator $\partial\psi(x,\mathbf{u})/\partial t$ is subject to the analogue of conditions (6.3), (6.4)

$$\frac{dE}{dt} \leq \left(\frac{\partial\psi(x,\mathbf{u})}{\partial t}, \mathbf{u}\right)_\Omega, \quad \|\mathbf{u}(\cdot,t)\|^{1+\gamma}_{L^{1+\gamma}(\Omega)} \leq E(t), \quad \gamma > 0.$$

It will also be assumed that the solution of (6.14) satisfies the boundary condition

$$\int_\Gamma \mathbf{n}\cdot\mathbf{g}\,\mathbf{u}\,d\Gamma \le 0.$$

Proceeding by analogy with Lemma 2.1 one may derive the scalar energy relation

$$\frac{dE}{dt} + (\mathbf{A},\, D\mathbf{u})_\Omega - (\mathbf{B},\, \mathbf{u})_\Omega - (\mathbf{C},\, \mathbf{u})_\Omega \le (\mathbf{f},\, \mathbf{u})_\Omega - (\mathbf{g},\, D\mathbf{u})_\Omega\,.$$

By properties (5.5a)–(5.5c) of Chapter 1, Section 3, this inequality can be written in the form

$$\begin{aligned}\frac{dE}{dt} &+ C_2\|D\mathbf{u}\|^p_{L^p(\Omega)} + C_4\,\|\mathbf{u}\|^{1+\sigma}_{L^{1+\sigma}(\Omega)} - C_3\left(|\mathbf{u}|^{1+\alpha},\, |D\mathbf{u}|^\beta\right)_\Omega \\ &\le \|\mathbf{f}\|_{L^{(1+\gamma)/\gamma}(\Omega)}\,\|\mathbf{u}\|_{L^{1+\gamma}(\Omega)} + \|\mathbf{g}\|_{L^{p/(p-1)}(\Omega)}\,\|D\mathbf{u}\|_{L^p(\Omega)}.\end{aligned}$$

and then studied as in Theorem 2.1.

7 Higher-order parabolic equations and other applications

7.1 Higher-order parabolic equations. Operator **L** in (6.1) may contain derivatives of orders higher than two. Let us consider a few examples. To begin with, we consider the following initial and boundary-value problem

$$\begin{aligned}&\frac{\partial u}{\partial t} + \Delta(|\Delta u|^{p-2}\,\Delta u) = 0 \quad \text{in } Q = \Omega\times(0,T), \quad p\in(1,2),\\ &u|_{\Sigma_T} = 0, \quad \left.\frac{\partial u}{\partial \nu}\right|_{\Sigma_T} = 0, \quad u(x,0) = u_0(x)\in L^2(\Omega),\end{aligned}\tag{7.1}$$

where ν is the unit normal outer vector to $\Gamma = \partial\Omega$, and $\Sigma_T = \Gamma\times(0,T)$. The existence of weak solution $u \in L^\infty(0,T;\, L^2(\Omega)) \cap L^p(0,T;\, W^{2,p}_0(\Omega))$ was proved in [243]. In the present case conditions (6.3)–(6.6), due to the embedding $W^{2,2}_0(\Omega)\subset L^2(\Omega)$, take the form

$$\frac{dE}{dt} \equiv \frac{d}{dt}\left(\frac{1}{2}\|u\|^2_{L^2(\Omega)}\right) = (u_t,\, u)_\Omega,$$

$$-(\Delta(|\Delta|^{p-2}\,\Delta\, u),\, u)_\Omega = -\|\Delta u\|^p_{L^p(\Omega)} \le -C\,\|u\|^p_{L^2(\Omega)}. \tag{7.2}$$

It follows then that the nonlinear differential equation

$$\frac{d\,E}{dt} + C\,E^{p/2} \le 0, \quad p\in(1,2)$$

holds, which provides the finite time extinction of the weak solutions.

In the same way, we can study the initial and boundary-value problem

$$\frac{\partial}{\partial t}(u|u|^{(\gamma-1)}) + Lu = f(x,t) \quad \text{in } Q, \tag{7.3}$$

$$\left.\frac{\partial^k u}{\partial^k \nu}\right|_{\Sigma_T} = 0, \quad k = 0, \ldots, m-1; \quad u(x,0) = u_0(x) \quad \text{in } \Omega, \tag{7.4}$$

with the operator

$$Lu = (-1)^m \sum_{|\alpha|=m\geq 1} D^\alpha(|D^\alpha u|^{(p-2)} D^\alpha u) + C_4 u|u|^{\sigma-1}.$$

The existence of weak solutions of problem (7.3), (7.4) was proved in [79].

The qualitative properties of the weak solutions such as extinction in a finite time were studied with the energy method in [75], [77].

We present a scheme of application of the energy method to this problem. The formal multiplication of equation (7.3) by $u(x,t)$ and integration by parts lead to the energy relation

$$\frac{\gamma}{1+\gamma}\frac{dE(t)}{dt} + \int_\Omega \left(|D^m u|^p + |u|^{\sigma+1} dx\right) = \int_\Omega f u dx \tag{7.5}$$

where

$$E(t) = \int_\Omega |u(x,t)|^{1+\gamma} dx, \quad |D^m u| = \left(\sum_{|\alpha|\leq m} |D^\alpha u|^p\right)^{1/p}.$$

Using Lemma 3.5 of the appendix, we can write the inequality

$$\begin{aligned} E^{1/(1+\gamma)}(t) &\leq K \|D^m\|^\theta_{L^p(\Omega)} \|u\|^{1-\theta}_{L^{1+\sigma}(\Omega)} \\ &\leq \left(\int_\Omega \left(|D^m|^p + |u|^{1+\sigma}\right) dx\right)^{\frac{\theta}{p} + \frac{(1-\theta)}{(1+\sigma)}}, \end{aligned} \tag{7.6}$$

where $K = K(p, \sigma, N)$ and

$$\frac{1}{1+\gamma} = \theta\left(\frac{1}{p} - \frac{m}{N}\right) + \frac{1-\theta}{1+\sigma}.$$

Let us assume that either

$$\sigma < \gamma \quad \text{and} \quad \frac{1}{1+\gamma} \geq \frac{1}{p} - \frac{m}{N},$$

or

$$\sigma > \gamma \quad \text{and} \quad \frac{1}{1+\gamma} \leq \frac{1}{p} - \frac{m}{N},$$

which provides the inequality

$$\nu = \frac{1+\gamma}{(\theta/p + (1-\theta)/(1+\sigma))} = \frac{1}{(1+\theta(1+\gamma)m/N)} < 1.$$

Gathering (7.5) with $f = 0$ and (7.6), we come to the ordinary differential inequality

$$\frac{dE(t)}{dt} + CE^{\nu} \leq 0$$

with the parameters $\nu < 1$, and $C = \frac{1+\gamma}{\gamma K^{\nu}}$.

7.2 Nonlocal operators. The effect of finite time extinction may also be caused by the nonlocal structure of the differential operators. Let us consider the equation

$$\frac{\partial u}{\partial t} = L(x, t, Du) - u\|u\|^{-\alpha}_{L^p(\Omega)} \quad \text{in } Q \tag{7.7}$$

with some positive constants α, $p \in (1, \infty)$. Let us assume that we are given a weak solution of some initial and boundary-value problem for equation (7.7) such that

$$(L(x, t, u, Du), u|u|^{p-2}))_{\Omega} \leq 0.$$

The formal multiplication of equation (7.7) by $u|u|^{p-2}$ and integration over Ω lead to the ordinary inequality

$$\frac{dE}{dt} + pE^{(1-\alpha/p)} \leq 0, \quad E(t) = \|u(\cdot, t)\|^p_{L^p(\Omega)},$$

and, correspondingly, to the estimate

$$E^{\alpha/p}(t) \leq E^{\alpha/p}(0) - \alpha t.$$

A similar analysis is applicable to the solutions of the initial and boundary-value problem

$$\frac{\partial u}{\partial t} = \operatorname{div}(\|u\|^{-\alpha}_{L^q(\Omega)}|Du|^{(p-2)}Du) \quad \text{in } Q, \tag{7.8}$$

$$u|_{\Sigma_T} = 0, \qquad u(x, 0) = u_0(x) \quad \text{in } \Omega \tag{7.9}$$

with constants $q \in (1, \frac{Np}{N-p})$, $2 \leq p < 2 + \alpha < \infty$. Indeed, the formal multiplication of equation (7.8) by u and integration by parts give the inequality

$$\frac{dE}{dt} + C\, E^{(p-\alpha)/2} \leq 0, \quad E(t) = \|u(\cdot, t)\|^2_{L^2(\Omega)}$$

with a constant $C = C(K_1, p, \alpha, q, N, \operatorname{meas}\Omega)$, and the constant K_1 defined in (2.14). This inequality yields the finite time of extinction of $u(x, t)$.

In the same way, we can study a more complicated equation of the form

$$\frac{\partial u}{\partial t} = \operatorname{div}\left(\|u\|^{-\alpha}_{L^q(\Omega)}\|Du\|^{\gamma}_{L^m(\Omega)}|Du|^{(p-2)}Du\right)$$

with some constants α, q, γ, m, p.

7.3 Equations with singular "superabsorption" terms. The energy method is well-settled for the study of equations involving "superabsorption" terms,

$$\frac{\partial u}{\partial t} = L(x,\, t,\, Du) - u\,|u|^{-\sigma}, \qquad 1 < \sigma < \infty. \tag{7.10}$$

Indeed, let us assume the existence of a weak solution $u(x,t)$ of equation (7.10) such that

$$(L(x,\, t\,u,\, Du),\, u|u|^{\gamma-2})_\Omega \leq 0, \qquad E(t) = \int_\Omega |u|^\gamma\, dx < \infty$$

with some constant $1 < \gamma < \sigma$. Multiplying equation (7.10) by $u|u|^{\gamma-2}$, integrating by parts and taking into account (7.3), we obtain the inequality

$$\frac{dE}{dt} + \gamma \int_\Omega |u|^{\gamma-\sigma}\, d\Omega \leq 0.$$

Next, using the inequality

$$|\Omega| = \int_\Omega dx \leq E^{\nu/\sigma} \left(\int_\Omega |u|^{-\nu}\, dx \right)^{\gamma/\sigma}$$

with $\nu = \sigma - \gamma > 0$, we obtain the standard differential inequality

$$\frac{dE}{dt} + C\,E^{-\nu} \leq 0, \quad C = \gamma |\Omega|^{\sigma/\gamma},$$

whence the estimate

$$E^{1+\nu}(t) \leq E^{1+\nu}(0) - C\,(1+\nu)\,t.$$

7.4 Finite time stabilization to a nonzero state. Let us consider the initial and boundary-value problem

$$u_t = \operatorname{div}(|u(x,t) - l(x)|^\alpha Du) \quad \text{in } Q, \tag{7.11}$$

$$u|_{\Sigma_T} = l(x), \quad u(x,0) = u_0(x) \geq l(x), \quad \text{in } \Omega \tag{7.12}$$

with some constant

$$\alpha \in \left(-\min\left(2, \frac{4}{N}\right), 0 \right).$$

This is an example in which $u_\infty \not\equiv 0$. Specifically, we shall see that $u_\infty = l(x)$.

We shall study the weak solution $u(x,t) \geq l(x)$ of problem (7.11), (7.12) such that

$$\int_\Omega |u|^2 dx + \int_0^t \int_\Omega (u-l)^\alpha |Du|^2 dx\,dt \leq E_0 < \infty.$$

The existence of such solutions can be proved by the methods of [243, 233]. Introduce the new function

$$v(x,t) = u(x,t) - l(x) \geq 0, \quad u_t = v_t, \quad v|_\Gamma = 0.$$

Using the formula of integration by parts, for the function $v(x,t)$, we obtain the relation

$$\frac{1}{2}\frac{d}{dt}\int_\Omega |v|^2 dx + \int_\Omega v^\alpha |Dv|^2 dx = \frac{1}{1+\alpha}\int_\Omega v^{1+\alpha}\Delta l(x)\, dx dt. \tag{7.13}$$

Let us additionally assume that the prescribed function $l(x)$ and the constant α satisfy the conditions

$$\Delta l \leq 0, \qquad \frac{4}{2+\alpha} \leq \frac{2N}{N-2}. \tag{7.14}$$

Introducing the function

$$w = v^{1+\alpha/2}, \qquad v^\alpha |Dv|^2 = \frac{4}{2+\alpha}|Dw|^2,$$

using inequality (2.14) with $p = 2$ and $q = \frac{4}{2+\alpha} \leq \frac{2N}{N-2}$, and gathering (7.13) with (7.14) we obtain the inequality

$$\frac{dE}{dt} + C\, E^\nu \leq 0, \quad E(t) = \int_\Omega w^{4/(2+\alpha)} d\,\Omega, \quad \nu = 1 + \alpha/2 < 1,$$

whence the finite time extinction of the solution, i.e.,

$$u(x,t) \equiv l(x) \quad \text{in } \Omega \text{ for all } t \geq t^*.$$

It is worth noting the following peculiar phenomenon. Let the function $l(x)$ satisfy the condition

$$-\Delta l \geq \delta = \text{const} > 0.$$

Then in problem (7.11), (7.12), the effect of finite time stabilization occurs if $0 < \alpha < 1$.

We infer from the energy relation (7.13) that

$$\frac{1}{2}\frac{d}{dt}\int_\Omega v^2 dx + \int_\Omega v^\alpha |Dv|^2 dx + \frac{\delta}{1+\alpha}\int_\Omega v^{1+\alpha} dx \leq 0, \tag{7.15}$$

$$\frac{1}{2}\frac{d}{dt}\int_\Omega w^{\frac{4}{2+\alpha}} dx + \int_\Omega \left(\frac{4}{2+\alpha}|Dw|^2 dx + \frac{\delta}{1+\alpha}|w|^{\frac{2(1+\alpha)}{2+\alpha}}\right) dx \leq 0, \tag{7.16}$$

$$\frac{dE}{dt} + C_0 \int_\Omega \left(|Dw|^2 dx + |w|^{\frac{2(1+\alpha)}{2+\alpha}}\right) dx, \quad E(t) = \int_\Omega w^{\frac{4}{2+\alpha}} dx \tag{7.17}$$

with

$$C_0 = 2\min\left(\frac{4}{2+\alpha}, \frac{\delta}{1+\alpha}\right).$$

Using inequality (2.15) with

$$q = \frac{4}{2+\alpha}, \quad p = 2, \quad r = \frac{2(1+\alpha)}{2+\alpha},$$
$$\delta = \frac{1/r - 1/q}{1/r - (N-p)/Np} = \frac{(2+\alpha)(1-\alpha)}{2+4(1+\alpha)/N}, \quad \lambda = \frac{2+\alpha}{4(1+\sigma)}$$

and Young's inequality, we obtain the estimate

$$(E)^{\lambda}(t) \le \frac{1}{2C}\int_{\Omega}\left(\frac{4}{2+\alpha}|Dw|^2 + \frac{\delta}{1+\alpha}w^{2(1+\alpha)}\right)dx. \tag{7.18}$$

Gathering (7.17) and (7.18), we come to the ordinary differential inequality

$$\frac{dE}{dt} + CE^{\lambda} \le 0, \quad \lambda < 1,$$

whence the effect of finite time stabilization.

The explanation of this effect lies in the fact that the equation for $w = (u - l)$ has the form

$$w_t = \operatorname{div}\left(w^{\alpha}\nabla w\right) + \alpha w^{\alpha-1}\nabla w \cdot \nabla l - |\Delta l| w^{\alpha}.$$

The last term on the right-hand side of this equation plays the role of the strong absorption term (see Chapter 3) and the equation is of the type "weak diffusion/strong absorption." The finite time extinction is an intrinsic property of solutions to such equations.

In [140] the energy method was applied to study the asymptotic behavior in time and, in particular, the finite time stabilization, of the weak solutions of the problem

$$\frac{\partial w}{\partial t} = \operatorname{div}(|Dw|^{p-2}Dw) \quad \text{in } Q = \Omega \times \mathbb{R}_+, \tag{7.19}$$

$$\left.\frac{\partial w}{\partial \nu}\right|_{\Sigma_N} = 0, \quad w^m\big|_{\Sigma_D} = h = \text{const}, \tag{7.20}$$

$$w(x,0) = w_0(x) \ge h, \quad \text{in } \Omega \subset \mathbb{R}^N \tag{7.21}$$

where $1 < p < \infty$, $\Gamma_D \cup \Gamma_N = \Gamma = \partial\Omega$, and $\Sigma_D = \Gamma_D \times (0,T)$, $\Sigma_N = \Gamma_N \times (0,T)$. In this problem $u_{\infty}(x) = h \not\equiv 0$.

Problem (7.19)–(7.21) arises from the mathematical modeling of the motion of a turbulent gas in a lengthy pipe [140].

In [41, 42], the energy method was applied to the study of properties of solutions of the equation

$$(u - l(x))^{\gamma}\, u_t = \operatorname{div}\left((u - l(x))^{\alpha}\, |\nabla u|^{\beta} \nabla u\right),$$

where $l(x)$ is a given surface. These works contain conditions which provide either the finite time of stabilization of the solution $u(x,t)$ to the stationary state $l(x)$, or the effect of finite speed of propagation of disturbances from the data (in the sense of Definition 1.1 of Chapter 3).

8 Bibliographical notes and open problems

Most results in the literature on *finite time stabilization to a stationary state* concern the case where the stationary state is identically zero, i.e., they deal with the property of *finite time extinction.* A few problems in which the stationary solution is nonzero are presented in Section 7. Some general references on the asymptotic behavior of solutions to the infinite-dimensional dynamical systems are given in Hale [189] and Temam [300].

It should be noted that the examples given in Subsection 1.1 frequently appear in the undergraduate ODE courses and serve an illustration of the fact that the solution need not be *unique* if the Lipschitz conditions fail. Nonetheless, the uniqueness takes place in the class of nonnegative solutions.

The property of finite time extinction also holds for some discrete dynamical systems, Le Roux [239], and for the functional differential equations; see Wu [309] for a general reference on such equations.

The first result in the literature on the finite time extinction in the solutions of the fast diffusion equation is apparently due to Sabinina [283]. She used a comparison method to study the one-dimensional equation with the exponents $p = 2$ and $\gamma < 1$. The same equation but in an open bounded set $\Omega \subset \mathbb{R}^N$, $N \geq 1$, was treated by Berryman and Holland [81, 82] by using some energy arguments. The sharp results for the equation with a nonlinear function φ, not necessarily homogeneous, are due to G. Díaz and J. I. Díaz [122]. The case $\Omega = \mathbb{R}^N$ was analyzed by Benilan, Brezis, and Crandall [66]. Some special examples with $p \neq 2$ are given in Bamberger [60]. Herrero and Vazquez [193] extended the result of Benilan, Brezis, and Crandall to the case $\Omega = \mathbb{R}^N$ and $p \neq 2$.

The property of finite time extinction for solutions of a quasilinear problem arising from image processing (with the limit exponent $p = 1$) was established in Andreu, Caselles, Díaz, and Mazón [6].

For the semilinear parabolic equations with strong absorption, $p = 2$, $\gamma = 1$, and $\sigma < 1$, results on the finite time extinction can be found in Kalashnikov [204] (see also Kersner [217]). We recall that in Chapter 2 we are dealing with the property of *global* extinction and that *local vanishing* of solutions will be studied in Chapter 3.

For the study of the finite time localization property, the energy method in the form we use it in this chapter was proposed in [15].

An energy method was used in Knops and Straughan [224] (see also Straughan [298]). An unpublished result by Luc Tartar on semilinear parabolic equations was later generalized for quasilinear equations by Evans and Knerr [155]. A doubly nonlinear equation with strong absorption was studied in Tsutsumi [301]. See alsothe result by Veron [306] in the framework of m-accretive operators in $L^1(\Omega)$ and its extension in Pazy [268]. For the parabolic obstacle problem (which formally corresponds to the case $\sigma = 0$), previous results were obtained by Brezis and Friedman [99] and Bensoussan and Lions [68]. Generalizations in the framework of m-accretive operators in $L^s(\Omega)$ were done by Brezis [94] ($s = 2$) and Díaz [125] ($s = \infty$). A sharper result on the finite extinction time for the obstacle problem via symmetrization can be found in Díaz and Mossino [141]. For the effect of finite time extinction of solutions to multivalued parabolic equations appearing in the study of *null controllability* of parabolic equations in which the multivalued term corresponds to the feedback control, see Slemrod [294], Barbu [63], and Díaz and Ramos [144].

The results on the instantaneous extinction time (for equations with sufficiently flat right-hand sides) are due to Antontsev and Díaz [27].

The asymptotic behaviour of the solution near the extinction instant was studied by several authors: Berryman and Holland [82], Kwong [232], Díaz and Liñan [140], and Andreu, Caselles, Díaz, and Mazón [6].

For the stabilization in a finite time to a nonzero stationary state in the case of Neumann boundary conditions, see Andreu, Caselles, Díaz, and Mazón [6]. The presence of an absorption term implies that this stationary state becomes zero (see Lair and Oxley [238]). The effect of finite time extinction on the boundary for linear equations but with Signorini boundary conditions was proved in Díaz and Jimenez [139].

The negative answer to the question about the existence of the property of finite time extinction can be given by using very different methods such as the Nirenberg strong maximum principle, the backward unique continuation (see Mizohata [253]), or by showing that certain norm of the solution is preserved in time. (This is true for most hyperbolic equations and the Schrödinger equation.) For sharper results concerning the case of parabolic equations with absorption, see Kersner [217], Lair [236], and Lair and Oxley [237].

The exponential decay of the solutions of parabolic reaction-diffusion equations with strong absorption but under nonhomogeneous boundary conditions was proved in Ricci [276]. For doubly nonlinear equations with nonhomogeneous boundary data, see Ricci and Tarzia [277] and Bandle, Nanbu, and Stakgold [61] (see Díaz and Liñan [140]).

For the study of numerical schemes employed to approximate the extinction instant, see Le Roux [239] for the case of $p = 2$ and $m \in (0, 1)$, unpublished results by E. Menendez concerning the case of doubly nonlinear equations and Nakaki [258], and their bibliography for equations with strong absorption terms.

A description of various weighted energy methods can be found in Straughan [298].

Results on the behaviour of solutions to the Cauchy problem can be found in Sabinina [283], Kalashnikov [204], Kersner [218], Evans and Knerr [155], Benilan, Brezis, and Crandall [66], and DiBenedetto [150], among others. The results relying on the use of Orlicz spaces are taken from Rykov [281].

For the effect of finite time localization for solutions of certain systems arising in chemical engineering, see Bandle and Stakgold [62], Díaz and Hernández [134], Stakgold [296], and Díaz and Stakgold [146]. Our presentation is original and new.

For the case of higher-order equations, see Bernis [77]. The finite time stabilization to a nonzero stationary solution was also proved in Haraux [190] for the wave equation with a multivalued friction (the finite time extinction for a suitable linear wave equation with friction was established by Majda [247]; see also Díaz and Liñán [140] for the case of non-Lipschitz perturbations of an ordinary second order equation). A pioneering result on the finite time extinction for the first-order hyperbolic equations is due to Murray [257]; see also Díaz and Veron [148] for a more general class of first-order quasilinear equations including the associate obstacle problem.

A different approach to a nonlocal problem related to that discussed in Subsection 7.2 was proposed in Pazy [268].

There is a very large literature on *singular superabsorption problems* related to equation (7.10). This is the so-called *quenching problem* in which the singularity is usually taken at the level $u = 1$:

$$u_t - L(x, t, u, Du, D^2u) = \frac{1}{(1-u)^k}$$

with some $k > 0$. For more details, see [160, 162, 107, 212, 308, 108, 109, 121, 120, 161, 213, 269]. Notice that the singular term blows up on the subset where the solution vanishes. Nonetheless, if the equation is written in the form

$$u_t - L(x, t, u, Du, D^2u) = \frac{1}{(1-u)^k}\chi_{\{u\neq 1\}},$$

where $\chi_{\{u\neq 1\}}$ is the characteristic function of the set $\{(x,t) : u \neq 1\}$, under suitable assumptions one can prove the existence of global solutions defined for every $t \in \mathbb{R}^+$ (see Phillips [269] and Levine [240] for the case $L = \Delta$, and Kawohl and Kersner [213] and Deng [120] for $Lu = \Delta u^m$). The local vanishing phenomena for the solutions of this class of equations can be proved by the tools we present in the next chapter. For the study of the first-order singular terms (singular nonlinearities in the convection term) see Galaktionov–Kersner [170].

Kawohl and Peletier [214] discuss the relation between the properties of the finite time extinction and finite time blowup for solutions of different nonlinear parabolic equations connected by a suitable transformation. The series of papers by Galaktionov and Vazquez [177, 178, 169, 175, 176] (see also the references therein) is devoted to the study of the fine structure of solutions to parabolic equations near the extinction or the blowup instant.

The decay of solutions to nonlinear parabolic equations with singular diffusion exponents was studied by several authors: see Herrero [192], Esteban, Rodriguez, and Vazquez [154], Rodriguez and Vazquez [279], Zhang [312].

Results on the nonzero stationary state, generalizing the results of Subsection 7.4 can be found in Antontsev and Kashevarov [41, 42].

A few open problems are in order.

1. To study the property of finite time extinction when the initial energy is infinite and there is no regularizing effect.

2. To study the effect of finite time extinction for the class of viscosity solutions for nondivergent parabolic equations (a partial result can be found in G. Díaz and J. I. Díaz [122]).

3. To study the continuous dependence of extinction instant on the data.

4. To construct the schemes of numerical approximation well-adapted to the presence of the extinction instant for higher-order equations and systems.

3

Space and Time Localization in Nonlinear Evolution Problems

1 Introduction

In this chapter the energy method is applied to study the space and time localization of weak solutions for nonlinear degenerate parabolic equations and systems of such equations.

1.1 Definitions. Let us first describe the types of localization we are going to study. Let $u(x,t)$ be a real-valued function of $N+1$ variables $t \in \mathbb{R}^+$ and $x \in \mathbb{R}^N$. It is assumed that $u(x,t)$ is defined in a cylinder $Q = \Omega \times (0,T)$ for some $\Omega \subseteq \mathbb{R}^N$, $T > 0$, and that for every $t \in [0,T]$ $u(x,t)$ is a measurable function of the variables $x \in \Omega$ and

$$u \in C\left([0,T];\ L^1_{\text{loc}}(\Omega)\right).$$

Let us take $x_0 \in \Omega$ and assume that

$$u(x,0) = 0 \quad \text{a.e. in a ball } B_{\rho_0}(x_0) = \{x : |x - x_0| < \rho_0\} \subset \Omega.$$

Definition 1.1. (i) A function $\rho(t) : [0, t^*) \mapsto [0, \infty)$, $\rho(0) \le \rho_0$, is called a *rate at the point* x_0 if

$$\forall t \in [0, t^*), \quad u(x,t) = 0 \quad \text{a.e. in } B_{\rho(t)}(x_0).$$

(ii) A function $u(x,t)$ is said to possess the property of *finite speed of propagation* (from nonzero disturbances) if for some $x_0 \in \Omega$ and $t^* > 0$ there exists a strictly positive rate at the point x_0.

(iii) Let $u(x,t)$ be defined on $Q = \Omega \times [0,\infty)$. We say that $u(x,t)$ possesses the property of *stable localization* if it has the property of finite speed of propagation with a rate $\rho(t)$ defined on the whole of $[0,\infty)$ such that

$$\liminf_{t\to\infty} \rho(t) > 0.$$

(iv) Given $x_0 \in \Omega$, let

$$\rho_0 = \sup\{\rho > 0 : u(x,0) = 0 \text{ a.e. in } B_\rho(x_0) \subset \Omega\}.$$

We say that $u(x,t)$ possesses the *generalized waiting time property* if for some $t^* > 0$ the function $\rho(t) \equiv \rho_0$ is a rate at the point x_0 on the interval $[0, t^*]$.

Let $x_* \in \overline{\operatorname{supp} u(x,0)}$. The instant

$$t_w(x_*) = \sup_{x_0\in\Omega} \{t^* : |x_* - x_0| = \rho_0 \text{ and } \rho_0 \text{ is a rate on } (0,t^*) \text{ at the point } x_0\}$$

is called the *waiting time* at the point x_*.

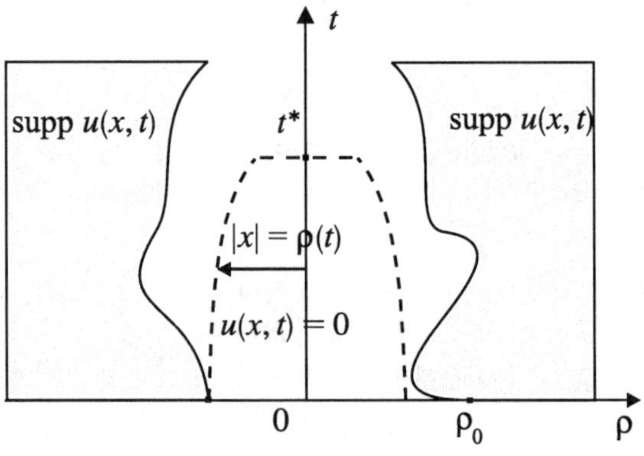

Figure 1.1: Finite speed of propagation, $x_0 = 0$.

This definition needs several comments. First of all, it is clear that the rate is not unique: given a rate $\rho(t)$ at the point $x_0 \in \Omega$, any positive function $\delta(t) \in (0, \rho(t)]$ is also a rate at the point x_0. The optimal rate on an interval $(0, t^*)$ can be defined by

$$\eta(t) = \sup\{\rho(t) : \rho(t) \text{ is a rate at the point } x_0, t \in (0,t^*)\}. \qquad (1.1)$$

For functions of two variables, the function $\eta(t)$ defined in this way coincides with the traditional definition of the free boundary or interface occurring in nonnegative solutions of nonlinear parabolic equations. (By the free boundary or the interface we mean a curve in the (x,t)-plain separating the regions where the solution is positive or zero). We abstain here from any detailed discussion of the questions related to the qualitative properties of $\eta(t)$, its asymptotic behavior for long times,

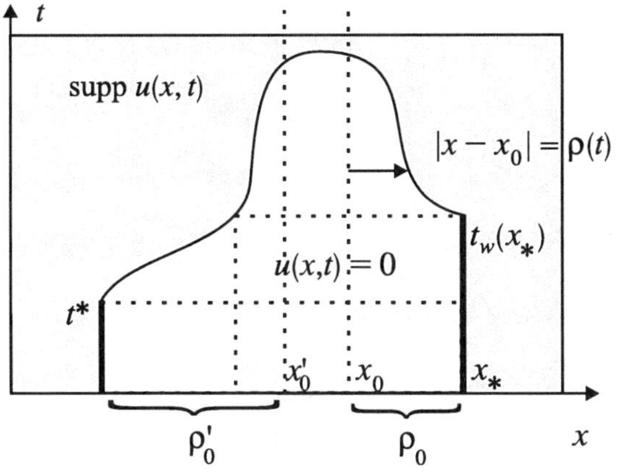

Figure 1.2: The generalized waiting time property.

regularity, etc. The reader is referred to the papers quoted in Section 9 for further information on this subject.

Definition 1.1 says, in essence, that a function possesses the property of finite speed of propagation of disturbances if the "zero caverns" take time to disappear. This is what is guaranteed by the existence of a rate. The instant speed of propagation need not be finite, however. Let $N = 1$. Given the optimal rate of propagation $\eta(t)$, we can introduce the functions

$$V^+(t) = \limsup_{\Delta t \to 0} \frac{\eta(t + \Delta t) - \eta(t)}{\Delta t}, \qquad V^-(t) = \liminf_{\Delta t \to 0} \frac{\eta(t + \Delta t) - \eta(t)}{\Delta t},$$

which are of dimension LT^{-1} (length/time) and can be interpreted as upper and lower bounds for the instant velocity of propagation of nonzero disturbances (the existence of $\eta'(t)$ is not assumed here). In the next subsection we present examples of explicit solutions to nonlinear parabolic equations which possess the property of finite speed of propagation in the sense of Definition 1.1 but with the instant velocity of propagation infinite at certain points. It is worth noting that finite speed of propagation is not displayed by the solutions of any linear parabolic problem but is typical for solutions of linear hyperbolic equations. Say in the simplest case, where $u(x, t)$ is a solution of the Cauchy problem for the linear hyperbolic equation

$$cu_{tt} - u_{xx} = 0,$$

it is given by d'Alembert's formula, and in the above notation we have $V^+(t) = V^-(t) = c$ for all $t > 0$.

For the sake of convenience, we have introduced the concept of finite speed of propagation of disturbances with respect to the zero-level of $u(x, t)$. Considering the function $u(x, t) - s$ with $s \neq 0$, we can extend these concepts in a natural way to define the finite speed of propagation of disturbances with respect to s-level.

In the next definition we describe the situation when the function $u(x,t)$ admits a strictly increasing rate $\rho(t)$ at a point x_0.

Definition 1.2. Let $x_0 \in \Omega$ be a given point, and

$$\rho_0 = \sup\{\rho > 0 : u(x,0) = 0 \text{ a.e. in } B_\rho(x_0) \subset \Omega\}.$$

(i) Let $\rho_0 > 0$. We say that $u(x,t)$ possesses the property of *support shrinking* if on some interval $(0,t^*) \neq \emptyset$ there exists a monotone increasing rate $\rho(t)$ at the point x_0 such that $\rho(0) = \rho_0$.

(ii) Let $\rho_0 = 0$. The function $u(x,t)$ is said to possess the property of a *dead core formation* if on some interval $(0,t^*) \neq \emptyset$ it admits a strictly positive rate $\rho(t)$ at the point x_0.

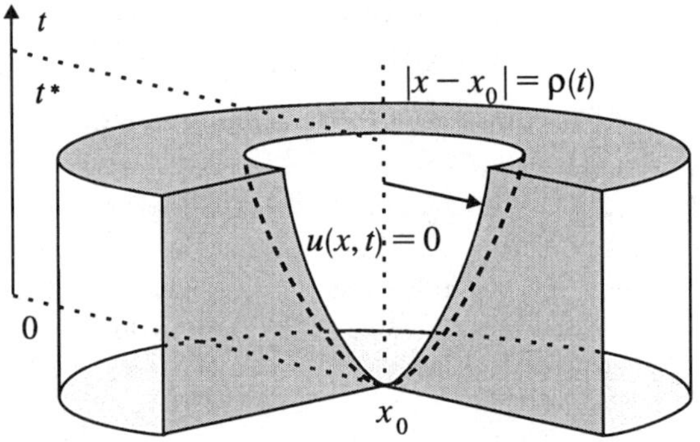

Figure 1.3: Formation of a dead core

1.2 Examples. For illustration, we present here a few simple examples of explicit solutions to nonlinear parabolic equations that admit explicit formulas.

(1) The first example is furnished by the self-similar solution for the so-called *porous medium equation*

$$u_t = \Delta u^m, \quad (x,t) \in \mathbb{R}^N \times \mathbb{R}^+ \tag{1.2}$$

with the parameter of nonlinearity $m > 1$. This name is due to one of the most natural interpretations of this equation. If we describe the motion of a polytropic gas with density u, pressure $p = \lambda u^{m-1}$, and velocity $v = -\kappa \nabla p$ (Darcy law) through a porous medium, equation (1.2) expresses the mass balance law of the motion (up to a constant which we scale out to unit). If $m = 2$, equation (1.2) becomes the Boussinesq equation in filtration theory.

Equation (1.2) admits the class of explicit self-similar solutions constructed in [311, 64] and then rediscovered in [267]:

$$U(x,t) = t^{-\alpha} f(\xi), \quad \xi = \frac{x}{t^\beta}, \tag{1.3}$$

where

$$f(\xi) = \left[A - B|\xi|^2\right]_+^{1/(m-1)}$$

and the constants B, α, β are defined as

$$\alpha = \frac{1}{m - 1 + \frac{2}{N}}, \qquad \beta = \frac{\alpha}{N}, \qquad B = \frac{m-1}{m}\beta.$$

The constant A is arbitrary. It distinguishes a concrete solution of this family. The evolution of the space profile of $U(x, t + \varepsilon)$, $\varepsilon > 0$, is shown in Figure 1.4.

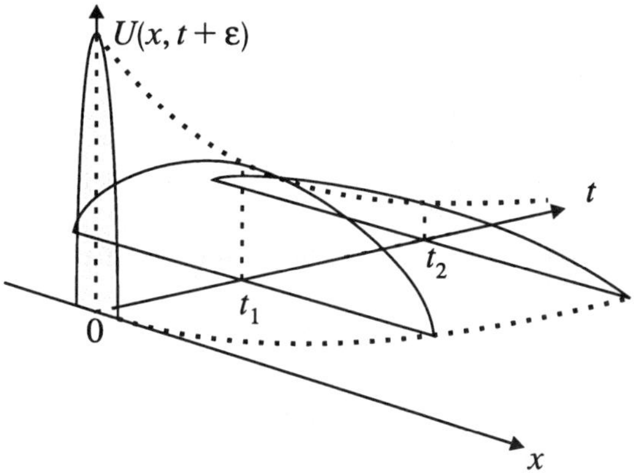

Figure 1.4: Self-similar solution $U(x, t + \varepsilon)$ of equation (1.2).

The solution $U(x, t)$ takes the Dirac mass as initial data

$$U(x, t) \to M\delta(x) \quad \text{as } t \to 0+,$$

where the constant M depends on A and can be found from the relation

$$\int_{\mathbb{R}^N} U(x, t)\, dx = M \quad \text{for all } t > 0.$$

The interface between the regions where the solution $U(x, t)$ is positive or is equal identically to zero is given by the exact formula

$$|x| = \sqrt{A/B}\, t^{\beta} \tag{1.4}$$

(the dotted line on Figure 1.4). The velocity of propagation of disturbances from the initial data is equal to $v_n = \beta\sqrt{A/B}\, t^{\beta-1}$, where v_n is the derivative in the direction of outer unit normal to the surface (1.4). Obviously, the velocity v_n is infinite at the instant $t = 0$.

(2) The next example is the stationary solution of equation (1.2):

$$u^m(x) = C\left[x' - x\right]_+, \quad C = \text{const}.$$

The interface of this solution is given by the formula $x = x'$ and so is the optimal rate at every point $x_0 > x'$: $\eta(t) = x' - x_0$ for all $t \geq 0$.

(3) Let us consider now the so-called "diffusion-absorption" equation

$$u_t = \left(u^m\right)_{xx} - u^p, \quad (x, t) \in \mathbb{R} \times \mathbb{R}^+, \tag{1.5}$$

with the parameters of nonlinearity $m > 1$ and $p \in (0, 1)$. If $m + p = 2$, this equation admits an explicit solution constructed by R. Kersner[219],

$$u(x,t)^{m-1} = \frac{m-1}{2m(m+1)t}\left[Ct^{\frac{2}{m+1}} - (m+1)^2t^2 - x^2\right]_+, \quad C = \text{const.} \tag{1.6}$$

The interface of this solution consists of two plane curves which meet within a finite time (defined by the choice of the parameter C). This solution illustrates the properties of support shrinking and dead core formation. The solution (1.6) corresponding to the initial profile $u(x, \varepsilon)$, $\varepsilon > 0$, is presented in Figure 1.5. The solution vanishes at some instant t_*. It is to be noted that in this example the interface velocity need not be positive (in accordance with the values of ε and C) so that the set $\operatorname{supp} u$ may expand and shrink. Moreover, the velocity is infinite at the extinction moment.

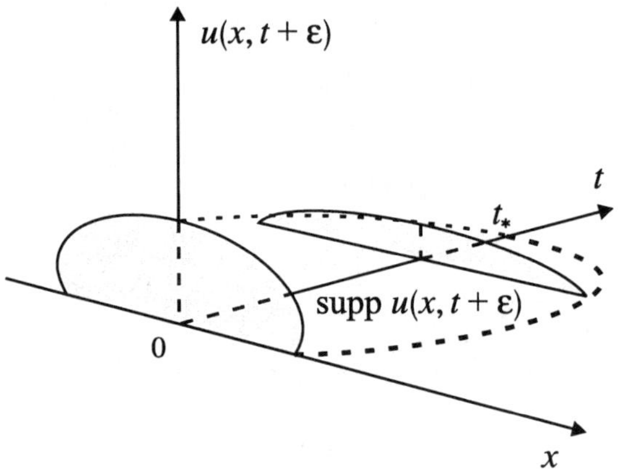

Figure 1.5: Self-similar solution to (1.5) with $m + p = 2$.

1.3 Scheme of the method. Once an equation is proven to possess a class of exact solutions (or sub-/supersolutions) which display one of the localization properties, a typical argument which allows one to extend this property to every admissible solution is to apply the maximum principle for parabolic equations and to compare the solution with an exact solution (or sub-/supersolution) through the input data. In so doing one has to impose certain restrictions on both the structure of the equation under study and the input data. Application of the local energy method makes it possible to avoid most of the difficulties of this sort. The idea consists in describing the evolution of the null set of the solution using only nonlinearity of the equation and a threshold value of the "total energy" of the solution. This value accumulates all the information required to perform a local study of a "zero cavern" in the space-time domain.

Let us present the scheme of the method by considering the one-dimensional equation

$$\left(u|u|^{\gamma-1}\right)_t - u_{xx} = f(x,t), \quad \gamma \in (0,1). \tag{1.7}$$

Equation (1.7) is considered in the rectangle

$$Q = (-L, L) \times (0, T) \quad \text{with } 0 < L < \infty, T > 0,$$

and is endowed with the initial condition

$$u(x,0) = u_0(x), \quad x \in (-L, L). \tag{1.8}$$

The solution of problem (1.7)–(1.8) is understood in the following sense.

Definition 1.3. A measurable-in-Q function $u(x,t)$ is called a *local weak solution* of problem (1.7), (1.8) if

(i) $u \in V(Q) = L^\infty(0,T; L^{1+\gamma}(-L,L)) \cap L^2(0,T; W^{1,2}(-L,L))$;

(ii) $\lim_{t\to 0} \|u(x,t) - u_0(x)\|_{L^{1+\gamma}(-L,L)} = 0$;

(iii) for every test function $\varphi \in C^\infty(0,T; C_0^\infty(-L,L))$, vanishing at $t = T$, the integral identity

$$\iint_Q \left\{u|u|^{\gamma-1}\varphi_t + u_x\varphi_x + f\varphi\right\} dxdt + \int_{-L}^{L} u_0|u_0|^{\gamma+1}\varphi(x,0)\,dx = 0$$

holds.

The application of the method can be conventionally divided into three steps:

(1) the choice of appropriate local energy functions;

(2) derivation of the differential inequality for the local energy functions;

(3) analysis of the differential inequality and interpretation of results.

Given a local weak solution $u(x,t)$ of problem (1.7), (1.8), let us define the *local energy functions*

$$\begin{aligned}
E(\rho,t) &= \int_0^t \int_{B_\rho} |u_x|^2 dxd\tau \equiv \int_0^t \|u_x(\cdot,\tau)\|^2_{L^p(B_\rho)} d\tau,\\
b(\rho,t) &= \int_{B_\rho} |u|^{1+\gamma} dx \equiv \|u(\cdot,t)\|^{1+\gamma}_{L^{1+\gamma}(B_\rho)},\\
\overline{b}(\rho,t) &= \operatorname*{ess\,sup}_{0\le\tau\le t} b(\rho,\tau), \quad B_\rho = \{x \in (-L,L) : |x| < \rho\}.
\end{aligned} \tag{1.9}$$

According to these definitions, the *local energy set* is a cylinder of radius ρ and height t; ρ is viewed as the independent variable while t serves as a parameter. This simplest choice is convenient to reveal the property of finite speed of propagation. More sophisticated choices of the new variable are discussed in the next sections.

We shall consider only those solutions to problem (1.7), (1.8) that satisfy the condition

$$D(u, L, T) \equiv \overline{b}(L, T) + E(L, T) \le D_0 < \infty. \tag{1.10}$$

The quantity $D(u, L)$ is called the *total energy* of the solution $u(x, t)$ in the domain Q. The constant D_0 (the upper estimate on the total energy) absorbs all the global information on the data of the problem under study.

The derivatives of $E(\rho, t)$ in ρ and t are given by the formulas

$$\begin{aligned}
\frac{\partial E(\rho, t)}{\partial \rho} &= \int_0^t \left(|u_x(\rho, \tau)|^2 + |u_x(-\rho, \tau)|^2\right) d\tau \\
&= \frac{\partial}{\partial \rho}\left(\sup_{0\le\tau\le t} E(\rho, \tau)\right) = \sup_{0\le\tau\le t} \frac{\partial E(\rho, \tau)}{\partial \rho}, \\
\frac{\partial^2 E(\rho, t)}{\partial \rho \partial t} &= (|u_x(\rho, t)|^2 + |u_x(-\rho, t)|^2).
\end{aligned}$$

Let us multiply equation (1.7) by $u(x, t)$ and then formally integrate the resulting equality over the domain $B_\rho \times (0, t) \subset Q$. This leads to the energy relation

$$\frac{\gamma}{1+\gamma} b(\rho, \tau)\Big|_{\tau=0}^{\tau=t} + E(\rho, t) = I(f) + I(\rho), \tag{1.11}$$

where

$$I(f) = \int_0^t \int_{B_\rho} f\, u\, dx d\tau, \quad I(\rho) = \int_0^t u_x u\, d\tau \Big|_{x=-\rho}^{x=\rho}.$$

Relation (1.11) is derived in full rigor in Section 2, where it is obtained at once for parabolic equations of general form.

1.3.1 *Finite speed of propagation of disturbances.* To simplify matters, we first assume that equation (1.7) is homogeneous:

$$f(x, t) = 0. \tag{1.12}$$

Let

$$u_0(x) = 0, \quad x \in B_{\rho_0} \subset (-L, L).$$

Assumption (1.12) allows us to rewrite (1.11) in the following form: for every $\rho \le \rho_0$,

$$\frac{\gamma}{1+\gamma} b(\rho, \tau)\Big|_{\tau=0}^{\tau=t} + E(\rho, t) \le I(\rho) \tag{1.13}$$

Note that $b(\rho, 0) = 0$ if $\rho \le \rho_0$. The term $I(\rho)$ on the right-hand side of (1.13) can be estimated by means of the second method proposed in the introduction to Chapter 1: we shall make use of the interpolation-trace inequality (1.12). In the notation of (1.9), this estimate takes the form

$$\begin{aligned}|I(\rho)| &\le C \int_0^t \left(\frac{\partial^2 E}{\partial\rho\,\partial t}\right)^{1/2} \left(E_t^{1/2} + \rho^{-\delta} b^{1/(1+\gamma)}\right)^{\theta} b^{(1-\theta)/(1+\gamma)} d\tau \\ &\le C_1 \rho^{-\delta\theta} \overline{b}^{(1-\theta)/(1+\gamma)} \int_0^t \left(\frac{\partial^2 E}{\partial\rho\,\partial t}\right)^{(1/2} (E_t + b)^{\theta/p} d\tau,\end{aligned} \tag{1.14}$$

where

$$\theta = \frac{2}{3+\gamma}, \quad \delta = \frac{1}{1+\gamma} + \frac{1}{2} = \frac{1}{\theta(1+\gamma)},$$
$$C_1 = C \max\left(\rho_0^{\delta\theta}, \overline{b}(\rho_0, T)^{\theta(1-\gamma)/(1+\gamma)}\right), \quad C = C(\gamma).$$

Using Hölder's inequality and the formulas

$$\int_0^t E_t(\rho, \tau) d\tau = E(\rho, t), \qquad \int_0^t E_{\rho t}(\rho, \tau) d\tau = E_\rho(\rho, t),$$

in (1.14), we arrive at the estimate

$$|I(\rho)| \le C_2 \rho^{-\delta\theta} t^{(1-\theta)/2} \left(E + \overline{b}\right)^{\theta/2 + (1-\theta)/(1+\gamma)} \left(\frac{\partial E}{\partial \rho}\right)^{1/2} \tag{1.15}$$

with $C_2 = C_1 \max(1, T)^{\theta/2}$. Returning to (1.13), we get the inequality

$$\begin{aligned}&\frac{\gamma}{1+\gamma} b(\rho, t) + E(\rho, t) \\ &\quad \le C_2 \rho^{-\delta\theta} t^{(1-\theta)/2} \left(E + \overline{b}\right)^{\theta/2 + (1-\theta)/(1+\gamma)} \left(\frac{\partial E}{\partial \rho}\right)^{1/2}.\end{aligned}$$

Since the right-hand side of this inequality is nondecreasing in t, the inequality holds if we replace $b(\rho, t)$ by $\overline{b}(\rho, t)$, and C_2 by $2C_2$. Thus

$$\begin{aligned}&\frac{\gamma}{1+\gamma} \overline{b}(\rho, t) + E(\rho, t) \\ &\quad \le 2C_2 \rho^{-\delta\theta} t^{(1-\theta)/2} \left(E + \overline{b}\right)^{\theta/p + (1-\theta)/(1+\gamma)} \left(\frac{\partial E}{\partial \rho}\right)^{1/2}.\end{aligned} \tag{1.16}$$

Next, applying Young's inequality, we derive from (1.16) the inequality

$$E^{\nu}(\rho, t) \le \left(E + \overline{b}\right)^{\nu} \le C_3 \rho^{-\alpha} t^{\beta} \frac{\partial E}{\partial \rho} \tag{1.17}$$

with the parameters

$$\alpha = 2\delta\theta, \quad \beta = 1 - \theta,$$
$$C_3 = \left(2C_2 \max\left(1, \frac{\gamma}{1+\gamma}\right)\right)^2,$$
$$\nu = 2\left(1 - \frac{\theta}{2} - \frac{1-\theta}{1+\gamma}\right) = \frac{2(1+\gamma)}{3+\gamma}.$$

Inequality (1.17) is the ordinary differential inequality for the function $E(\rho, t)$ that we consider depending on t as a parameter. Integrating (1.17) in ρ leads to the estimate

$$\begin{aligned} E^{1-\nu}(\rho, t) &\le E^{1-\nu}(\rho_0, t) - \frac{1-\nu}{C_3(1+\alpha)} t^{-\beta}\left(\rho_0^{1+\alpha} - \rho^{1+\alpha}\right) \\ &< D_0^{1-\nu} - \frac{1-\nu}{C_3(1+\alpha)} t^{-\beta}\left(\rho_0^{1+\alpha} - \rho^{1+\alpha}\right) \end{aligned}$$

with the "threshold" value of the total energy D_0 from (1.10). Since $E(\rho, t)$ is a nondecreasing nonnegative function of ρ, we conclude that

$$E(\rho, t) = b(\rho, t) = 0 \quad \text{for all } \rho \in (0, \rho(t))$$

with

$$\rho^{1+\alpha}(t) = \rho_0^{1+\alpha} - \frac{C_3(1+\alpha)t^\beta}{1-\nu} D_0^{1-\nu}.$$

That is,

$$u(x, t) = 0 \quad \text{for a.a. } |x| \le \rho(t).$$

Our conclusion is meaningful because

$$\rho(0) = \rho_0 \quad \text{and} \quad \rho(t) > 0 \text{ for } 0 < t < t^* = \left(\frac{1-\nu}{C_3(1+\alpha)} \rho_0^{1+\alpha} D_0^{\nu-1}\right)^{1/\beta}.$$

Proposition 1.1. *Let $u(x, t)$ be a weak solution of problem* (1.7), (1.8) *in the sense of Definition* 1.3. *Let $u(x, 0) = 0$ in $B_{\rho_0} \subset (-L, L)$. If $D_0 < \infty$, there always exists $t^* > 0$ such that the solution $u(x, t)$ has a nonzero rate $\rho(t)$ at the point $x = 0$. An admissible rate is given by the formula*

$$\rho(t) = \left(\rho_0^{1+\alpha} - \frac{C_3(1+\alpha)t_*^\beta}{1-\nu} D_0^{1-\nu}\right)^{1/(1+\alpha)}.$$

1.3.2 *The waiting time effect.* Let us consider the energy relation (1.11) for $\rho \in (\rho_0, L)$. We have the inequality

$$\frac{\gamma}{1+\gamma} b(\rho, t) + E(\rho, t) \leq I(\rho) + \frac{\gamma}{1+\gamma} b(\rho, 0). \tag{1.18}$$

Let us assume that

$$b(\rho, 0) = \int_{B_\rho} |u_0(x)|^{1+\gamma}\, dx \leq \varepsilon(\rho - \rho_0)_+^{1/(1-\nu)} \tag{1.19}$$

for some constant $\varepsilon > 0$, and with the exponent ν defined in (1.17). Making use of (1.15), (1.19), we arrive at the following inequality which generalizes (1.17):

$$E^\nu(\rho, t) \leq \left(E + \overline{b}\right)^\nu \leq C_3 \rho_0^{-\alpha} t^\beta \frac{\partial E}{\partial \rho} + C_4 \varepsilon^\nu (\rho - \rho_0)_+^{\nu/(1-\nu)} \tag{1.20}$$

with a constant $C_4 = C_4(\gamma)$. It is easy to check that if t^* satisfies the relation

$$A^\nu = C_3 \rho_0^{-\alpha} (t^*)^\beta A \frac{1}{1-\nu} + C_4 \varepsilon^\nu$$

with $A > A_0 = \max(C_4 \varepsilon^\nu,\ D_0 (L - \rho_0)^{1/(\nu-1)})$, then the function

$$z(\rho,\ t^*) \equiv z(\rho) = A(\rho - \rho_0)_+^{1/(1-\nu)},$$

solves the problem

$$\begin{cases} z^\nu(\rho) = C_3 \rho_0^{-\alpha} (t^*)^\beta \dfrac{dz}{d\rho} + C_4 \varepsilon^\nu (\rho - \rho_0)_+^{\nu/(1-\nu)}, \\ z(L) \geq D_0. \end{cases}$$

On the other hand, $z(\rho)$ is a majorant for the function $E(\rho,\ t^*)$. The monotonicity of $E(\rho,\ t)$ in t yields

$$E(\rho,\ t) \leq E(\rho,\ t^*) \leq z(\rho) = A(\rho - \rho_0)_+^{1/(1-\nu)} \quad \text{for all } t \leq t^* \text{ and } \rho \in (\rho_0,\ L).$$

Since $E(\rho,\ t)$ is also monotone with respect to ρ, we conclude that

$$E(\rho,\ t) = 0 \quad \text{for all } \rho \leq \rho_0, 0 \leq t \leq t^*,$$

which means that the solution of problem (1.7), (1.8) possesses the *generalized waiting time property* in the sense of Definition 1.1.

The same arguments apply to the case $f(x, t) \not\equiv 0$. Instead of (1.19), let us assume that

$$\left(\int_0^T \int_{B_\rho} |f|^{(1+\gamma)/\gamma} dx d\tau + b(\rho, 0)\right) \leq \varepsilon(\rho - \rho_0)_+^{1/(1-\nu)} \quad (\rho_0 \leq \rho \leq L). \tag{1.21}$$

The energy relation (1.11) becomes

$$\frac{\gamma}{1+\gamma}b(\rho,t)+E(\rho,t)+\lambda C(\rho,t)\le I(\rho)+\frac{\gamma}{1+\gamma}b(\rho,0)+I(f).$$

The last term on the right-hand side of this inequality is estimated as

$$\begin{aligned}|I(f)| &\le \int_0^T\int_{B_\rho}|u||f|dxd\tau\\ &\le \delta T\overline{b}(\rho)+C(\delta)\int_0^T\int_{B_\rho}|f(x,t)|^{(1+\gamma)/\gamma}dxd\tau, \qquad \delta\in(0,1),\end{aligned}$$

so that we arrive once again at the already-studied differential inequality of the type (1.20).

Proposition 1.2. *Let $u(x,t)$ be a weak solution of problem* (1.7), (1.8) *in the sense of Definition* 1.3. *If conditions* (1.19), (1.21) *are fulfilled, $u(x,t)$ possesses the* generalized waiting time property *in the sense of Definition* 1.1.

Conditions (1.19), (1.21) impose a constraint on the vanishing rate of the initial function $u_0(x)$ and the source function $f(x,t)$ when $\rho\to\rho_0+0$.

One can show that condition (1.19) is sharp and cannot be improved. Let $v(x,t)$ be a bounded nonnegative solution of the problem

$$v_t=\left(v^m\right)_{xx}, \qquad m>1 \quad \text{in } Q=(-L,L)\times(0,T),$$

$$v(x,0)=v_0(x)\begin{cases}=0 & \text{for } |x|\le\rho_0<L,\\ >0 & \text{in } (-L,\,L)\setminus(-\rho_0,\rho_0).\end{cases}$$

It is known [204, 223] that the condition

$$v_0(x)\le c|x-\rho_0|^{2/(m-1)} \quad \text{in a neighborhood of the point } x=\rho_0, \tag{1.22}$$

$c=\text{const}>0$, is sufficient and necessary for the solution to possess the waiting time effect of the interface $x=\eta(t)$ defined by (1.1). If we recalculate condition (1.22) for the solution $u(x,t)$ of problem (1.7)–(1.8), it will take the form

$$|u_0(x)|\le C|x-\rho_0|^{2m/(m-1)}, \qquad C=\text{const}.$$

This shows the necessity of (1.19) because

$$\int_{\rho_0}^{\rho}|u_0(x)|^{1+\gamma}dx\le C^{1+\gamma}(1-\mu)(\rho-\rho_0)^{1/(1-\nu)}, \qquad \rho\in(\rho_0,\,L).$$

2 General second-order equations

2.1 Finite speed of propagation. In this section, we study the property of finite speed of propagation of disturbances from the initial data for weak solutions of second-order parabolic equations. The energy estimates derived here will serve the analytic framework for all further considerations of this chapter.

We consider the parabolic equations

$$\frac{\partial\psi(x,u)}{\partial t} - \operatorname{div}\mathbf{A}(x,t,u,Du) + B(x,t,u,Du) + C(x,t,u) = f(x,t), \tag{2.1}$$

where

$$Du = \nabla u, \quad \mathbf{A} = (A_1,\dots,A_N), \quad \operatorname{div}\mathbf{A} = \sum_{i=1}^{N} \frac{d}{d x_i} A_i(x,t,u,Du).$$

Equation (2.1) is considered in a cylinder $Q = \Omega \times (0,T)$, $T \in \mathbb{R}^+$, where Ω is an open subset of $\mathbb{R}^N$, $N \geq 1$. The case $\Omega = \mathbb{R}^N$ will be treated separately. It is assumed that the coefficients of equation (2.1) satisfy the structural conditions

$$\forall\,(t,r,\mathbf{q}) \in \mathbb{R}^+ \times \mathbf{R} \times \mathbb{R}^N \quad \text{and} \quad \text{a.e. } x \in \Omega,$$

$$|\mathbf{A}(x,t,r,\mathbf{q})| \leq C_1\,|\mathbf{q}|^{p-1}, \tag{2.2a}$$

$$C_2\,|\mathbf{q}|^p \leq \mathbf{A}(x,t,r,\mathbf{q})\cdot\mathbf{q}, \tag{2.2b}$$

$$|B(x,t,r,\mathbf{q})| \leq C_3\,|r|^{\alpha}\,|\mathbf{q}|^{\beta}, \tag{2.2c}$$

$$C_4\,|r|^{1+\sigma} \leq C(x,t,r)\,r, \tag{2.2d}$$

$$C_6\,|r|^{\gamma+1} \leq G(x,r) \leq C_5\,|r|^{\gamma+1}, \tag{2.2e}$$

$$G(x,r) = \psi(x,r)\,r - \int_0^r \psi(x,\tau)\,d\tau \equiv \psi(x,r) - j(r).$$

Here $C_1 - C_6$, p, α, β, σ, γ, k are positive constants which will be specified later on.

With respect to the function $\psi(x,r)$ we assume the following:

(i) $\psi(x,r)$ is a Caratheodory function (measurable in x for all $r \in \mathbb{R}$ and continuous in r for almost all $x \in \Omega$);

(ii) $\psi(x,r)$ is nondecreasing in r for almost all $x \in \Omega$;

(iii) $\psi(x,r)$ satisfies conditions (2.18) of Chapter 2.

We will consider the weak solutions of equation (2.1) satisfying the initial condition

$$u(x,0) = u_0(x), \quad x \in \Omega. \tag{2.3}$$

Definition 2.1. A measurable-in-Q function $u(x,t)$ is called a weak solution of problem (2.1), (2.3) if

(i) $u \in L^{\infty}(0,T;\,L^{\gamma+1}(\Omega')) \cap L^p(0,T;\,W^{1,p}(\Omega'))$, $\overline{\Omega'} \subset \Omega$;

(ii) $\mathbf{A}(\cdot,\cdot,u,Du),\ B(\cdot,\cdot,u,Du),\ C(\cdot,\cdot,u)\in L^1(Q)$;

(iii) $\liminf_{t\to 0} G(x,u(\cdot,t)) = G(x,u_0)$ in $L^1(\Omega)$;

(iv) for every test function

$$\varphi \in L^\infty\left(0,T;\, W^{1,p}(\Omega)\right)\cap W^{1,2}(0,T;\, L^\infty(\Omega)),$$

$\varphi = 0$ on $\partial\Omega\times(0,T)$ in the sense of traces, the identity

$$\begin{aligned}\int_Q \{\psi(x,u)\varphi_t - \mathbf{A}\cdot D\varphi - B\,\varphi - C\,\varphi\}\,dx\,dt \\ - \int_\Omega \psi(x,u)\varphi\,dx\Big|_{t=0}^{t=T} = -\int_Q f\,\varphi\,dx\,dt\end{aligned} \tag{2.4}$$

holds.

Let us introduce the energy functions

$$\begin{aligned}
E(\rho,t) &= \int_0^t\int_{B_\rho} \mathbf{A}(x,\tau,u,Du)\cdot Du\,dxd\tau,\\
C(\rho,t) &= \int_0^t\int_{B_\rho} |u(x,\tau)|^{1+\sigma}\,dxd\tau,\\
b(\rho,t) &= \int_{B_\rho} |u(x,t)|^{1+\gamma}\,dx \equiv \|u(\cdot,t)\|^{1+\gamma}_{L^{1+\gamma}(B_\rho)},\\
\overline{b}(\rho,t) &= \operatorname*{ess\,sup}_{0\le\tau\le t} b(\rho,\tau), \qquad B_\rho = \{x\in\Omega : |x-x_0|<\rho\}.
\end{aligned} \tag{2.5}$$

The following equalities hold:

$$\begin{aligned}
\frac{\partial E(\rho,t)}{\partial\rho} &= \int_0^t\int_{S_\rho} \mathbf{A}(x,t,u,Du)\cdot Du\,dSdt, \qquad S_\rho = \partial B_\rho(x_0),\\
\frac{\partial E(\rho,t)}{\partial t} &= \int_{B_\rho(x_0)} \mathbf{A}(x,t,u,Du)\cdot Du\,dx,\\
\frac{\partial^2 E(\rho,t)}{\partial\rho\partial t} &= \int_{S_\rho} \mathbf{A}\cdot Du\,dS,\\
C_1\int_0^t\int_{S_\rho} |Du|^p\,dSd\tau \le \frac{\partial E(\rho,t)}{\partial\rho} &\le C_2\int_0^t\int_{S_\rho}|Du|^p\,dS\,d\tau,\\
C_5 b(\rho,t) \le \int_{B_\rho(x_0)} G(u(\cdot,t))dx &\le C_6 b(\rho,t),\\
C_2\|Du\|^p_{L^p((0,t)\times B_\rho)} \le E(\rho,t) &\le C_1\|Du\|^p_{L^p((0,t)\times B_\rho)}.
\end{aligned} \tag{2.6}$$

We will assume that

$$u_0(x) = 0 \quad \text{for } x\in B_{\rho_0}(x_0) \text{ with } \rho_0\in(0,\operatorname{dist}(x_0,\partial\Omega)). \tag{2.7}$$

Theorem 2.1 (Finite speed of propagation). *Assume that conditions* (2.2a)–(2.2e), (2.7) *and conditions* (i)–(iii) *on* ψ *are fulfilled. Let the structural constants satisfy the inequalities*

$$C_2 > 0,\ C_3 \geq 0, \quad C_4 \geq 0, \quad C_5 > 0$$

and

$$1+\gamma < p, \quad 0 \leq \beta \leq p, \quad \alpha = \gamma - \frac{(1+\gamma)\beta}{p}.$$

Let T^* *be defined by the formulas*

$$T^* = \begin{cases} (pC_5/2(p-\beta))\, C_3^{-p/(p-\beta)}\, (C_2 p/\beta)^{\beta/(p-\beta)} & \text{if } 0 < \beta < p, \\ C_3/C_2 & \text{if } \beta = 0, \\ \infty & \text{if } \beta = p \text{ and } C_3 < C_2. \end{cases}$$

Then every weak solution of problem (2.1), (2.3) *with* $f(x,t) \equiv 0$ *in* $B_{\rho_0}(x_0) \times (0, T^*)$ *possesses the finite speed of propagation property,*

$$u(x,t) = 0 \quad \text{in } x \in B_{\rho(t)}(x_0), \quad 0 \leq t \leq T^*$$

with $\rho(t)$ *given by the formula*

$$\rho^{\nu}(t) = \rho_0 - Ct^{\lambda} \min_{(\gamma+1)/p<\tau\leq 1} \left(E^{\varepsilon}(\rho_0,t) M(\rho_0,t)\right), \tag{2.8}$$

$$M = \frac{1}{\tau p - 1 - \gamma} \max(1, \rho_0^{\nu-1}) \max(b^{\mu}(\rho_0,t),\ b^{\eta}(\rho_0,t)),$$

with some constant $C = C(C_1, C_2, C_3, C_5, N, p, \gamma, \beta, T)$ *and*

$$\nu = \frac{\gamma k}{(p-1)(\gamma+1)}, \quad \lambda = \frac{\gamma+1}{\gamma k}, \quad \mu = \frac{p(1-\tau)}{\gamma k},$$
$$\eta = \frac{p-1-\gamma}{(p-1)(\gamma+1)} - \frac{1+\gamma-\tau p}{\gamma k}, \quad \varepsilon = \frac{\tau p - \gamma - 1}{\gamma k},$$
$$k = N(p-1-\gamma)/\gamma + p(1+\gamma)/\gamma.$$

Remark 2.1. Since $\rho(t)$ is a monotone decreasing function with $\rho(0) > 0$, the set $B_{\rho(t)}(x_0)$ is nonempty for small t.

The proof of Theorem 2.1 is split into two steps. The first step is to derive the energy relation valid for every weak solution $u(x,t)$ of problem (2.1), (2.3). The second step is to study the properties of the functions satisfying the ordinary nonlinear differential inequality which follows from the energy relation.

Lemma 2.1. *Under the hypotheses of Theorem* 2.1

$$\mathbf{A}(\cdot,\cdot,u,Du)\cdot Du,\ B(\cdot,\cdot,u,Du)\,u,\ C(\cdot,\cdot,u)\,u \in L^1(0,T; B_{\rho_0}(x_0)),$$

and for almost all $\rho \in (0, \rho_0)$ *and* $t \in (0, T^*) \subset (0, T)$, *the integration-by-parts inequality*

$$\int_{B_\rho(x_0)} G(x, u(x, \tau))dx \Bigg|_{\tau=0}^{\tau=t}$$
$$+ \int_0^t \int_{B_\rho(x_0)} (\mathbf{A}(x, \tau, u, Du) \cdot Du + B(x, \tau, u, Du)u + C(x, \tau, u)u + fu)\, dxd\tau$$
$$\leq \int_0^t \int_{S_\rho(x_0)} \mathbf{A}(x, \tau, u, Du) \cdot \mathbf{n}\, u\, dSd\tau \equiv I \tag{2.9}$$

holds, where $\mathbf{n}$ *is the unit outer normal vector to* $S_\rho(x_0)$.

Proof. Introduce the function

$$\varphi(x, \tau) \equiv \varphi_{n,l,k,h}(x, \tau) = \zeta_n(|x - x_0|)\chi_k(\tau)\frac{1}{h}\int_\tau^{\tau+h} T_l(u(x, s))ds,$$

where $h \in (0, T^* - t)$

$$T_l(u) = \min(|u|, l)\,\text{sign}\, u, \quad l \in \mathbb{N},$$
$$\chi_k(\tau) = \begin{cases} 1 & \text{if } \tau \in [0, \tau - 1/k], \\ k(t - \tau) & \text{if } \tau \in [\tau - 1/k, t], \\ 0 & \text{if } \tau \in [t, T^*], k \in \mathbb{N}, \end{cases} \tag{2.10}$$

and $\zeta_n(r)$ is given by formula (2.22). The function $\varphi(x, \tau)$ is an admissible test function. Substituting it into (2.4), we get

$$\int_0^T \int_{B_\rho(x_0)} (\mathbf{A}(x, \tau, u, Du) \cdot D\varphi + B(x, t, u, Du)\,\varphi + C(x, t, u)\,\varphi)\, dxd\tau$$
$$= \int_0^T \int_{B_\rho(x_0)} \psi(u(x, \tau))\frac{\partial \varphi}{\partial \tau}\, dxd\tau + \int_{B_\rho(x_0)} \psi(u(x, 0))\varphi(x, 0)\, dx. \tag{2.11}$$

Using conditions (iv)–(vii) of Definition 2.1, let us pass in (2.10) to the limit as $k \to \infty$. This procedure leads to the equality

$$\int_{B_\rho(x_0)} \psi(u(x, \tau))\widetilde{\varphi}(x, \tau)dx \Bigg|_{\tau=0}^{\tau=t} + \int_0^t \int_{B_\rho(x_0)} (\mathbf{A} \cdot D\widetilde{\varphi} + B\,\widetilde{\varphi} + C\,\widetilde{\varphi})\, dxd\tau$$
$$= \int_0^t \int_{B_\rho(x_0)} \psi(u(x, \tau))\frac{\partial \widetilde{\varphi}}{\partial \tau} dxd\tau, \tag{2.12}$$

where

$$\widetilde{\varphi} = \widetilde{\varphi}_{n,l,h}(x,\tau) = \zeta(|x-x_0|)\frac{1}{h}\int_{\tau}^{\tau+h} T_l(u(x,s))\,ds.$$

Since j is convex and increasing in R^+ (decreasing in R^-), we obtain the inequality

$$\begin{aligned} &\psi(u(x,\tau))\,(T_l(u(x,\tau+h)) - T_l(u(x,\tau))) \\ &\quad \le j(T_l(u(x,\tau+h))) - j(T_l(u(x,\tau))). \end{aligned} \tag{2.13}$$

Calculating the derivative $\partial\widetilde{\varphi}/\partial\tau$, using (2.13), and changing the limits of integration in t, we have

$$\begin{aligned} &\int_0^t \int_{B_{\rho_0}} \zeta(x)\psi(u(x,\tau))\frac{\partial\widetilde{\varphi}}{\partial\tau}dx d\tau \\ &\quad = \int_0^t \int_{B_{\rho_0}} \zeta(x)\psi(u)\frac{T_l(u(x,\tau+h)) - T_l(u(x,\tau))}{h}dx d\tau \\ &\quad \le \int_t^{t-h}\int_{B_{\rho_0}} \zeta_n T_l(u(x,\tau))dx d\tau - \int_0^h \int_{B_{\rho_0}} \zeta_n T_l(u(x,\tau))dx d\tau \\ &\quad \equiv F_h(T_l(u(x,\tau)). \end{aligned} \tag{2.14}$$

It follows from the properties of $j(u)$ and conditions (iv)–(vii) of Definition 2.1 that for almost all t,

$$\lim_{h\to 0} F_h = \int_{B_{\rho_0}} (j(T_l(u(x,t))) - j(T_l(u(x,0))))\,\zeta(x)dx, \tag{2.15}$$

$$\begin{aligned} &\lim_{h\to 0}\int_{B_\rho(x_0)} \psi(u(x,\tau))\widetilde{\varphi}(x,\tau)dx\Bigg|_{\tau=0}^{\tau=t} \\ &\quad = \int_{B_\rho(x_0)} \psi(u(x,\tau))\zeta_n T_l(u(x,\tau))dx\Bigg|_{\tau=0}^{\tau=t}, \end{aligned} \tag{2.16}$$

$$\lim_{h\to 0} I_{h,l,n} = \int_0^t \int_{B_\rho} (\mathbf{A}\cdot D(\zeta_n T_l(u)) + B\,\zeta_n T_l(u) + C\,\zeta_n T_l(u))\,dx d\tau. \tag{2.17}$$

Gathering (2.14)–(2.16), we come to the inequality

$$\begin{aligned} &\int_0^t \int_{B_{\rho_0}(x_0)} (\mathbf{A}\cdot DT_l(u) + B\,T_l(u) + C\,T_l(u))\,\zeta_n dx d\tau \\ &\quad \le \int_{B_\rho(x_0)} (j(T_l(u(x,\tau))) - \psi(u(x,\tau))T_l(u(x,\tau)))\,\zeta_n dx\Bigg|_{\tau=0}^{\tau=t} \\ &\qquad - \int_0^t \int_{B_{\rho_0(x_0)}} \mathbf{A}\cdot T_l(u)\,D\zeta_n dx d\tau. \end{aligned} \tag{2.18}$$

The restrictions imposed in Definition 2.1 allow us to perform in the last inequality the limit passage as $l \to \infty$ and thus to substitute $T_l(u(x,\tau))$ by $u(x,\tau)$. The last limit passage as $n \to \infty$ is performed by analogy with (2.23), which leads to the final inequality

$$\begin{aligned}\int_{B_\rho} G(x,u(x,\tau))dx\Big|_{\tau=0}^{\tau=t} &+ \int_0^t \int_{B_\rho} (\mathbf{A}\cdot Du + B\,u + C\,u)\,dxd\tau \\ &\le \int_0^t \int_{S_\rho} u\,\mathbf{A}\cdot\mathbf{n}\,dSd\tau \equiv I.\end{aligned} \tag{2.19}$$

□

It is worth noting that derivation of the energy relation (2.9) lends itself to different methods. However, all of them have in common two essential points: the justification of the first term on the left-hand side of (2.9), (the integration-by-parts formula with respect to the variable t), and of the term $I(\rho)$ on the right-hand side of (2.9) (integration by parts in x). These are typical difficulties already discussed in Chapter 1.

Additionally to (2.2d), let us assume that

$$C(x,t,r) \le C_4^*|r|^{1+\sigma}. \tag{2.20}$$

Proceeding in the same way as in Lemma 2.1 of Chapter 2, we derive the integration-by-parts formula in t,

$$\int_\Omega G(x,u(x,\tau))\zeta(x)dx\Big|_{\tau=0}^{\tau=t} = \int_0^t \int_\Omega \frac{\partial\psi(u(x,\tau))}{\partial t} u(x,\tau)\zeta(x)\,dxdt. \tag{2.21}$$

This formula is true for every $\zeta \in C_0^1(\Omega)$ and every weak solution $u(x,t)$. Therefore, the integral identity

$$\begin{aligned}\int_\Omega G(x,u(x,\tau))\zeta(x)dx\Big|_{\tau=0}^{\tau=t} &+ \int_0^t \int_\Omega (\mathbf{A}\cdot Du + B\,u + C\,u)\,\zeta\,dxdt \\ &+ \int_0^t \int_\Omega u\,\mathbf{A}\cdot D\zeta\,dxdt = 0\end{aligned} \tag{2.22}$$

holds. Here $\zeta(x) = \zeta_n(x)$, with $\zeta_n(x)$ defined by the formula

$$\zeta_n(r) = \begin{cases} 1 & \text{if } r \in [0, \rho - 1/n], \\ n(\rho - r) & \text{if } r \in [\rho - 1/n, \rho], \\ 0 & \text{if } r \in [\rho, \rho_0],\ n \in \mathbb{N}. \end{cases} \tag{2.23}$$

Using formula (2.16) of Chapter 1,

$$\lim_{n\to\infty} \int_0^t \int_{B_\rho} u\,\mathbf{A}\cdot D\zeta_n\,dxdt = \int_0^t \int_{S_\rho} u\,\mathbf{A}\boldsymbol{\nu}\,dS, \tag{2.24}$$

we come to the integral identity (2.9).

Let us proceed to derive from (2.18) an ordinary differential inequality for the energy function E. Fix some $T, 0 < t \le T < T^*$. We shall use the relations

$$\int_{B_\rho(x_0)} G(x, u(x,0))dx = 0 \quad \text{for } \rho \le \rho_0,$$

$$C_5 b(\rho, t) \le \int_{B_\rho(x_0)} G(x, u(x,t))dx, \tag{2.25}$$

$$\begin{aligned}\left|\int_0^t \int_{B_{\rho(x_0)}} B(x,\tau,u,Du)\,u\,dx d\tau\right| &\le \varepsilon C_3 \frac{p-\beta}{p} \int_0^t \int_{B_{\rho(x_0)}} |u|^{(\gamma+1)/\gamma} dx d\tau \\ &\quad + \frac{\beta C_3}{pC_2}\varepsilon^{-(p-\beta)/\beta} E(\rho, t) \\ &\le \varepsilon C_3 \frac{p-\beta}{p}\, t\, \overline{b}(\rho, t) \\ &\quad + \frac{\beta C_3}{pC_2}\varepsilon^{-(p-\beta)/\beta} E(\rho, t),\end{aligned} \tag{2.26}$$

$$\begin{aligned}|I| &\le \int_0^t \int_{S_{\rho(x_0)}} |u||A| dS d\tau \\ &\le C_1 \left(\int_0^t \int_{S_{\rho(x_0)}} |Du|^p\, dS d\tau\right)^{(p-1)/p} \left(\int_0^t \int_{S_{\rho(x_0)}} |u|^p\, dS d\tau\right)^{1/p} \\ &\le C_1 1 C_2^{-(p-1)/p} \left(\frac{\partial E}{\partial \rho}\right)^{(p-1)/p} \left(\int_0^t \int_{S_\rho} |u|^p dS d\tau\right)^{1/p},\end{aligned} \tag{2.27}$$

which follow from (2.2a)–(2.2e), (2.6), (2.7). Gathering (2.19)–(2.27), we come to the inequality

$$\begin{aligned}&C_5 b(\rho, t) + E(\rho, t) + C_4 C(\rho, t) \\ &\quad \le \varepsilon C_3 \frac{p-\beta}{p}\, t\, \overline{b}(\rho, t) + \frac{\beta C_3}{pC_2}\varepsilon^{-(p-\beta)/\beta} E(\rho, t) \\ &\quad + C_1 C_2^{-(p-1)/p} \left(\frac{\partial E}{\partial \rho}\right)^{(p-1)/p} \left(\int_0^t \int_{S_\rho} |u|^p dS d\tau\right)^{1/p}.\end{aligned} \tag{2.28}$$

Since the right-hand side of the last inequality is nondecreasing in t, we can replace $b(\rho, t)$ by $\overline{b}(\rho, t)$ and C_5 by $C_5/2$.

Since $t < T^*$, one may take in (2.26) ε so that

$$\varepsilon C_3 t \frac{p-\beta}{p} < C_5/2 \quad \text{and} \quad \frac{\beta C_3}{pC_2}\varepsilon^{-(p-\beta)/p}.$$

If we set

$$K = \min\left(C_5/2 - \varepsilon C_3 t \frac{p-\beta}{p},\ 1 - \frac{\beta C_3}{pC_2}\varepsilon^{-(p-\beta)/p}\right), \tag{2.29}$$

(2.28) yields the inequality

$$\overline{b}(\rho,t)+E(\rho,t)\le K_1\left(\frac{\partial E}{\partial \rho}\right)^{(p-1)/p}\left(\int_0^t\int_{S_\rho}|u|^p dS d\tau\right)^{1/p} \tag{2.30}$$

with $K_1 = C_1/KC_2^{(p-1)/p}$.

To estimate the integral on the right-hand side of (2.30), we proceed in the usual way and apply the inequality

$$\|u\|_{L^p(S_\rho)} \le C(p,q)\left(\|Du\|_{L^p(B_\rho)}+\rho^\delta\|u\|_{L^q(B_\rho)}\right)^\theta\left(\|u\|_{L^q(B_\rho)}\right)^{1-\theta} \tag{2.31}$$

with the exponents

$$\theta=\frac{N(p-q)+q}{N(p-q)+pq},\qquad \delta=-\frac{N(p-q)+qp}{pq},\qquad q=1+\gamma.$$

Let us raise both sides of (2.31) to the power p and then integrate over the interval $t\in(0,t)$. Applying Hölder's inequality and formulas (2.6), we have that

$$\left(\int_0^t\int_{S_\rho}|u|^p dS d\tau\right)^{1/p} \le 2Ct^{(1-\theta)/p}\overline{b}^{(1-\theta)/(1+\gamma)}\left(\|Du\|_{L^p((0,t)\times(B_\rho)}+\rho^\delta t^{1/p}\overline{b}^{1/(1+\gamma)}\right)^\theta. \tag{2.32}$$

Using (2.6), from (2.30), (2.32) we obtain

$$\overline{b}(\rho,t)+E(\rho,t) \le K_2\left(\frac{\partial E}{\partial\rho}\right)^{\frac{p-1}{p}} t^{(1-\theta)/p}\overline{b}^{(1-\theta)/(1+\gamma)}\left(E^{1/p}+\rho^\delta t^{1/p}\overline{b}^{1/(1+\gamma)}\right)^\theta \tag{2.33}$$

with a constant $K_2 = K_2(K_1, C, p, q)$. Let us make use of the following identity, which is valid for every $\tau\in(0,1)$, and $\kappa = 1/\theta(1+\gamma)$:

$$\begin{aligned}E^{1/p}\overline{b}^{(1-\theta)\kappa}&+\rho^\delta t^{1/p}\overline{b}^{\kappa}\\ &=E^{1/p}\overline{b}^{\tau(1-\theta)\kappa}\overline{b}^{(1-\tau)(1-\theta)\kappa}\\ &\quad+\rho^\delta t^{1/p}\overline{b}^{-1/p+\tau(1-\theta)\kappa}\overline{b}^{\kappa+1/p+\tau(1-\theta)/(1+\gamma)}.\end{aligned}$$

We have

$$\overline{b}^{(1-\theta)/(1+\gamma)}\left(E^{1/p}+\rho^\delta t^{1/p}b^{-1/(1+\gamma)}\right)^\theta \le K_3\,\rho^{\delta\theta}\left(\overline{b}+E\right)^{\theta/p+\tau(1-\theta)/(1+\gamma)} \tag{2.34}$$

with the constant K_3 given by the formula

$$K_3 = 2^\theta \max\left(1,\ \rho_0^{-\delta\theta}\right) \max\left(1,\ T^{\theta/p}\right) \max\left(\overline{b}^\mu,\ \overline{b}^\eta\right).$$

Then

$$\overline{b} + E \le K_2\, K_3\, t^{(1-\theta)/p} \rho^{\delta\theta} \left(\frac{\partial E}{\partial \rho}\right)^{(p-1)/p} \left(\overline{b} + E\right)^{\theta/p + \tau(1-\theta)(1+\gamma)}. \tag{2.35}$$

Raising both sides of the last inequality to the power $p/(p-1)$ and dropping

$$(\overline{b} + E)^{\theta/(p-1) + p\tau(1-\theta)/(1+\gamma)(p-1)},$$

we transform (2.35) to the form

$$E^{1-\varepsilon} \le (\overline{b} + E)^{1-\varepsilon} \le K\, t^\lambda \rho^{1-\nu} E_\rho, \tag{2.36}$$

with the constants

$$K = C \max\left(1,\ \rho_0^{\nu-1}\right) \max\left(\overline{b}^\mu(\rho_0, T), \overline{b}^\eta(\rho_0, T)\right),$$
$$C = C\,(C_1, \ldots, C_5, N, p, \gamma, \beta, T).$$

The exponents

$$\varepsilon = 1 - \frac{p}{p-1}\left(1 - \frac{\theta}{p} - \frac{\tau(1-\theta)}{1+\gamma}\right), \quad \lambda = \frac{1-\theta}{p-1}, \quad \nu = 1 - \frac{\delta\theta p}{p-1}$$

coincide with the constants defined in (2.8). Integrating (2.36), we come to the estimate for the energy function $E(\rho, t)$,

$$E^\varepsilon(\rho, t) \le E^\varepsilon(\rho_0, t) - \frac{\varepsilon}{\nu} \left(K\, t^\lambda\right)^{-1} \left(\rho_0^\nu - \rho^\nu\right), \tag{2.37}$$

and, correspondingly, to the equality that defines the function $\rho(t)$:

$$\rho^\nu(t) = \rho_0^\nu - \frac{\nu}{\varepsilon} K\, t^\lambda E^\varepsilon(\rho_0, t). \tag{2.38}$$

Since ν is independent of τ, we obtain (2.8) just taking the maximum in τ of the right-hand side of (2.38). This completes the proof.

Remark 2.2. As in Theorem 2.1, the hypotheses on α and β can be relaxed if we know a priori that $u \in L^\infty((0, T) \times B_{\rho_0}(x_0))$. To this end, we only have to assume that $\alpha \ge 0$, $0 \le \beta \le p$ and that the constant C_3 is sufficiently small (see [306] for some specific L^∞ estimates).

Corollary 2.1. *Let u be a weak solution of the equation*

$$\frac{\partial \psi(x, u)}{\partial t} - \operatorname{div} \mathbf{A}(x, t, u, D\,u) + B(x, t, u, D\,u) + C(x, t, u) = 0 \tag{2.39}$$

in $\mathbb{R}^+ \times \mathbb{R}^N$ *such that* $j(u) \in C^0(\mathbb{R}^+; L^1_{\mathrm{loc}}(\mathbb{R}^N))$ *and that for any* $\rho > 0$, $t > 0$ *there exists* $K = K(\rho, t)$ *such that for any* $y \in \mathbb{R}^N$

$$\operatorname*{ess\,sup}_{0 \le \tau \le t} \int_{B_\rho(y)} G(x, u(x,\tau))dx + \int_0^t \int_{B_\rho(y)} \mathbf{A}(x, \tau, u, Du) \cdot Du\, dx d\tau \le K \tag{2.40}$$

and assume also that the structural hypotheses of Theorem 2.1 *are fulfilled (with* $C_3 < C_2$ *if* $\beta = p$*). If the initial data* u_0 *vanishes outside* $B_R(0)$*, there exists a nondecreasing function* $t \mapsto R(t)$ *defined on* $\mathbb{R}^+$ *such that* $R(0) = R_0$ *and* $\operatorname{supp} u(\cdot, t) \subset B_{R(t)}(0)$ *for any* $t \ge 0$*. If* $\beta = 0$ *or* $\beta = p$*, then*

$$R(t) = R_0 + C \max(t^\lambda, t^{1/(p-1)}) \tag{2.41}$$

with λ *given in* (2.8) *and* C *depending on the structural constants and the norm of* u_0 *in* $L^{(1+\gamma)/\gamma}(\mathbb{R}^N)$.

Proof. **Step 1.** It is asserted: there exists $T > 0$ such that for any $t \in [0, T]$ the support of $u(\cdot, t)$ is compact. Fix $T' < T^*$, $\rho > 0$, and $R + \rho$, and then apply Theorem 2.1 in $B_\rho(x_0)$. By (2.40), there exists a constant M depending on the structural constants, T' and ρ, but not on x_0, such that if $\rho^\nu \ge M t^\lambda \max(1, \rho^{\nu-1})$ and $t < T'$, then

$$u(x,t) = 0 \quad \text{a.e in } B_{\rho_1}(x_0) \quad \text{with} \quad \rho_1^\nu(t) = \rho^\nu - M t^\lambda \max(1, \rho^{\nu-1}).$$

If we set $T^\lambda = \min((T')^\lambda, \min(\rho^{\nu,\rho})/M)$ and make x_0 run over all the complement of $B_{R+\rho}(0)$, we deduce that for any $t \le T$, $u(x,t)$ vanishes for a.a. $|x| > R + \rho - \rho_1$.

Step 2. We assert that the support of $u(\cdot, t)$ is compact every any $t > 0$. Let us argue by contradiction: assume that the set

$$\{t \in \mathbb{R}^+ : \operatorname{supp} u(\cdot, t) \text{ is compact throughout } (0, t)\}$$

admits an upper bound $t^* < \infty$. From (2.40), we have

$$\begin{aligned} &\operatorname*{ess\,sup}_{\tau \le 2t^*} \int_{B_\rho(y)} G(x, u(x,\tau))dx \\ &\quad + \int_0^{2t^*} \int_{B_\rho(y)} \mathbf{A}(x, \tau, u, Du) \cdot Du\, dx d\tau \le K(t^*, \rho) \end{aligned} \tag{2.42}$$

for every $y \in \mathbb{R}^N$. For any $t < t^*$ the support of $u(\cdot, t)$ is included within a ball $B_{R(t)}(0)$ so that we apply Theorem 2.1 on $[t, \infty) \times \mathbb{R}^N$ (if we set $s = \tau - t$ and $v(x,s) = u(x,\tau)$ the function v satisfies (2.39) in $\mathbb{R}^+ \times \mathbb{R}^N$ with $u(\cdot, t)$ as initial data). Proceeding as in Step 1, we see that there exists $M > 0$ such that for any $|y| > R(t) + \rho$ and $(\tau - t)^\lambda \le \min(t^{*\lambda}, \min(\rho^\nu, \rho)/M)$,

$$u(x,t) = 0 \quad \text{a.e. in } B_{\rho_\tau}(y) \quad \text{with} \quad \rho_1^\nu(\tau) = \rho^\nu - M(\tau - t)^\lambda \max(1, \rho^{\nu-1}).$$

Moreover, it follows from (2.42) and the definition of v that the constant M is independent of $t < t^*$ and $y \in \mathbb{R}^N - B_{R(t)+\rho}$, whence $u(x,\tau) = 0$ for a.a. $|x| > R + \rho - \rho(\tau)$. In particular, for

$$\tau = \min\left(t^* + t, t + (1/m \min(\rho^\nu, \rho))^{1/\lambda}\right),$$

$u(\cdot,\tau)$ vanishes a.e. in $\mathbb{R}^N - B_{R+\rho-\rho(\tau)}(0)$. If we take t close enough to t^*, we have a contradiction, so that $t^* = +\infty$. Moreover, by construction there exists a nondecreasing function R defined on $\mathbb{R}^+$ such that $R(0) = R_0$ and $\operatorname{supp} u(\cdot,t) \subset B_{R(t)}(0)$.

Step 3. *End of the proof.* If we apply Lemma 2.1 in $[0,t] \times B_{2R(t)}(0)$, we get for $t \geq 0$,

$$\int_{\mathbb{R}^N} G(x,u(x,t))dx + \int_0^t \int_{\mathbb{R}^N} (\mathbf{A}(x,\tau,u,Du)\cdot Du + B(x,\tau,u,Du)u)\,dxd\tau \tag{2.43}$$
$$\leq \int_{\mathbb{R}^N} G(x,u(x,0))dx.$$

If $\beta = 0$ or $\beta = p$ and $C_3 < C_2$, we deduce from (2.43)

$$\operatorname*{ess\,sup}_{0\leq t} \int_{\mathbb{R}^N} G(u(x,t))dx + \int_0^\infty \int_{\mathbb{R}^N} (\mathbf{A}(x,\tau,u,Du)\cdot Du)\,dxd\tau \tag{2.44}$$
$$\leq K \int_{\mathbf{R}^N} G(u(x,0))dx,$$

where K depends only on the structural constants. If we fix y outside $B_{R+1}(0)$ and set $\rho_0 = |y| - R$, we get as in Step 1

$$B_{\rho_1}(y) \subset \{x : u(x,t) = 0\} \quad \forall t \in \left[0, \min\{T_\rho, T\}\right]$$

with $\rho_1^\nu(t) = \rho_0^\nu - M\max(t^\lambda, t^\lambda T^{1/(p-1)-\lambda})\rho_0^\nu$ where M depends only on the structural constants and $\|u_0\|_{L^{(1+\gamma)/\gamma}(\mathbb{R}^N)}$, and an arbitrary T. By the mean value theorem

$$\rho_0^\nu - \rho_1^\nu(t) = \nu\widetilde{\rho}^{\nu-1}(\rho_0 - \rho_1(t)) = M, \max(t^\lambda, t^\lambda T^{1/(p-1)-\lambda})\rho_0^\nu$$

where $\widetilde{\rho} \in (\rho_1(t), \rho_0)$, so that

$$\rho_0 - \rho_1(t) = \frac{M}{\nu}\max(t^\lambda, t^\lambda T^{1/(p-1)-\lambda})\left(\frac{\rho_0}{\widetilde{\rho}}\right)^\nu.$$

Moreover, $\operatorname{supp} u(\cdot,t) \in B_{R+\rho_0-\rho_1(t)}(0)$. As

$$1 - \left(\frac{\rho_1}{\rho_0}\right)^\nu = \frac{M}{\nu\rho_0}\max(t^\lambda,\ t^\lambda T^{1/(p-1)-\lambda}),$$

we conclude that

$$\lim_{y\to\infty}(\rho_0/\rho_1) = \lim_{y\to\infty}(\rho_0/\widetilde{\rho}) = 1$$

and

$$\operatorname{supp} u(\cdot, t) \subset \{x : |x| \geq R - \frac{M}{\nu} \max(t^{\lambda}, t^{\lambda} T^{1/(p-1)-\lambda})\}$$

for $t \leq \lim_{\rho\to\infty} T_\rho = +\infty$. In particular, to get (2.41) we can take $t = T$. □

Remark 2.3. Estimate (2.44) ceases to be true if $0 < \beta < p$ so that (2.41) is only valid for $t \leq T < T^*$ with a constant M depending on T. Moreover, we do not know whether relation (2.40) (which says that the energy of the solution is locally uniform in $\mathbb{R}^N$) is necessary to assure the finite speed of propagation for $u(x, t)$.

2.2 Stable localization. In the next theorem we establish the property of localization of the support of $u(x, t)$ independently of t.

Theorem 2.2 (Stable localization). *Let us assume that* $C_2 > 0$, $C_4 > 0$, $C_5 > 0$, $\sigma \geq 0$, $p > 1$, $0 < \gamma < \infty$, $0 \leq \beta \leq p$, $\alpha = \sigma - \beta(1+\sigma)/p$, *and* $C_3 < C_4$ *if* $\beta = 0$ ($C_3 < C_2$ *if* $\beta = p$), *or*

$$C_3 < \left(C_4 \frac{p}{p-\beta}\right)^{(p-\beta)/p} \left(C_2 \frac{p}{\beta}\right)^{\beta/p} \tag{2.45}$$

if $0 < \beta < p$. *Moreover, let* $\max(\sigma+1, \gamma+1) < p$. *If* $u(x, t)$ *is a weak solution of* (2.39) *in* $\mathbb{R}^+ \times \Omega$ *with the initial data* u_0 *vanishing in* $B_{\rho_0}(x_0)$, $x_0 \in \Omega$, $\rho_0 < \operatorname{dist}(x_0, \partial\Omega)$ *and such that* $u \in L^\infty(\mathbb{R}^+, L^{1+\gamma}(B_{\rho_0}(x_0)) \cap L^{1+\sigma}(\mathbb{R}^+ \times (B_{\rho_0}(x_0))$, *and* $Du \in L^p(\mathbb{R}^+ \times (B_{\rho_0}(x_0))$, *then*

$$u(x, t) = 0 \quad \textit{for a.a. } (x, t) \in \mathbb{R}^+ \times B_\rho(x_0)$$

where ρ *is defined by the formula*

$$\begin{aligned} \rho^\nu &= \rho_0^\nu - C \min_{\kappa/p<\tau\leq 1} \left(E^\varepsilon(\rho_0) M(\rho_0)\right), \\ M &= \max(1, {\rho_0}^{\nu-1}) \max\left(b^\mu(\rho_0), b^\eta(\rho_0)\right)/(\tau p - \kappa), \end{aligned} \tag{2.46}$$

with a constant $C = C(C_1 - C_5, N, p, \gamma, \beta, \sigma)$ *and*

$$\begin{aligned} E(\rho) &= E(\rho, \infty) = \int_0^\infty \int_{B_\rho} \mathbf{A}(x, \tau, u, Du) \cdot Du \, dx d\tau, \\ C(\rho) &= C(\rho, \infty) = \int_0^\infty \int_{B_\rho} |Du(x, \tau)|^{1+\sigma} dx d\tau, \\ b(\rho) &= \overline{b}(\rho, \infty) = \sup_{0\leq\tau\leq\infty} \operatorname{ess} \int_{B_\rho} G(u(x, t)) dx, \\ \varepsilon &= \frac{(\tau p - \kappa))(1+\gamma)(p-1-\gamma)}{(p-\kappa))(N(p-1-\sigma) + p(1+\gamma 1))}, \\ \nu &= 1 + \frac{N(p-1-\sigma) + 1 + \gamma)}{(p-1)(1+\gamma)}, \end{aligned}$$

$$\mu = \frac{(p-1-\sigma)(1+\gamma)}{N(p-1-\sigma)+p(1+\gamma)} - \varepsilon,$$

$$\eta = 1 + \frac{p-1-\sigma)}{p-1} - \varepsilon, \qquad \kappa = \max(\sigma+1,\ \gamma+1).$$

As a preliminary step, we prove the following proposition. By convention, throughout this section we denote by K different constants depending on N and the constants from conditions (2.2a)–(2.2e).

Lemma 2.2. *Let* $u \in V(B_\rho(x_0) \times (0,T))$, $\rho \le \rho_0$,

$$V = L^\infty(0,T; L^{1+\gamma}(B_\rho(x_0))) \cap L^{1+\sigma}(0,T; L^{1+\sigma}(B_\rho(x_0))) \\ \cap L^p(0,T; W^{1,p}(B_\rho(x_0))),$$

and $Du \in L^p(0,T; L^p(B_\rho(x_0)))$. *Then*

$$\left(\int_0^t \int_{S_\rho} |u|^p dS d\tau\right)^{1/p} \\ \le K\rho^{\theta\delta}\max(1,\rho^{-\delta\theta})\left(\int_0^t\int_{B_\rho}\left(|Du|^p + |u|^{1+\sigma}\right)dxd\tau\right)^{1/p} \\ \times \max\left(\sup_{0\le\tau\le t}\text{ess}\,\|u(\cdot,\tau)\|_{L^{1+\gamma}(B_\rho(x_0))}^{(1-\theta)(p-1-\sigma)/p},\ \sup_{0\le\tau\le t}\text{ess}\,\|u(\cdot,\tau)\|_{L^{1+\gamma}(B_\rho(x_0))}^{(p-1-\sigma)/p}\right) \tag{2.47}$$

where

$$\theta = \frac{N(p-q)+q}{N(p-q)+pq}, \quad \delta = -\frac{N(p-q)+q}{+pq}, \quad q = \frac{p(1+\gamma)}{p+\gamma-\sigma} < p.$$

Proof. To begin with, note that in the notation used for the energy functions (2.47) is equivalent to

$$\left(\int_0^t\int_{S_\rho}|u|^p dSd\tau\right)^{1/p} \le K\rho^{\delta\theta}\left(E(\rho,t)+C(\rho,t)\right)^{1/p}\max(1,\rho^{-\delta\theta}) \\ \times\max\left(\overline{b}^{(p-1-\sigma)(1\theta)/p}(\rho,t),\ \overline{b}^{(p-1-\sigma)/p}(\rho,t)\right). \tag{2.48}$$

Let us make use of inequality (2.31) with $q = p(1+\gamma)/(p+\gamma-\sigma) < p$. We have

$$\|u\|_{L^p(S_\rho)} \le C\left(\|Du\|_{L^p(B_\rho)} + \rho^\delta\|u\|_{L^{1+\sigma}(B_\rho)}^{(1+\sigma)/p}\|u\|_{L^{1+\gamma}(B_\rho)}^{(p-1-\sigma)/p}\right)^\theta \\ \times\left(\|u\|_{L^{1+\sigma}(B_\rho)}^{(1+\sigma)/p}\|u\|_{L^{1+\gamma}(B_\rho)}^{(p-1-\sigma)/p}\right)^{1-\theta}, \tag{2.49}$$

with $C = C(p, q), \theta$, and δ defined in (2.47). In our notation (2.49) takes the form

$$\|u\|_{L^p(S_\rho)} \le K \left(E_t^{1/p}(\rho, t) + \rho^\delta b^{(p-1-\sigma)/p(1+\gamma)}(\rho, t) C_t^{1/p}(\rho, t)\right)^\theta \times \left(b^{(p-1-\gamma)/p(1+\gamma)}(\rho, t) C_t^{1/p}(\rho, t)\right)^{1-\theta}. \tag{2.50}$$

Raising both sides of (2.50) to the power p, we obtain the inequality

$$\begin{aligned}\|u(\cdot, \tau)\|^p_{L^p(S_\rho)} &\le K\rho^{\delta\theta p} \max(1, \rho^{-p\delta\theta}) \max(b^{\kappa(1-\theta)}, b^\kappa) (E_t + C_t)^\theta C_t^{1-\theta} \\ &\le K\rho^{\delta\theta p} \max(1, \rho^{-p\delta\theta}) \max(b^{\kappa(1-\theta)}, b^\kappa) (E_t(\rho, \tau) + C_t(\rho, \tau))\end{aligned} \tag{2.51}$$

with $\kappa = (p - 1 - \sigma)/(1 + \gamma)$.

Integrating both sides of (2.51) in $\tau \in (0, t)$, using the inequality $b(\rho, \tau) \le \overline{b}(\rho, t)$ if $\tau \le t$, and raising to the power $1/p$, we come to the desired inequality (2.48). □

Observe that in the notation of Theorem 2.2, inequality (2.48) yields

$$\left(\int_0^t \int_{S_\rho} |u|^p dS d\tau\right)^{1/p} \le K\rho^{\delta\theta p} (E + C)^{1/p + (p-1)\varepsilon/p} \times \left(\max(b^\mu(\rho_0), b^\eta(\rho_0))\right)^{(p-1)/p} \max(1, \rho^{\delta\theta}). \tag{2.52}$$

Proof of Theorem 2.2. First, proceeding by analogy with the proof of Lemma 2.1, we derive the energy relation

$$\begin{aligned}\Lambda \equiv \int_{B_\rho(x_0)} G(x, u(x, t)) dx + \int_0^t \int_{B_\rho(x_0)} (\mathbf{A} \cdot Du + B\,u + C\,u)\, dx d\tau \\ \le \int_0^t \int_{S_{\rho(x_0)}} u\,\mathbf{A} \cdot \mathbf{n}\, dS d\tau \equiv I.\end{aligned} \tag{2.53}$$

Then we use the following inequalities, similar to (2.26), (2.27):

$$\left|\int_0^t \int_{B_\rho} B(x, \tau, u, Du)\, u dx d\tau\right| \le \varepsilon C_3 \frac{p - \beta}{p} C(\rho, t) + \frac{\beta C_3}{p C_2} \varepsilon^{-(p-\beta)/\beta} E(\rho, t), \tag{2.54}$$

$$|I| \le C_1 C_2^{-(p-1)/p} \left(\frac{\partial E}{\partial \rho}\right)^{(p-1)/p} \left(\int_0^t \int_{S_\rho} |u|^p dS d\tau\right)^{1/p}. \tag{2.55}$$

Gathering (2.53)–(2.55) and taking into account (2.25), (2.45) we arrive at the inequality

$$K_0\left(b(\rho,t)+E(\rho,t)+C(\rho,t)\right) \leq C_1 C_2^{-(p-1)/p}\left(\frac{\partial E}{\partial\rho}\right)^{(p-1)/p}\left(\int_0^t\int_{S_\rho}|u|^p dS d\tau\right)^{1/p}, \tag{2.56}$$

where

$$K_0=\min\left(C_4-\varepsilon C_3\frac{p-\beta}{p},\ 1-\frac{\beta C_3}{pC_2}\varepsilon^{(p-\beta)/\beta},\ C_5\right)>0.$$

The right-hand side of the last inequality is nondecreasing in t; therefore $b(\rho,t)$ can be replaced by $\overline{b}(\rho,t)$ and K by $K/2$. Thus we have

$$K_0/2\left(\overline{b}(\rho,t)+E(\rho,t)+C(\rho,t)\right) \leq C_1 C_2^{-(p-1)/p}\left(\frac{\partial E}{\partial\rho}\right)^{(p-1)/p}\left(\int_0^t\int_{S_\rho}|u|^p dS d\tau\right)^{1/p}. \tag{2.57}$$

Substituting (2.48) into (2.57) and calculating the exponents, we come to the inequality

$$\begin{aligned}&\left(\overline{b}(\rho,t)+E(\rho,t)+C(\rho,t)\right)\\ &\quad\leq K\left(\frac{\partial E}{\partial\rho}\right)^{(p-1)/p}\left(\overline{b}(\rho,t)+E(\rho,t)+C(\rho,t)\right)^{1/p+\varepsilon(p-1)/p}\\ &\quad\times\rho^{\delta\theta}\left(\max(1,\rho_0^{\nu-1})\max(b^{\mu}(\rho_0),\ b^{\eta}(\rho_0))\right)^{(p-1)/p}.\end{aligned}$$

Raising both sides to the power $p/(p-1)$, we have

$$\begin{aligned}E^{1-\varepsilon}(\rho,\tau)&\leq\left(\overline{b}(\rho,\tau)+E(\rho,\tau)+C(\rho,\tau)\right)^{1-\varepsilon}\\ &\leq K\rho^{1-\nu}\max(1,\rho_0^{\nu-1})\max(b^{\mu}(\rho_0),\ b^{\eta}(\rho_0))\left(\frac{\partial E}{\partial\rho}\right).\end{aligned} \tag{2.58}$$

Integrating (2.58) with respect to $\tau\in(0,t)$, we get the inequality

$$\rho_0^{\nu}-\rho^{\nu}(t)\leq\overline{K}\left(E^{\varepsilon}(\rho_0)-E^{\varepsilon}(\rho,t)\right) \tag{2.59}$$

with

$$\overline{K}=\frac{\nu}{\varepsilon}\max(b^{\mu}(\rho_0),\ b^{\eta}(\rho_0))$$

and, correspondingly, the equation which defines the function $\rho(t)$:

$$\rho^{\nu}(t)=\rho_0^{\nu}-\overline{K}E^{\varepsilon}(\rho_0). \tag{2.60}$$

Taking the maximum of the right-hand side of (2.60), we obtain (2.46). □

Remark 2.4. If the energy $b + E + C$ of u is not large, then $\rho(t) > 0$ and there indeed exists a cylinder $\mathbb{R} \times B_{\rho(t)}(x_0)$ where u is zero a.e. On the other hand, if u does not have a finite energy in $\mathbb{R} \times B_{\rho_0}(x_0)$, we just obtain a finite speed of propagation for the nonzero set of u in $\mathbb{R}^+ \times B_{\rho_0}(x_0)$. Moreover, if we know a priori that u is bounded in $\mathbb{R}^+ \times B_{\rho_0}(x_0)$, the hypotheses on α and β can be relaxed as in Remark 2.2.

Corollary 2.2. *Let us assume that $u(x,t)$ is a weak solution of* (2.39) *satisfying* (2.40) *such that $j(u) \in C^0(\mathbb{R}^+; L^1_{\text{loc}}(\mathbb{R}^N))$ and that $C_2 > 0$, $C_4 > 0$, $C_5 > 0$, $0 \le \sigma$, $0 < q$,$0 < \gamma < \infty$, $0 \le \beta \le p$, $\alpha = -\beta(1+\sigma)/p$ and $C_3 < C_4$ if $\beta = 0$ (resp., $C_3 < C_2$ if $\beta = p$) or 2.45) if $0 < \beta < p$, and assume also that $\max(\sigma, \gamma) < p - 1$. If the initial function u_0 vanishes outside $B_{R_0}(0)$, then there exists $R \ge R_0$ depending on the structural constants and $\|u_0\|_{L^{1+\gamma}(\mathbb{R}^N)}$ such that $u(\cdot, t)$ vanishes a.e. outside $B_R(0)$ for any $t \ge 0$.*

Proof. As in Corollary 2.1, we first notice that the support of u has a finite speed of propagation. If we apply Lemma 2.1 in $B_\rho(0) \times (0,t)$, we get (2.19) and when $\rho \to \infty$ we obtain

$$\begin{aligned} &\int_0^t \int_{\mathbb{R}^N} (\mathbf{A}(x,\tau,u,Du) \cdot Du + B(x,\tau,u,Du)u)\, dxd\tau \\ &\qquad \le \int_{\mathbb{R}^N} G(x,u(x,0))dx - \int_{\mathbb{R}^N} G(x,u(x,t))dx, \end{aligned} \tag{2.61}$$

whence (with (2.45)) the energy estimate

$$\begin{aligned} &\int_{\mathbb{R}^N} G(x,u(x,t))dx + \int_0^t \int_{\mathbb{R}^N} \left(\mathbf{A}(x,\tau,u,Du) \cdot Du + |u|^{1+\sigma}\right) dxd\tau \\ &\qquad \le K \int_{\mathbb{R}^N} G(x,u(x,0))dx, \end{aligned} \tag{2.62}$$

where K depends on the structural constants. We now fix x_0 outside $B_{R+1}(0)$, $\rho_0 = |x_0| - R$ and apply Theorem 2.1 in $B_{\rho_0}(x_0)$: there exists a constant L depending on K and $\|u_0\|_{L^{1+\gamma}(\mathbb{R}_N)}$ such that $u(x,t)$ is zero a.e. in $\mathbb{R}^+ \times B_\rho(x_0)$, where

$$\rho^\nu = \rho_0^\nu - L\rho_0^{\nu-1}.$$

Correspondingly, $\rho \ge 0$ provided $\rho_0 \ge L$. Hence $u(\cdot,t)$ vanishes a.e. outside $B_{R+L}(0)$. □

3 The waiting time property

In this section we study the generalized waiting time property for weak solutions to equation (2.1) understood in the sense of Definition 1.2.

We again assume that conditions (2.2a)–(2.2e), (i), and (ii) are fulfilled and that the energy functions possess properties (2.5), (2.6). Let

$$u_0(x) \equiv 0 \quad x \in B_{\rho_0}(x_0) \quad \text{for some } x_0 \in \Omega \text{ and } \rho_0 > 0 \tag{3.1}$$

and

$$f(x,t) \equiv 0 \quad (x,t) \in P(0,\rho_0) = B_{\rho_0}(x_0) \times (0,T). \tag{3.2}$$

It is assumed that the data $u_0(x)$ and $f(x,t)$ are sufficiently "flat" near the boundaries of their supports. These conditions are formulated as follows: either

$$F \equiv F(u_0, f, \rho, T) = \left(\|u_0\|_{L^{1+\gamma}(B_\rho(x_0))} + \|f(\cdot,t)\|_{L^{(q+1)/q}(B_\rho(x_0))}\right) \\ \le \varepsilon(\rho-\rho_0)_+^{1/(1-\nu)} \tag{3.3}$$

or

$$I(\rho) := \int_{\rho_0}^{\rho} (s-\rho_0)^\nu F^\nu ds < \infty, \quad B_{\rho_0} \subset B_{\rho_*}, \quad \overline{B}_{\rho_*} \subset \Omega, \tag{3.4}$$

where

$$q = 1+\gamma, \quad \nu \frac{N(p-1-\gamma)+\gamma(p+1)+1}{N(p-1-\gamma)+p(1+\gamma)} \quad \text{if} \quad C_4 = 0$$

and

$$q = 1+\sigma, \quad \nu \frac{(p-1-\sigma)+(1+\gamma)}{N(p-1-\sigma)+p(1+\gamma)} \quad \text{if } C_4 > 0$$

(cf. with (1.19), (1.21)).

We assume that the global energy of the solution $u(x,t)$,

$$D(u,\Omega,T) = \operatorname*{ess\,sup}_{0\le\tau\le t} \int_\Omega |u(\cdot,t)|^{1+\gamma} dx \\ + \int_0^T \int_\Omega \left(|Du(x,t)|^p + C_4|u(x,t)|^{1+\sigma}\right) dxdt,$$

is finite.

Theorem 3.1 (The waiting time property). *Let $u(x,t)$ be a weak solution of problem* (2.1), (2.3) *and u_0, f satisfy* (3.1), (3.2), *and* (3.3) (*or* (3.4)). *Let the global energy be bounded: $D(u,\Omega,T) \le D_0 < \infty$. Then*

(a) *under the conditions of Theorem* 2.1, *there exists a positive constant $t^* \le T^*$ depending only on the constants in* (2.2a)–(2.2e), N, ρ_0, $\operatorname{dist}(x_0, \partial\Omega)$, *and* $D(u,\Omega,T)$, *such that*

$$u(x,t) = 0 \quad \text{in } B_{\rho_0}(x_0) \times (0,t^*);$$

(b) *under the conditions of Theorem* 2.2, *there exists a positive constant E_* depending only on the constants in* (2.2a)–(2.2e), N, ρ_0, $\operatorname{dist}(x_0,\partial\Omega)$, *such that*

$$u(x,t) = 0 \quad \text{in } P(0,\rho_0) = B_{\rho_0}(x_0) \times (0,T).$$

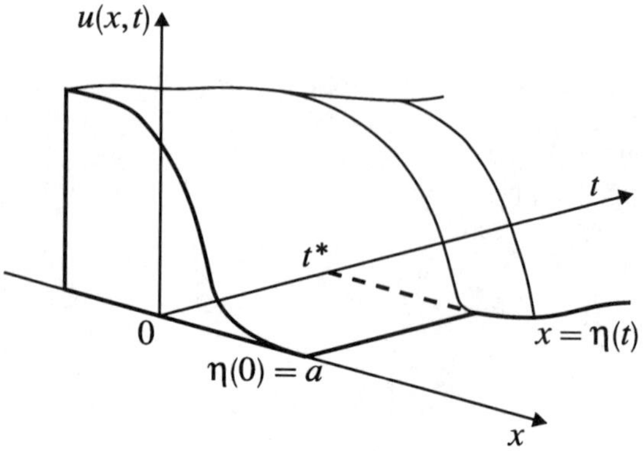

Figure 3.1: The waiting time property: if the initial data are sufficiently flat, the disturbances do not propagate across the endpoint $x = \eta(0)$ of the initial support.

Proof. According to Lemma 2.1 and (2.9), (2.2a)–(2.2e), (2.27), for each weak solution of problem (2.1), (2.3) the inequality

$$
\begin{aligned}
b(\rho,t) &+ E(\rho,t) + C_4 C(\rho,t) \\
&\le K\left(E_\rho^{(p-1)/p}\right)\left(\int_0^t \int_{S_\rho} |u|^p dS d\tau\right)^{1/p} \\
&\quad + K\left(b(0,t) + \left|\int_0^t \int_{B_\rho} Budxd\tau\right| + \left|\int_0^t \int_{B_\rho} fudxd\tau\right|\right) \\
&\equiv K\,(I_1 + I_2 + I_3 + I_4)\,, \quad B_{\rho_0} \subset B_\rho \subset \Omega
\end{aligned}
\tag{3.5}
$$

holds. To estimate the terms I_1, I_3, and I_4 in (3.5), we distinguish cases (a) and (b). We begin with the case (a) where $C_4 = 0$. Using (2.32) and Hölder's and Young's inequalities analogously to the second step of the proof of Theorem 2.1, we can deduce the estimates

$$
\begin{aligned}
|I_1| &\le K\,L\,t^{(1-\theta)/p}\left(\overline{b} + E\right)^{\theta/p+(1-\theta)/(1+\gamma)} \\
&\le \delta\left(\overline{b} + E\right) + C(\delta, p, \gamma)\left(KLt^{(1-\theta)/p}E_\rho^{(p-1)/p}\right)^{p/\nu(p-1)},
\end{aligned}
\tag{3.6}
$$

$$
|I_3| \le \varepsilon C_3 \frac{p-\beta}{p}\, t\,\overline{b}(\rho,t) + \frac{\beta C_3}{p C_2}\varepsilon^{-(p-\beta)/\beta} E(\rho,t),
\tag{3.7}
$$

$$
|I_4| \le \delta t \overline{b}(\rho,t) + C(\delta,\gamma)\int_0^t \int_{B_\rho} |f|^{(1+\gamma)/\gamma} dxd\tau,
\tag{3.8}
$$

where $(\delta, \varepsilon) \in (0,1)$ and $L = \max(1, \rho_0^{\delta\theta})\max(1, T^{\theta/p})$. Choosing δ small enough, ε as in (2.29), and replacing b by $\overline{b}$, from (3.5)–(3.8) we obtain the inequality

$$
\overline{b} + E \le K\left(\left(LE_\rho^{(p-1)/p}t^{(1-\theta)/p}\right)^{p/\nu(p-1)} + F\right).
$$

Raising both sides of this inequality to the power ν, we come to the final inequality

$$E^{\nu} \le \left(\overline{b} + E\right)^{\nu} \le K\left(L^{p/(p-1)}t^{(1-\theta)/(p-1)}E_{\rho} + F^{\nu}\right), \tag{3.9}$$

with $\rho_0 \le \rho$. Inequality (3.9) can be given the form

$$E^{\nu} \le \left(\overline{b} + E\right)^{\nu} \le K L^{p/(p-1)}t^{(1-\theta)/(p-1)}E_{\rho} + K\varepsilon^{\nu}\left(\rho - \rho_0\right)^{\nu/(1-\nu)}, \tag{3.10}$$

which is analogous to (1.20) if condition (3.3) holds. Then we can repeat the analysis given earlier to inequality (1.20). According to it, there exists $t^* = t^*(K, L, \varepsilon, \theta, p) > 0$ such that

$$u(x,t) = 0 \quad \text{for } x \in B_{\rho_0}(x_0), 0 \le t \le t^*.$$

Now let condition (3.4) hold with $q = 1 + \gamma$. Then we use Lemma 2.4, whence

$$\Lambda = K\, L^{p/(p-1)}(t^*)^{(1-\theta)/(p-1)} \equiv \Lambda_0(t^*)^{(1-\theta)/(p-1)}, \qquad F = K\, F^{\nu}.$$

The constant t^* will be defined later on. The energy function $E(\rho, t^*)$ obeys the estimate

$$\begin{aligned} E(\rho_0, t^*) &\le \omega(s) \\ &\equiv E(\rho_*, t^*) - (s - \rho_0)^{1/\nu}(t^*)^{(1-\theta)/p}\left(\left(\frac{\nu}{\Lambda_0}\right)^{1/\nu} - \frac{I(s)}{\Lambda_0}\right), \end{aligned} \tag{3.11}$$

with $s \in (\rho_0, \rho_*)$, $0 < \theta < 1$, $0 < \nu < 1$. Let us fix some $s_* \in (\rho_0, \rho_*)$. It is easy to see that there exists $t^* > 0$ such that $\omega(s_*, t^*) = 0$. Correspondingly, we have $E(\rho_0, t^*)$. This completes the proof in case (a).

Let us pass to case (b). Now the terms I_1, I_3, I_4 on the right-hand side of (3.5) can be estimated in the following way:

$$\begin{aligned} |I_1| &\le K\, L \max(1, \rho_0^{\delta\theta}) \max(\overline{b}^{\mu}(\rho,t), \overline{b}^{\eta}(\rho,t)) E_{\rho}^{(p-1)/p}\left(\overline{b} + E\right)^{1/p} \\ &\le K\, L E_{\rho}^{(p-1)/p}\left(\overline{b} + E + C\right)^{1/p+\nu(p-1)/p} \\ &\le \delta\left(\overline{b} + E + C\right) + C(\delta, p, \sigma)\left(K\, L\, E_{\rho}^{(p-1)/p}\right)^{p/\nu(p-1)}, \end{aligned} \tag{3.12}$$

$$|I_3| \le \varepsilon C_3 \frac{p-\beta}{p} C(\rho,t) + \frac{\beta C_3}{p C_2}\varepsilon^{-(p-\beta)/\beta} E(\rho,t), \tag{3.13}$$

$$|I_4| \le \delta C(\rho,t) + C(\sigma,\gamma)\int_0^t \int_{B_\rho} |f|^{(1+\sigma)/\sigma} dx d\tau, \tag{3.14}$$

where

$$\begin{aligned} L &= \max(1, \rho_0^{\theta\delta}) \max(1, \overline{b}^{\eta-\mu}), \quad \delta \in (0,1), \\ \nu &= \frac{p\mu}{p-1} = \frac{(1-\theta)(p-1-\sigma)}{p-1} = \frac{\eta(1-\theta)p}{p-1}. \end{aligned}$$

To derive (3.12)–(3.14) we also made use of analogues of inequalities (2.47), (2.54). Choosing δ sufficiently small, ε as in (2.54), and replacing b by $\overline{b}$, from (3.5), (3.12)–(3.14) we get the inequality

$$\overline{b} + E + C \leq K\left(L^{p/\nu(p-1)} E_\rho^{1/\nu} + F\right),$$

whence

$$E^\nu \leq \left(\overline{b} + E + C\right)^\nu \leq K\left(L^{p/(p-1)} E_\rho + F^\nu\right). \tag{3.15}$$

Let us consider now the case of condition (3.4). To analyze (3.15), we make use of Lemma 2.4, which gives the following estimate on the energy function $E(\rho, T)$:

$$E \leq \omega(s, D_0) \equiv D_0 - (s - \rho_0)^{1/\nu}\left(\left(\frac{\nu}{\Lambda}\right)^{1/\nu} - \frac{I(s)}{\Lambda}\right). \tag{3.16}$$

Taking into account (3.4), it is easy to see that there exist $s_* \in (\rho_0, \rho_*)$ and S_0 such that $\omega(s_*, D_0) = 0$. Correspondingly, we obtain $E(\rho_0, T) = 0$. The case of condition (3.3) can be considered by analogy. □

4 Shrinking of supports and formation of a dead core

In this section, we deal with the effects of support shrinking and a dead core formation, which are intrinsic for weak solutions of problem (2.1), (2.3). These effects are due to the presence in equation (2.1) of the term responsible for "strong absorption."

We assume that conditions (2.2a)–(2.2e) are fulfilled (now with $C_4 > 0$) and, in addition,

$$\sigma < \gamma, \quad 1 + \sigma \leq \frac{\gamma p}{p-1}, \tag{4.1}$$

$$C_3 < \left(C_4 \frac{p}{p-1}\right)^{(p-\beta)/p} \left(C_2 \frac{p}{\beta}\right)^{\beta/p} \tag{4.2}$$

if $0 < \beta < p$, and $C_3 < C_4$ if $\beta = 0$, (respectively, $C_3 < C_2$ if $\beta = p$).

The results we obtain below may be given the following simplified description. Let $v(x, t)$ be a local weak solution of the model equation

$$v_t = \Delta_p(|v|^{m-1}v) - |v|^{\lambda-1}v + f(x, t),$$

where $\Delta_p(\cdot)$ denotes the p-Laplace operator

$$\Delta_p v \equiv \operatorname{div}\left(|\nabla v|^{p-2}\nabla v\right), \quad p > 1.$$

Set $\gamma = 1/m$, $\sigma = \lambda/m$.

(i) If γ, σ, and p satisfy relation (4.1), the initial support shrinks, i.e., the inclusion below is strict:

$$\operatorname{supp} v(\cdot, t) \subset\subset \operatorname{supp} v(\cdot, 0) \quad \text{for } t > 0 \text{ small enough;}$$

(ii) under the assumptions of item (i) on the exponents, but without any assumption on the initial function, a null set with nonempty interior, alias a *dead core*, is formed:

$$\exists t^* > 0 : \quad \forall t > t^* \quad \overline{\Omega} \setminus \{\operatorname{supp} v(\cdot, t)\} \neq \emptyset.$$

In contrast to considerations of the previous sections, we shall use the energy functions defined on domains of a special form. Let us introduce the following notation: given $T > 0$, $t \in [0, T)$, $x_0 \in \Omega$, $\rho \geq 0$, and nonnegative parameters ϑ and υ, we define the local energy set

$$\begin{aligned} P(t, \rho) &\equiv P(t, \rho; \vartheta, \upsilon) \\ &= \left\{(x, s) \in Q : |x - x_0| < \rho(s) \equiv \rho + \vartheta(s - t)^{\upsilon}, \ s \in (t, T)\right\}. \end{aligned}$$

The shape of $P(t, \rho)$ is determined by the choice of the parameters $\vartheta, \upsilon, \rho, T$. We will distinguish three cases:

(a) $\vartheta = 0$, $\upsilon = 0$, $\rho > 0$; in this case $P(t, \rho)$ is the cylinder $B_\rho(x_0) \times (t, T)$;

(b) $\vartheta > 0$, $\upsilon = 1$, $\rho > 0$; $P(0, \rho)$ becomes a truncated cone centered at the point $x_0 \in \Omega$ and with the base $B_\rho(x_0) := \{x \in \Omega : |x - x_0| < \rho\}$ on the plane $t = 0$;

(c) $\vartheta > 0$, $0 < \upsilon < 1$, $\rho = 0$; then $P(t, 0)$ becomes a paraboloid.

Figure 4.1: The set $P(t, \rho; \theta, \upsilon)$.

Case (a) is already studied. That was the cylinder $B_\rho(x_0) \times (t, T)$ that was chosen for the local energy set in the study of space localization properties in Sections 2.1 and 2.2). Cases (b) and (c) allow us to study more sophisticated localization properties.

To simplify the notation, we will omit the arguments of P wherever possible. Treating cases (b) and (c) separately, we specially indicate which of the parameters are essential and which are not.

Choosing $P(t,\rho)$ for the *local energy set*, we define the local energy functions

$$E(P) := \int_{P(t,\rho)} |\nabla u(x,\tau)|^p \, dxd\tau,$$

$$C(P) := \int_{P(t,\rho)} |u(x,\tau)|^{\sigma+1} \, dxd\tau, \tag{4.3}$$

$$b(T) := \operatorname*{ess\,sup}_{s\in(t,T)} \int_{|x-x_0|<\rho+\vartheta(s-t)^\nu} |u(x,s)|^{\gamma+1} \, dx,$$

associated with any local weak solutions of problem (2.1). This choice of the set $P(t,\rho)$ is explained by the convenience of having at our disposal the domains of variable form but in fact depending only on a single variable: on ρ in cases (a) and (b) and on t in case (c).

Let us pass to the precise statement of our results. The required global information about the solution under study is formulated in terms of the *global energy function*

$$D(u(\cdot,\cdot)) := b(T,\Omega) + \int_Q \left(|\nabla u|^p + |u|^{\sigma+1}\right) dxdt, \tag{4.4}$$

where

$$b(T,\Omega) := \operatorname*{ess\,sup}_{t\in(0,T)} \int_\Omega |u(x,t)|^{\gamma+1} dx.$$

It will be assumed that the data u_0 and f are "flat" enough near the boundary of their supports. Suppose that

$$u_0 \equiv 0 \quad \text{in} \quad B_{\rho_0}(x_0) \quad \text{for some } x_0 \in \Omega \text{ and } \rho_0 > 0, \tag{4.5}$$

and

$$f \equiv 0 \quad \text{in the cylinder } P = P(0,\rho_0), \tag{4.6}$$

$P(0,\rho_0) = P(0,\rho_0 : 0, 0) \quad (= B_{\rho_0}(x_0) \times (0,T))$. Then the *flatness condition* is stated by claiming convergence of the auxiliary integral

$$I = \int_{\rho_0+0} (\rho-\rho_0)^\beta \left(\|u_0\|^{\gamma+1}_{L^{\gamma+1}(B_\rho(x_0))} + \|f\|^{(1+\sigma)/\sigma}_{L^{(1+\sigma)/\sigma}(P(0,\rho))} \right)^{p/(p-1)} d\rho, \tag{4.7}$$

where

$$\beta = (1-\delta\widetilde{\theta})(1+\kappa), \quad \delta = -\left(1 + \frac{p-1-\sigma}{p(1+\sigma)} N\right), \tag{4.8}$$

with some

$$\kappa \in \left(0, \frac{p(1+\gamma)}{(p-1-\sigma)(1-\widetilde{\theta})}\right), \quad \widetilde{\theta} = \frac{pN-(1+\sigma)(N-1)}{(N+1)p-N(1+\sigma)}.$$

Theorem 4.1 (Shrinking of support). *Assume that conditions* (2.2a)–(2.2e) *hold with* $\alpha = \gamma - (1+\gamma)\beta/p$, *and that* (4.1) *is true. Let* (4.2) *be fulfilled. Let* u_0 *satisfy* (4.5). *Assume that*

$$f \equiv 0 \quad \textit{in the truncated cone } P \equiv P(0, \rho_0 : \vartheta, 1) \tag{4.9}$$

for some $\vartheta > 0$ *and let* (4.7) *be true. Then there exist positive constants* M *and* t^* *such that each weak solution of problem* (2.1) *with global energy satisfying the inequality* $D(u) \leq M$ *possesses the property*

$$u(x,t) \equiv 0 \quad \textit{in } P(0, \rho_0 : \vartheta, 1) \cap \{t \leq t^*\}.$$

The next result applies to the case when the initial function need not vanish, namely when the parameter ρ_0 in the conditions of Theorem 4.1 equals zero. Assuming that $f \equiv 0$ (which is convenient but not necessary), we show how the strong absorption term causes formation of the dead core of the solution.

Theorem 4.2 (Dead core formation). *Let us assume that conditions* (2.2a)–(2.2e) *hold with* $\alpha = \gamma - (1+\gamma)\beta/p$, (4.1), (4.2) *are fulfilled, and that* $\gamma \leq p - 1$. *Let* $f \equiv 0$. *Then there exist positive constants* M, t^*, *and* $\upsilon \in (0, 1)$ *such that any weak solution of problem* (2.1) *satisfying the inequality* $D(u) \leq M$ *possesses the property*

$$u(x,t) \equiv 0 \quad \textit{in } P(t^*, 0 : 1, \upsilon).$$

The proof of Theorem 4.1 splits into several steps.

4.1 The energy relation: Integration-by-parts formula. Given $x_0 \in \Omega$, $t \in [0, T]$, $\vartheta \geq 0$ and $\upsilon \in (0, 1)$, we define the cutting function on the set $P(t, \rho)$,

$$\zeta(x,\theta) := \psi_\varepsilon\left(|x - x_0|, \theta\right) \xi_k(\theta) \frac{1}{h} \int_\theta^{\theta+h} T_m\left(u(x,s)\right) ds, \qquad h > 0,$$

where

$$T_m\left(u(x,t)\right) := \operatorname{sign} u(x,s) \min\left\{m, |u(x,s)|\right\}, \qquad m \in \mathbb{N},$$

and

$$\xi_k(\theta) := \begin{cases} 1 & \text{if } \theta \in \left[t, T - \frac{1}{k}\right], \\ k(T-\theta) & \text{for } \theta \in \left[T - \frac{1}{k}, T\right], \\ 0 & \text{otherwise}, \quad k \in \mathbb{N}, \end{cases}$$

$$\psi_\varepsilon\left(|x - x_0|, \theta\right) := \begin{cases} 1 & \text{if } d > \varepsilon, \\ \frac{1}{\varepsilon} d & \text{if } d < \varepsilon, \\ 0 & \text{if } (x,\theta) \in Q \setminus P(t,\rho). \end{cases}$$

Here $d = \operatorname{dist}((x,\theta), \partial_l P(t,\rho))$, $\varepsilon > 0$, and in what follows $\partial_l P$ denotes the lateral boundary of P, i.e.,

$$\partial_l P = \{(x,s) : \; |x - x_0| = \rho + \vartheta(s-t)^{\upsilon}, \; s \in (t,T)\}.$$

By construction we have $\operatorname{supp}\zeta(x,\theta) \equiv P(t,\rho)$. It is easy to verify that for every natural m and k and positive real numbers h and ε,

$$\zeta, \frac{\partial \zeta}{\partial t} \in L^{\infty}((0,T)\times\Omega), \qquad \frac{\partial \zeta}{\partial x_i} \in L^{p}((0,T)\times\Omega).$$

These properties allow one to substitute $\zeta(x,\theta)$ into the integral identity (2.4) as a test function. Passing to the limits in the appearing equality, analogously by Lemma 2.1, one has

$$\begin{aligned} & i_1 + i_2 + i_3 + i_4 \\ & =: \int_{P\cap\{t=T\}} G(x,u(x,t))dx + \int_P \mathbf{A}\cdot Du\,dxd\theta \\ & \quad + \int_P B\,udxd\theta + \int_P C\,udxd\theta \\ & = \int_{\partial_l P} \boldsymbol{n}_x,\cdot\mathbf{A}\,u\,d\Gamma d\theta + \int_{\partial_l P} \boldsymbol{n}_\tau G(u(x,t))d\Gamma d\theta \\ & \quad + \int_{P\cap\{t=0\}} G(x,u(x,t))dx + \int_P uf\,dxd\theta \\ & := j_1 + j_2 + j_3 + j_4. \end{aligned} \tag{4.10}$$

Here $d\Gamma$ is the differential form on the hypersurface $\partial_l P \cap \{t = \text{const}\}$, $\boldsymbol{n}_x$ and $\boldsymbol{n}_\tau$ are the components of the unit normal vector to $\partial_l P$.

4.2 Differential inequalities. We begin with the most complicated case (c), where the domain P is a paraboloid determined by the parameters $\upsilon \in (0,1)$, $\vartheta > 0$, and $t \in (0,T)$:

$$P = P(t) = \{(x,\tau) : |x - x_0| \equiv \rho(\tau) \le \vartheta(\tau - t)^{\upsilon}, \quad \tau \in (t,T)\}.$$

We assume that $f \equiv 0$ and that P does not touch the initial plane $\{t = 0\}$. These assumptions simplify the basic energy inequality (4.10):

$$i_1 + i_2 + i_3 \le j_1 + j_2.$$

Let us estimate the term j_1. It is easy to see that

$$\boldsymbol{n} \equiv (\boldsymbol{n}_x, \boldsymbol{n}_\tau) = \frac{1}{\left(\vartheta^2\upsilon^2 + (\theta - t)^{2(1-\upsilon)}\right)^{1/2}} \left((\theta - t)^{1-\upsilon}\boldsymbol{e}_x - \upsilon\vartheta\boldsymbol{e}_\tau\right),$$

where $\boldsymbol{e}_\tau$ and $\boldsymbol{e}_x$ are unit vectors orthogonal to the hyperplane $t = 0$ and the axis t, respectively.

Let (ρ, ω), $\rho > 0$, $\omega \in \partial B_1$, be the spherical coordinate system in $\mathbb{R}^N$. Given an arbitrary function $F(x,t)$, we use the notation $x = (\rho, \boldsymbol{\omega})$ and $F(x,t) = \Phi(\rho, \boldsymbol{\omega}, t)$. There holds the equality

$$I(t) := \int_P F(x,\theta)dxd\theta \equiv \int_t^T d\theta \int_0^{\rho(\theta,t)} \rho^{N-1}d\rho \int_{\partial B_1} \Phi(\rho, \boldsymbol{\omega}, \theta)|J|d\boldsymbol{\omega},$$

where J is the Jacobi matrix and, due to the definition of P, $\rho(\theta,t) = \vartheta(\theta - t)^{\upsilon}$. It is easy to check that

$$\begin{aligned} \frac{dI(t)}{dt} &= -\int_0^{\rho(\theta,t)} \rho^{N-1}d\rho \int_{\partial B_1} \Phi(\rho, \boldsymbol{\omega}, \theta)|J|d\boldsymbol{\omega}\Big|_{\theta=t} \\ &\quad + \int_t^T \rho_t \rho^{N-1} d\theta \int_{\partial B_1} \Phi(\rho, \boldsymbol{\omega}, t)|J|d\boldsymbol{\omega} \\ &= \int_{\partial_l P} \rho_t F(x,\theta)d\Gamma d\theta. \end{aligned} \tag{4.11}$$

Viewing the energy function E as a function of t, with the use of (2.2a)–(2.2e) and Hölder's inequality, we obtain the relation

$$\begin{aligned} &\left|\int_{\partial_l P} \boldsymbol{n}_x \cdot \mathbf{A}\, u\, d\Gamma d\theta\right| \\ &\quad \le M_2 \int_{\partial_l P} |\boldsymbol{n}_x||\nabla u|^{p-1}|u| d\Gamma d\theta \\ &\quad \le M_2 \left(\int_{\partial_l P} |\rho_t||\nabla u|^p d\Gamma d\theta\right)^{(p-1)/p} \left(\int_{\partial_l P} \frac{|\boldsymbol{n}_x|^p}{|\rho_t|^{p-1}}|u|^p d\Gamma d\theta\right)^{1/p} \\ &\quad = M_2 \left(-\frac{dE}{dt}\right)^{(p-1)/p} \left(\int_t^T \frac{|\boldsymbol{n}_x|^p}{|\rho_t|^{p-1}} \left(\int_{\partial B_{\rho(\theta,t)}} |u|^p d\Gamma\right) d\theta\right)^{1/p}. \end{aligned} \tag{4.12}$$

To estimate the right-hand side of (4.12) we use the interpolation inequality: if $\sigma \le p-1$, then $\forall\, v \in W^{1,p}(B_\rho)$,

$$\|v\|_{p,S_\rho} \le L_0 \left(\|\nabla v\|_{p,B_\rho} + \rho^{\delta}\|v\|_{\sigma+1,B_\rho}\right)^{\widetilde{\theta}} \cdot \left(\|v\|_{r,B_\rho}\right)^{1-\widetilde{\theta}} \tag{4.13}$$

with a universal constant $L_0 > 0$ not depending on $v(x)$ and the exponents

$$r \in [1+\sigma, 1+\gamma], \quad \widetilde{\theta} = \frac{pN - r(N-1)}{(N+1)p - Nr}, \quad \delta = -\left(1 + \frac{p-1-\sigma}{p(1+\sigma)}N\right).$$

Let us introduce the notation

$$E_*(t,\rho) := \int_{B_\rho} |\nabla u|^p dx, \qquad C_*(t,\rho) := \int_{B_\rho} |u|^{\sigma+1} dx$$

so that

$$E = \int_t^T E_*(\theta, \rho(\theta, t))d\theta, \qquad C = \int_t^T C_*(\theta, \rho(\theta, t))d\theta$$

and make use of Hölder's inequality

$$\left(\int_{B_\rho} |u|^r dx\right)^{1/r} \le \left(\int_{B_\rho} |u|^{1+\sigma} dx\right)^{1/qr} \cdot \left(\int_{B_\rho} |u|^{\gamma+1} dx\right)^{(q-1)/qr},$$

where

$$q = \frac{\gamma - \sigma}{\gamma - r + 1}.$$

Then by virtue of (4.13),

$$\begin{aligned}
\int_{\partial B_\rho} |u|^p d\Gamma &\le L_0 \left(\int_{B_\rho} |\nabla u|^p dx + \rho^{\delta p} \left(\int_{B_\rho} |u|^{\sigma+1} dx\right)^{p/(\sigma+1)}\right)^{\widetilde{\theta}} \\
&\quad \times \left(\int_{B_\rho} |u|^r dx\right)^{p(1-\widetilde{\theta})/r} \\
&\le L_0 \rho^{\delta\widetilde{\theta} p} \left(\int_{B_\rho} |\nabla u|^p dx + \int_{B_\rho} |u|^{\sigma+1} dx\right)^{\widetilde{\theta}} \\
&\quad \times \left(\int_{B_\rho} |u|^{\sigma+1} dx\right)^{p(1-\widetilde{\theta})/qr} \left(\int_{B_\rho} |u|^{\gamma+1} dx\right)^{p(q-1)(1-\widetilde{\theta})/qr} \\
&\le K\rho^{\delta\widetilde{\theta} p} (E_* + C_*)^{\widetilde{\theta}} C_*^{(1-\widetilde{\theta})p/qr} b^{(q-1)(1-\widetilde{\theta})p/qr} \\
&\le K\rho^{\delta\widetilde{\theta} p} (E_* + C_*)^{\widetilde{\theta}+(1-\widetilde{\theta})p/qr} b^{(q-1)(1-\widetilde{\theta})p/qr}
\end{aligned} \tag{4.14}$$

with

$$\begin{aligned}
K &= L_0 \max\left(1, \left(\operatorname*{ess\,sup}_{(t,T)} \int_{B_{\rho(\theta)}} |u|^{\sigma+1} dx\right)^{\frac{p}{\sigma+1}-1}\right)^{\widetilde{\theta}/p} \\
&\le L_0 \max\left(1, (\text{meas}\, B_{\rho(T)})^{\frac{\gamma-\sigma}{\gamma+1}\left(\frac{p}{\sigma+1}-1\right)} (b(T))^{\frac{\sigma+1}{\gamma+1}\left(\frac{p}{\sigma+1}-1\right)}\right)^{\widetilde{\theta}/p}.
\end{aligned}$$

Let us assume that

$$\mu = \widetilde{\theta} + p\frac{1-\widetilde{\theta}}{qr} < 1. \tag{4.15}$$

Returning to (4.12) and applying once again Hölder's inequality with the exponent μ, we have from (4.14)

$$
\begin{aligned}
|j_1| &\le L\left(-\frac{dE}{dt}\right)^{(p-1)/p} \\
&\quad\times\left(\int_t^T \frac{|\vec{n}_x|^p}{|\rho_t|^{p-1}} K\rho^{\delta\widetilde{\theta}p}\,(E_*+C_*)^{\mu}\, b^{(q-1)(1-\widetilde{\theta})p/qr}\,d\tau\right)^{1/p} \\
&\le L\left(-\frac{dE}{dt}\right)^{(p-1)/p} b^{(q-1)(1-\widetilde{\theta})/qr} \\
&\quad\times\left(\int_t^T (E_*+C_*)\,d\tau\right)^{\frac{\mu}{p}}\left(\int_t^T\left(\frac{|\vec{n}_x|^p}{|\rho_t|^{p-1}}\rho^{\delta\widetilde{\theta}p}(\tau)\right)^{\frac{1}{1-\mu}}d\tau\right)^{\frac{1-\mu}{p}} \\
&\le L\sigma(t)\left(-\frac{d\,(E+C)}{dt}\right)^{(p-1)/p}\cdot b^{(q-1)(1-\widetilde{\theta})/qr}\,(E+C)^{\frac{\widetilde{\theta}}{p}+\frac{1-\widetilde{\theta}}{qr}}
\end{aligned}
\tag{4.16}
$$

for a suitable positive constant L. To obtain (4.16), we have assumed (4.15) and

$$
\frac{p}{\sigma+1} > 1, \tag{4.17}
$$

$$
\sigma(t) := \left(\int_t^T\left(\frac{1}{|\rho_t|^{p-1}}\rho^{\delta\widetilde{\theta}p}(\tau)\right)^{\frac{1}{1-\mu}}d\tau\right)^{\frac{1-\mu}{p}} < \infty. \tag{4.18}
$$

Inequality (4.15) is fulfilled if $p < qr$, which is true under an extra restriction on r. We have

$$
p < qr \iff (\gamma-\sigma)r > p(\gamma-r+1) \iff r > \frac{1+\gamma}{1+\frac{\gamma-\sigma}{p}}.
$$

The last inequality must be consistent with the previous choice of r: $r \in [1+\sigma, 1+\gamma]$. Thus we have to claim first that

$$
\gamma+1 > \frac{\gamma+1}{1+\frac{\gamma-\sigma}{p}} = \frac{p(\gamma+1)}{p+\gamma-\sigma} \iff p+\gamma-\sigma > p \iff \gamma > \sigma,
$$

which is true because of (4.1), and then to choose

$$
r \in \left[\frac{p(\gamma+1)}{p+\gamma-\sigma}, \gamma+1\right].
$$

Inequality (4.17) coincides with (4.1). To satisfy (4.18), one only has to choose $\nu \in (0,1)$ sufficiently small because the condition of convergence of the integral $\sigma(t)$ has the form

$$
(1-\nu)(p-1)+\nu\delta\widetilde{\theta}p > -(1-\widetilde{\theta})\left(1-\frac{p}{qr}\right).
$$

Thus we have obtained the estimate of the type

$$|j_1| \le L_1\, \Lambda(t)\, D(u)^{(q-1)(1-\widetilde{\theta})/qr-\lambda} \times (E+C+b)^{1-\omega+\lambda}\left(-\frac{d(E+C)}{dt}\right)^{(p-1)/p}, \tag{4.19}$$

where L_1 is a universal positive constant, $D(u)$ is the *total energy* of the solution under investigation, $\lambda \in [0, (q-1)(1-\widetilde{\theta})/qr]$, and

$$\omega := 1 - \frac{\widetilde{\theta}}{p} - \frac{1-\widetilde{\theta}}{qr} \in \left(1-\frac{1}{p}, 1\right).$$

This allows us to choose λ so that

$$\frac{p(\omega-\lambda)}{p-1} \in (0,1).$$

Let us estimate j_2. Using (2.2e) and the expression for $\boldsymbol{n}_\tau$, we have

$$|j_2| \le C_5 \int_{\partial_l P} |u|^{1+\gamma}\, d\Gamma\, d\theta. \tag{4.20}$$

We apply then the interpolation inequality

$$\forall\, v \in W^{1,p}(B_\rho), \quad \|v\|_{\gamma+1,\partial B_\rho} \le L_0 \left(\|\nabla v\|_{p,B_\rho} + \rho^{\delta}\|v\|_{\sigma+1,B_\rho}\right)^{s} \cdot \|v\|_{r,B_\rho}^{1-s} \tag{4.21}$$

with a universal positive constant $L_0 > 0$ and the exponents

$$s = \frac{(\gamma+1)N - r(N-1)}{(N+r)p - Nr} \cdot \frac{p}{\gamma+1}, \qquad r \in [1+\sigma, 1+\gamma],$$

and with the constant δ from (4.13). By analogy with the previous estimate, we can write

$$\int_{\partial B_\rho} |u|^{\gamma+1} dx \le L^{1+\gamma} K^{s(\gamma+1)/\widetilde{\theta}p} \left(\int_{B_\rho} |\nabla u|^p dx + \int_{B_\rho} |u|^{\sigma+1} dx\right)^{s(\gamma+1)/p} \times \left[\left(\int_{B_\rho} |u|^{\sigma+1} dx\right)^{1/qr} \left(\int_{B_\rho} |u|^{\gamma+1} dx\right)^{(q-1)/qr}\right]^{(1-s)(\gamma+1)}. \tag{4.22}$$

Here K is the same as before. Let

$$\eta = \frac{s(\gamma+1)}{p} + \frac{(1-s)(\gamma+1)}{qr} < 1,$$

$$\pi = \frac{(q-1)(1-s)(\gamma+1)}{qr},$$

$$\eta + \pi \ge 1.$$

Then gathering (4.20), (4.22), we come to the inequality

$$
\begin{aligned}
|j_2| &= \left| \int_t^T d\tau \int_{\partial B_{\rho(\tau)}} |u|^{\gamma+1} d\Gamma \right| \\
&\le L\,(b(T))^{\pi} \left(\int_t^T K^{s(\gamma+1)/\widetilde{\theta}p} (E_* + C_*)^{\eta} |n_\tau| d\tau \right) \qquad (4.23)\\
&\le L\,(E + C + b(T,\Omega))\,(b(T,\Omega))^{\kappa} \left(\int_t^T \left(K^{s(\gamma+1)/\widetilde{\theta}p} \right)^{\varepsilon} d\tau \right)^{1/\varepsilon}
\end{aligned}
$$

with some $L = L(C_5, L_0)$ and the exponents

$$\kappa := \eta + \pi - 1, \qquad \varepsilon = 1/(1-\eta).$$

To perform this estimate, we have assumed that $\eta < 1$ and $\eta + \pi \ge 1$. The first of these inequalities is a simplified version of (4.15). As for the second one, a direct computation shows that it coincides with the second condition of (4.1).

We now turn to estimating the left-hand side of (4.10). By (2.2a)–(2.2e) and (4.2), we have at once that

$$C_5 \int_{P\cap\{t=T\}} |u|^{1+\gamma} dx + E + C_4 C \le i_1 + i_2 + i_3, \tag{4.24}$$

$$|i_4| \le \varepsilon C_3 \frac{p-\beta}{p} C(\rho,t) + \frac{\beta C_3}{p C_2} \varepsilon^{-(p-\beta)/\beta} E(\rho,t), \tag{4.25}$$

$$K \left(\int_{P\cap\{t=T\}} |u|^{1+\gamma} dx + E + C \right) \le i_1 + i_2 + i_3 + i_4 \tag{4.26}$$

where

$$K = \left(C_5,\ C_4 - \varepsilon C_3 \frac{p-\beta}{p},\ 1 - \frac{\beta C_3}{p C_2} \varepsilon^{-(p-\beta)/p} \right) > 0.$$

Since the right-hand side of (4.10) is an increasing function of t, we may always replace the first term on the left-hand side of (2.1) by $b(T)$. Now assuming $T - t$ and $D(u)$ so small that

$$L\,(b(T,\Omega))^{\kappa} \left(\int_t^T \left(K^{s(\gamma+1)/\widetilde{\theta}p} \right)^{\varepsilon} d\tau \right)^{1/\varepsilon} < \frac{K}{2},$$

we arrive at the inequality

$$
\begin{aligned}
E + C + b(T,\Omega) &\le L_1\, \Lambda(t)\, D(u)^{(q-1)(1-\widetilde{\theta})/qr-\lambda} \\
&\quad \times (E + C + b(T,\Omega))^{1-\omega+\lambda} \left(-\frac{d(E+C)}{dt} \right)^{(p-1)/p},
\end{aligned} \tag{4.27}
$$

whence the desired differential inequality for the energy function $Y(t) := E + C$:

$$Y^{(\omega-\lambda)p/(p-1)}(t) \le c(t)\,(-Y(t))', \tag{4.28}$$

where

$$c(t) = \left(L_1\,(D(u))^{(q-1)(1-\widetilde{\theta})/qr-\lambda}\,\sigma(t)\right)^{p/(p-1)}, \qquad L_1 = \text{const} > 0.$$

Note that $c(t) \to 0$ as $t \to T$. Moreover, the exponent $(\omega-\lambda)\frac{p}{p-1}$ belongs to the interval (0, 1).

We pass to the consideration of domains of type (b). The differential inequality for the energy function $E + C$ is now derived in much the same way as in case (c) but with certain simplification due to the choice of the domain P. Let

$$P = \{(x,t) : |x - x_0| < \rho + \vartheta\theta,\ \vartheta \in (0,T)\}, \quad \rho \ge \rho_0 > 0.$$

The unit outer normal to $\partial_l P$ has the form

$$\boldsymbol{n} = \frac{1}{\sqrt{1+\vartheta^2}}(1, -\vartheta)$$

and if the energy function $Y := E + C$ is considered as a function of ρ, we have

$$\begin{aligned}
\frac{dY(\rho)}{d\rho} &= \int_{\partial_l P}\left(|\nabla u|^p + |u|^{\sigma+1}\right) d\Gamma d\theta \\
&= \frac{d}{d\rho}\left\{\int_0^T d\theta \int_0^{\rho+\vartheta\theta} \tau^{N-1} d\tau \int_{\partial B_1} |J| \left(|\nabla u|^p + |u|^{\sigma+1}\right)\Big|_{x=(\tau,\omega)} d\omega\right\} \\
&= \int_0^T d\theta \int_{\partial B_1}\left\{(\rho+\vartheta\theta)^{N-1}|J|\left(|\nabla u|^p + |u|^{\sigma+1}\right)\Big|_{x=(\rho+\vartheta\theta,\omega)}\right\} d\omega.
\end{aligned} \tag{4.29}$$

Following the scheme above, we estimate the term j_1 in (4.10), and then applying (4.29), we arrive at the inequality

$$\begin{aligned}
|j_1| \le{}& \frac{K}{\sqrt{1+\vartheta^2}}\left(\frac{dE}{d\rho}\right)^{(p-1)/p} \rho^{\delta\widetilde{\theta}}\,(b(T))^{(q-1)(1-\widetilde{\theta})/qr} \\
&\times \left(\int_0^T (E_*) + C_*)^{\widetilde{\theta}+p(1-\widetilde{\theta})/qr} d\theta\right)^{1/p}.
\end{aligned}$$

Let r be such that

$$\widetilde{\theta} + \frac{(1-\widetilde{\theta})p}{qr} = 1.$$

Such a choice is always possible, since

$$\widetilde{\theta} + \frac{(1-\widetilde{\theta})p}{qr} = 1 \Leftrightarrow r = \frac{p(1+\gamma)}{p+\gamma-\sigma},$$

and the last equality is compatible with the choice of r. The estimate for j_1 then takes the form

$$|j_1| \leq \frac{K\rho^{\delta\widetilde{\theta}}}{\sqrt{1+\vartheta^2}} \left(\frac{dE}{d\rho}\right)^{(p-1)/p} (b(T))^{(q-1)(1-\widetilde{\theta})/qr-\varepsilon} (E+C)^{\varepsilon+1/p}$$

with an arbitrary $\varepsilon \in (0, (q-1)(1-\widetilde{\theta})/qr)$.

The estimate for j_2 is the same as that of case (c). The only difference is that now we need not claim the smallness of T. The value of the coefficient in the estimate for j_2 is controlled now by the choice of ϑ, since $n_\tau = -\vartheta/\sqrt{1+\vartheta^2}$. Due to (4.5) we have $j_3 = 0$. At last, we estimate j_4 with the help of (2.15) and the Young inequality

$$j_4 \leq \tau C + L(\tau, C_5) \int_P |f|^{(\sigma+1)/\sigma} dx d\theta.$$

The left-hand side of (4.10) is estimated in the same way as in case (c). Gathering these estimates, we arrive at the inequality

$$Y(\rho) \leq c(\rho) Y^{\varepsilon+1/p}(\rho) \left(Y'(\rho)\right)^{(p-1)/p} + F(\rho), \quad \rho > \rho_0$$

with the coefficient

$$c(\rho) = \rho^{\delta\widetilde{\theta}} K \, (d(u))^{(q-1)(1-\widetilde{\theta})/qr-\varepsilon}$$

and the right-hand side

$$F(\rho) = \int_{R_\rho(x_0)} |u_0|^{\gamma+1} dx + L(\tau) \int_P |f|^{(\sigma+1)/\sigma} dx d\theta.$$

It is easy to see now that the function

$$Z := Y^{p(1+\varepsilon)/(p-1)}(\rho)$$

satisfies the inequality

$$Z^\omega(\rho) \leq \frac{p-1}{p(1+\varepsilon)} c^{p/(p-1)} Z'(\rho) + F^{p/(p-1)}(\rho) \tag{4.30}$$

with $\rho > \rho_0$, $\omega = 1/(1+\varepsilon)$.

4.3 Analysis of the differential inequalities. First of all, we note that in case (b) the coefficient $c(\rho)$ admits the estimate $c \leq c(\rho_0)$. Then (4.30) transforms into inequality (3.15), which has already been studied.

In case (c), we assume that T is taken so as to provide the inclusion $P \subset Q$. Recall that the coefficient $c(t)$ in inequality (4.30)) can be estimated from above by $l := c(0)$. Introduce the function

$$z(T-t) := Y(t), \quad (z(t) \in [0, D(u)]).$$

It obeys the already studied differential inequality

$$z^{\gamma p/(p-1)}(t) \le l z'(t) \quad \text{as } t \in (0, T), \quad z(0) = 0, \tag{4.31}$$

and the proof of Theorems 4.1 and 4.2 is thus completed.

To make evident the dependence between the parameters t^*, T, and M, let us integrate inequality (4.31) over the interval $t \in (t_1, t_2) \subset (0, T)$. Then

$$z^{1-\kappa}(t_1) \le z^{1-\kappa}(t_2) - \frac{1-\kappa}{l}(t_2 - t_1),$$

where

$$\kappa = \frac{p\nu}{p-1}, \quad l = \left(L_1 \Lambda(0) M_*^{(q-1)(1-\theta)/qr}\right)^{p/(p-1)} \equiv l_0 M_*^{\kappa_0},$$
$$\kappa_0 = \frac{p}{p-1} \frac{(q-1)(1-\theta)}{qr}.$$

It follows that

$$z^{1-\kappa}(t_1) \le \frac{1-\kappa}{l_0 M_*^{\kappa_0}} \left(\frac{l_0}{1-\kappa} M_*^{1+\kappa+\kappa_0} - t_2 + t_1 \right)$$

and hence $z(t) = 0$ if

$$t \ge t^* \equiv \left(T - \frac{l_0}{1-\kappa} M_*^{1+\kappa_0-\kappa} \right), \quad 0 < t^* < T. \tag{4.32}$$

Therefore, for every $0 < T < \infty$ and M_* satisfying the condition

$$M_* < \left(\frac{1-\kappa}{l_0} T \right)^{1/(1+\kappa_0-\kappa)},$$

we have $u(x, t) \equiv 0$ in $P(t^*, 0)$ if $D(u(x, t)) \le M_*$, and t^* is given by (4.32).

Conversely, if $u(x, t)$ is a weak solution of equation (2.1), and

$$\sup_{0<t<\infty} D(u(., t)) \le M = \text{const} < \infty,$$

there always exists $t^* < \infty$ such that $u(x, t) \equiv 0$ in $P(t^*, 0)$.

As in the situation considered in Section 2, the condition $D(u) \le M$, (M is small, generally speaking), can be replaced by $C_4 \ge C_4^*$. The latter means that the intensity of the absorption term is supposed large.

4.4 Estimates on the total energy. Throughout this chapter we were assuming existence of a weak solution $u(x, t)$ whose total energy $D(u, \Omega, T) \equiv D(u)$ was bounded. Moreover, it was implicitly assumed that the value of the total energy might be somehow estimated. As was pointed out in Chapter 2, the conditions of the formulated theorems in most cases provide validity of such estimates. Let us present here a sketch of the derivation of energy estimates for the weak solutions of equation (2.1) both in a strictly interior subdomain of the cylinder Q and in its closure. We will also give a brief review of the works where the existence and uniqueness of weak solutions were studied and some of their properties were investigated.

Consider the boundary-value problem

$$\begin{aligned}\frac{\partial\psi(x,u)}{\partial t} &- \operatorname{div}\mathbf{A}(x,t,u,D\,u) + B(x,t,u,D\,u) + C(x,t,u)\\ &= f(x,t),\end{aligned} \tag{4.33}$$

$$u(x,0) = u_0(x) \quad \text{on } \Omega, \tag{4.34}$$

$$u(x,t) = \varphi(x,t) \quad \text{on } \Gamma_D \times (0,T), \tag{4.35}$$

$$\mathbf{A}(x,t,u,\nabla u)\cdot\vec{\nu} = 0 \quad \text{on } \Gamma_N \times (0,T), \tag{4.36}$$

where $\Gamma_D \cup \Gamma_N = \partial\Omega$. We still assume that conditions (2.2a)–(2.2e) and (3.13) are fulfilled with

$$\alpha = \gamma - \frac{(1+\gamma)\beta}{p}, \qquad 1 < p < \infty.$$

First, let in (4.33) $\varphi \equiv 0$. It follows then from Lemma 2.1 (see (2.9)) that

$$\begin{aligned}\int_\Omega G(x,u(x,\tau))dx\Big|_{\tau=0}^{\tau=t} &+ \int_0^t\int_\Omega (\mathbf{A}(x,\tau,u,Du)\cdot Du + B(x,\tau,u,Du)\,u\\ &+ C(x,\tau,u)\,u - f\,u)\,dxd\tau \le 0.\end{aligned} \tag{4.37}$$

With the use of (2.2a)–(2.2e), (3.13), (3.14) the last inequality can be written as

$$\begin{aligned}C_5\int_\Omega |u|^{1+\gamma}dx &+ \left(1 - \frac{\beta C_3}{pC_2}\epsilon^{-(p-\beta)/p}\right)E(\rho,t)\\ &+ \left(C_4 - \epsilon C_3\frac{p-\beta}{p}\right)C(\rho,t)\\ &\le C_6\int_\Omega |u_0|^{1+\gamma}dx\tau C(\rho,t) + L(\tau)\int_0^t\int_\Omega |f|^{(1+\sigma)/\sigma}.\end{aligned} \tag{4.38}$$

Choosing δ small enough and ε as in (2.54), we come to the desired estimate

$$\begin{aligned}D(u) &:= b(T,\Omega) + \int_Q \left(|\nabla u|^p + |u|^{\lambda+1}\right)dxdt\\ &\le K\left(\int_\Omega |u_0|^{1+\gamma}dx + \int_0^T\int_\Omega |f|^{(1+\sigma)/\sigma}dxd\tau\right),\end{aligned} \tag{4.39}$$

where $K = K(p,\sigma,\gamma,\beta,p,C_1 - C_6)$. Since the constant K does not depend on Ω, estimate (4.39) is valid as well for the weak solutions of the Cauchy problem with $\Omega = \mathbb{R}^N$.

Let us present an example of direct estimating of the total energy through the input data of the Dirichlet problem. We adopt the notation $\Gamma = \Gamma_D = \partial\Omega$ and assume, for simplicity, that in (2.2c) $C_3 = 0$. Let

$$u = \Phi(x,t) \text{ on the parabolic boundary of } Q. \tag{4.40}$$

In (2.4), set

$$\varphi(x,\theta) = \chi_k(\theta)\frac{1}{h}\int_\theta^{\theta+h} T_m\left(u(x,s) - G(x,u(x,s))\right) ds,$$

where χ_k, T_m are determined at the first step of the proof of Theorem 4.2, and $\Phi(x,t)$ is a function continuing the initial and boundary data into the interior of Q. Passing to the limits in the appearing inequality, as in Lemma 2.1, we come to the relations

$$\int_{B_\rho(x_0)} (G(x,u(x,\tau)) - \psi(x,u(x,\tau))\phi(x,t))\, dx \Bigg|_{\tau=0}^{\tau=t}$$
$$+ \int_0^t \int_{B_\rho(x_0)} (\mathbf{A}\cdot(Du - D\Phi) + C\,(u-\Phi) + \psi\Phi_t)\, dx d\tau \le 0.$$

Using (2.2a)–(2.2e) and Young's inequality, we get the final estimate

$$\begin{aligned} D(u(x,t)) &:= b(T,\Omega) + \int_Q \left(|\nabla u|^p + |u|^{\lambda+1}\right) dxdt \\ &\le K\Bigg[\sup_{t\le T}\int_\Omega |\Phi(x,t)|^{1+\gamma} dx \\ &\quad + \int_Q \left(|\nabla\Phi|^p + |\Phi|^{1+\sigma} + |f|^{(1+\sigma)/\sigma}\right) dxdt \\ &\quad + \left(\int_0^T \left(\int_\Omega |\Phi_t|^{1+\gamma} dx\right)^{1/(1+\gamma)} d\tau\right)^{1+\gamma}\Bigg], \end{aligned} \tag{4.41}$$

where $K \equiv K(\gamma,\sigma,p,C_1 - C_4, C_4^*)$, and, additionally to (2.2d),

$$\forall (x,t,s) \in \Omega\times\mathbb{R}^+\times\mathbb{R}, \quad |C(x,t,s)| \le C_4^*|s|^\lambda. \tag{4.42}$$

Let us pass to the *interior estimates*. Substitute into (2.4) the function

$$\varphi(x,\theta) = \xi(x)\chi_k(\theta)\frac{1}{h}\int_\theta^{\theta+h} T_m(u(x,s))ds$$

with $\xi \in C^1(\Omega)$, $\xi|_{\partial\Omega} = 0$, $\xi > 0$ and $|D\xi|\xi^{(1-p)/p} < \infty$. By analogy with the preceding estimates (4.39), (4.41), we then come to the inequality

$$\begin{aligned}
&\int_{B_\rho(x_0)} G(x, u(x,\tau))dx \Bigg|_{\tau=0}^{\tau=t} \\
&\quad + \int_0^t \int_{B_\rho(x_0)} (\xi\, (\mathbf{A}(x,\tau,u,Du)\cdot Du + B(x,\tau,u,Du)\, u \\
&\quad + C(x,\tau,u)\, u + f\, u) + u\,\mathbf{A}\cdot D\xi)\, dx d\tau \le 0.
\end{aligned} \tag{4.43}$$

Assuming conditions (2.2a)–(2.2e), (4.2),(4.42) and using Young's inequality, from (4.43) we obtain the estimate

$$\begin{aligned}
D(u(x,t)) &:= \sup_{t\le T}\int_\Omega \xi|u(x,t)|^{1+\gamma}dx + \int_Q \xi\left(|\nabla u|^p + |u|^{\sigma+1}\right)dxdt \\
&\le K\Bigg[\sup_{t\le T}\int_\Omega \xi|u_0(x,t)|^{1+\gamma}dx + \int_Q \Big(\xi|f|^{(1+\sigma)/\sigma} \\
&\quad + |u|^p|D\xi|^p\xi^{(-p+1)} + |\Phi|^{1+\sigma} + |f|^{(1+\sigma)/\sigma}\Big)dxdt \\
&\quad + \left(\int_0^T\left(\int_\Omega |\Phi_t|^{1+\gamma}dx\right)^{1/(1+\gamma)}d\tau\right)^{1+\gamma}\Bigg].
\end{aligned} \tag{4.44}$$

If the estimate on $\|u|D\xi|\xi^{(1-p)/p}\|_{L^p(Q)}$ is a priori known, (4.44) yields the estimate of the total energy in the domain $\overline{Q}' = \overline{\Omega}' \times [0,T]$, $\overline{\Omega}' \subset \Omega$. This property allows one to apply the local energy method to investigation of the weak solutions of (2.1), which need not be bounded on the boundary of domain Ω.

4.5 The one-directional phenomena: Localization via diffusion-convection balance. In this subsection we show how the anisotropy of an equation can lead to existence of localized solutions even if the equation under study has no absorption terms. This phenomenon is of one-dimensional nature and can be caused by two different reasons: (i) the degeneracy of the diffusion operator in some space directions x_i, (ii) a suitable diffusion-convection balance in the direction x_i. Examples of such situations are furnished by solutions of the equations (see Chapter 1, Subsection 4.2)

$$\frac{\partial u}{\partial t} - \sum_{i=1}^N \frac{\partial^2 u}{\partial x_i^2} - \frac{\partial}{\partial x_{N+1}}\left(|u|^\alpha\left|\frac{\partial u}{\partial x_{N+1}}\right|^{m-2}\frac{\partial u}{\partial x_{N+1}}\right) = 0, \tag{4.45}$$

$$\frac{\partial u}{\partial t} - \sum_{i=1}^{N+1}\frac{\partial^2 u}{\partial x_i^2} + B(x,t)\frac{\partial\left(|u|^{\sigma-1}u\right)}{\partial x_{N+1}} = 0, \tag{4.46}$$

(if, respectively, $m+\alpha > 2$ and $0 < \sigma < 1$).

Because of the one-dimensional character of this phenomenon, its analysis becomes the most transparent in the special case where the domain Ω is a cylinder: $\Omega = \mathcal{G} \times \mathbb{R}^+$ and $\mathcal{G}$ is an open bounded subset of $\mathbb{R}^N$. For the sake of presentation, in this section we restrict our study to the domains of this class. First, let us see how the degeneracy of the diffusion operator in one dimension can cause localization of solutions. We adopt the notation $x = (x_1, \dots, x_N)$, $y = x_{N+1}$,

$$|\nabla_x u|^2 = \sum_{i=1}^{N} \left(\frac{\partial u}{\partial x_i}\right)^2, \qquad |\nabla u|^2 = |\nabla_x u|^2 + |u_y|^2.$$

We will consider equation (4.45) under the following initial and boundary conditions:

$$\begin{aligned} u(x, y, 0) &= u_0(x, y) && \text{for } (x, y) \in \Omega, \\ u_0(x, y) &= 0 && \text{for } 0 < \rho_0 \le y, \end{aligned} \tag{4.47}$$

$$\begin{aligned} u(x, y, t) &= 0 && \text{when } x \in \Gamma_D, \\ \nabla_x u \cdot \mathbf{n} &= 0 && \text{when } x \in \Gamma_N, \quad y, t > 0, \end{aligned} \tag{4.48}$$

where $\partial\Omega = \Gamma = \Gamma_D \cup \Gamma_N$ and meas $\Gamma_D > 0$. Set

$$b(\rho, t) = \int_\rho^\infty \int_{\mathcal{G}} |u(x, y, t)|^{2k} \, dx \, dy \quad \text{with some } k \ge 1,$$

$$E(\rho, t) = \int_0^t \int_\rho^\infty \int_{\mathcal{G}} \left(|\nabla_x u|^2 + |u|^\alpha \left|\frac{\partial u}{\partial y}\right|^m\right) |u(x, y, \tau)|^{2k-2} \, dx \, dy \, d\tau,$$

$$\overline{b}(\rho) = \sup_{t \le t_0} b(\rho, t), \ \ \overline{E}(\rho) = \sup_{t \le t_0} E(\rho, t) = E(\rho, t_0).$$

It is easy to verify that

$$b_\rho = \frac{\partial b(\rho, t)}{\partial \rho} = -\int_{\mathcal{G}} |u(x, \rho, t)|^{2k} \, dx,$$

$$E_\rho = \frac{\partial E(\rho, t)}{\partial \rho} = -\int_0^t \int_{\mathcal{G}} \left(|\nabla_x u|^2 + |u|^\alpha \left|\frac{\partial u}{\partial y}\right|^m\right) |u(x, \rho, \tau)|^{2k-2} \, dx \, d\tau,$$

$$\begin{aligned} &\frac{\partial}{\partial \rho}\left(\sup_{t \le t_0}(-E(\rho, t))\right) \\ &\quad = -\frac{\partial \overline{E}(\rho)}{\partial \rho} = \sup_{t \le t_0}\left(\frac{\partial}{\partial \rho}(-E(\rho, t))\right) = -\frac{\partial E(\rho, t_0)}{\partial \rho}, \end{aligned}$$

$$E_t = \frac{\partial E(\rho, t)}{\partial t} = \int_0^\rho \int_{\mathcal{G}} \left(|\nabla_x u|^2 + |u|^\alpha \left|\frac{\partial u}{\partial y}\right|^m\right) |u(x, y, t)|^{2k-2} \, dx \, dy,$$

$$E_{t\rho} = \frac{\partial^2 E(\rho, t)}{\partial t \, \partial \rho} = -\int_{\mathcal{G}} \left(|\nabla_x u|^2 + |u|^\alpha \left|\frac{\partial u}{\partial y}\right|^m\right) |u(x, \rho, t)|^{2k-2} \, dx.$$

We will consider a weak solution of equation (4.45) such that

$$\overline{b}(\rho) + \overline{E}(\rho) \le M(t_0) < \infty. \tag{4.49}$$

Multiplying equation (4.45) by u^{2k-1} and integrating the result by parts in the domain $G \times (\rho, \infty)$ with $\rho \geq \rho_0$, we come to the energy relation

$$\frac{1}{2k} b(\rho, t) + (2k-1) E(\rho, t) = I(\rho, t), \tag{4.50}$$

$$I = -\int_0^t \int_{\mathcal{G}} |u|^\alpha u^{2k-1} \left| \frac{\partial u}{\partial y} \right|^{m-2} \frac{\partial u}{\partial y} \bigg|_{y=\rho} dx\, d\tau.$$

The parameter k will be defined later.

Using Hölder's inequality we obtain

$$|I| \leq \int_0^t \left(\int_{\mathcal{G}} |u|^{\alpha+2k-2} \left| \frac{\partial u}{\partial y} \right|^m dx \right)^{(m-1)/m} \left(\int_{\mathcal{G}} |u|^\lambda \, dx \right)^{1/m} d\tau,$$

or

$$|I| \leq \int_0^t (-E_{\rho t})^{(m-1)/m} \left(\int_{\mathcal{G}} |u|^\lambda \, dx \right)^{1/m} d\tau, \quad \lambda = \frac{\alpha + m - 2 + 2k}{m}, \tag{4.51}$$

and, in addition,

$$\left(\int_{\mathcal{G}} |u|^\lambda \, dx \right)^{1/m} \leq \left(\int_{\mathcal{G}} |u|^\gamma \, dx \right)^{1/pm} \left(\int_{\mathcal{G}} |u|^{(\lambda p - \gamma)/(p-1)} \, dx \right)^{(p-1)/pm} \tag{4.52}$$

with

$$\gamma \begin{cases} = \dfrac{\alpha + m - 2 + 2km}{m} & \text{if } 1 < p < \infty, \\ < \lambda & \text{if } \alpha + m > 2. \end{cases}$$

On the other hand,

$$\begin{aligned} \int_{\mathcal{G}} |u|^\gamma \, dx &\leq \gamma \int_\rho^\infty \int_{\mathcal{G}} |u|^{\gamma-1} |u_y| \, dx\, dy \\ &\leq \gamma \left(\int_\rho^\infty \int_{\mathcal{G}} |u|^{2k} dx\, dy \right)^{(m-1)/m} \left(\int_\rho^\infty \int_{\mathcal{G}} |u|^{\alpha+2k-2} |u_y|^m dx\, dy \right)^{1/m} \end{aligned}$$

or

$$\left(\int_{\mathcal{G}} |u|^\gamma \, dx \right)^{1/pm} \leq \left(\gamma b^{(m-1)/m} (E_t)^{1/m} \right)^{1/pm} \tag{4.53}$$

with

$$\gamma = \frac{\alpha + m - 2 + 2km}{m} < \lambda.$$

Assuming that

$$\frac{\lambda p-\gamma}{k(p-1)} \le \frac{2N}{N-2}, \qquad 1<p<\infty,$$

and applying the embedding theorems in the domain G, we get the inequality

$$\begin{aligned}\left(\int_{\mathcal{G}} |u|^{(\lambda p-\gamma)/(p-1)}\,dx\right)^{(p-1)/pm} &\le C\left(\int_{\mathcal{G}} |u|^{2k-2}\,|\nabla_x u|^2\,dx\right)^{(\lambda p-\gamma)/2kpm}\\ &\le C\left(-E_{\rho t}\right)^{(\lambda p-\gamma)/2kpm}.\end{aligned}$$

Gathering (4.51)–(4.53), we come to the inequality

$$\begin{aligned}|I| &\le C\int_0^t \left(-E_{\rho t}\right)^{(m-1)/m} b^{(m-1)/pm^2}\,(E_t)^{1/pm^2}\left(-E_{\rho t}\right)^{(\lambda p-\gamma)/2kpm}\,d\tau\\ &\le C\overline{b}^{(m-1)/pm^2}\int_0^t \left(-E_{\rho t}\right)^{(m-1)/m+(\lambda p-\gamma)/2kpm}(E_t)^{1/pm^2}\,d\tau, \qquad 0\le t\le t_0.\end{aligned}$$

Let us choose p and k according to the conditions

$$\frac{m-1}{m}+\frac{\lambda p-\gamma}{2kpm}+\frac{1}{pm^2}=1, \qquad \frac{\lambda p-\gamma}{k(p-1)} \le \frac{2N}{N-2},$$

that is,

$$\begin{aligned}p&=\frac{m\gamma-2k}{m(\lambda-2k)}=\frac{m+\alpha-2+2k(m-1)}{(m+\alpha-2)}=1+\frac{2k(m-1)}{(m+\alpha-2)}>1,\\ k&\ge \max\left(1,\frac{N(m+\alpha-2)}{4}\right),\end{aligned}$$

apply to (4.53) Hölder's inequality and make use of the conditions

$$E(\rho,0)=E_\rho(\rho,0)=0.$$

It follows that

$$|I| \le C\overline{b}^{(m-1)/pm^2}\left(-E_\rho\right)^{1-1/pm^2}(E)^{1/pm^2} \le C\left(\overline{b}+\overline{E}\right)^{1/p}\left(-E_\rho\right)^{1-1/pm^2},$$

or

$$|I| \le C\left(\varepsilon\left(\overline{b}+\overline{E}\right)+C_\varepsilon\left(-E_\rho\right)^{1/\nu}\right)$$

with

$$\nu=\frac{(p-1)m^2}{pm^2-1}<1, \qquad \varepsilon\in(0,1).$$

It follows from the last inequality and the energy relation (4.50) that either

$$\overline{E}^{\nu} \le \left(\overline{b}+\overline{E}\right)^{\nu} \le C\left(-\overline{E}_\rho\right),$$

or

$$\overline{E}^{1-\nu}(\rho) \le \overline{E}^{1-\nu}(\rho_0) - (\rho - \rho_0)\frac{1-\nu}{C}$$

and

$$\overline{E}^{1-\nu}(\rho) = 0, \qquad \frac{C}{1-\nu}\overline{E}^{1-\nu}(\rho_0) + \rho_0 \le \rho.$$

Theorem 4.3. *Let* $u(x, y, t)$ *be a weak solution of problem* (4.45), (4.47) *and* (4.48) *with* $m + \alpha > 2$ *satisfying* (4.49). *Then*

$$u(x, y, t) = 0, \qquad x \in \Omega, \tag{4.54}$$
$$t \le t_0, \quad \frac{CM(t_0)}{1-\nu} + \rho_0 \le y.$$

If $M(t_0) \le M_0 = \text{const}$, *then property* (4.54) *holds for*

$$t \in R^+, \quad \frac{CM_0}{1-\nu} + \rho_0 \le y.$$

Now we are going to show that the localization of solutions of parabolic equations of the type (4.46) can also be caused by the terms of the form

$$B(x, y, t)\frac{\partial}{\partial x_{N+1}}\left(|u|^{\sigma-1}u)\right),$$

which usually appear in modelling of physical processes of convective nature.

Let us consider equation (4.46) with the initial and boundary conditions (4.47), (4.48). We introduce the energy functions

$$b(\rho, t) = \int_\rho^\infty \int_G |u(x, y, t)|^2\, dx\, dy,$$
$$E(\rho, t) = \int_0^t \int_\rho^\infty \int_G \left(|\nabla u(x, y, \tau)|^2 + \left|\frac{\partial B}{\partial y}\right| \frac{\sigma}{\sigma+1}|u|^{1+\sigma}\right) dx\, dy d\tau,$$
$$\overline{b}(\rho) = \sup_{t \le t_0} b(\rho), \quad \overline{E}(\rho) = E(\rho, t_0),$$
$$\Lambda(\rho, t) = \int_0^t \int_G |B(x, \rho, \tau)|\frac{\sigma}{1+\sigma}|u|^{1+\sigma}\, dx\, d\tau, \quad \overline{\Lambda}(\rho) = \Lambda(\rho, t_0)$$

and assume that

$$C \le |B| \le \frac{1}{C}, \quad \max\left(B, \frac{\partial B}{\partial y}\right) \le 0, \quad 0 < \sigma < 1, \tag{4.55}$$
$$\overline{b}(\rho) + \overline{E}(\rho) + \overline{\Lambda}(\rho) \le M(t_0). \tag{4.56}$$

Theorem 4.4. *Let $u(x, y, t)$ be a weak solution of problem* (4.46), (4.47), *and* (4.48) *satisfying* (4.56) *and let* (4.55) *be fulfilled. Then*

$$u(x, y, t) = 0, \; x \in \Omega, \qquad t \le t_0, \quad \frac{CM(t_0)}{1-\nu} + \rho_0 \le y. \tag{4.57}$$

If $M(t_0) \le M_0 = \text{const}$, *then the property* (4.57) *holds for*

$$t \in R^+, \quad \frac{CM_0}{1-\nu} + \rho_0 \le y.$$

Proof. The following formula of integration by parts

$$\frac{1}{2} b(\rho, t) + E(\rho, t) + \Lambda(\rho, t) = I \tag{4.58}$$

holds with

$$I(\rho, t) = -\int_0^t \int_{\mathcal{G}} u_y u \, dx \, d\tau.$$

Making use of the inequalities

$$\|u(\cdot, \rho)\|_{L^2(G)} \le C \|\nabla u\|^{\theta}_{L^2(G)} \cdot \|u(\cdot, \rho)\|^{1-\theta}_{L^{1+\sigma}(G)} \le C \left(-E_{t\rho}\right)^{\theta/2} (\Lambda_t)^{(1-\theta)/(1+\sigma)}$$

with $\theta = \frac{N(1-\sigma)}{N(1-\sigma)+2(1+\sigma)}$ and

$$\int_{\mathcal{G}} u^2 \, dx \le 2 \int_{\rho}^{\infty} \int_{\mathcal{G}} |u_y u| \, dx \, d\rho \le 2 b^{1/2} (E_t)^{1/2},$$

we estimate I as

$$\begin{aligned} |I| &\le \int_0^t \|\nabla u\|_{L^2(G)} \|u\|^{1-\delta+\sigma}_{L^2(G)} \, d\tau \\ &\le C \int_0^t \left(-E_{\rho t}\right)^{1/2+\theta\delta/2} (\Lambda_t)^{(1-\theta)\delta/(1+\sigma)} b^{(1-\delta)/4} (E_t)^{(1-\delta)/4} \, d\tau. \end{aligned}$$

Choosing $\delta \in (0, 1)$ from the condition

$$\frac{1+\delta\theta}{2} + \frac{1-\delta}{4} + \frac{\delta(1-\theta)}{1+\sigma} = 1,$$

which is equivalent to

$$\delta = \frac{1+\sigma}{3-\sigma-2\theta(1-\sigma)} \in (0, 1),$$

and applying Hölder's inequality, we obtain that

$$\begin{aligned} |I| &\le C \bar{b}^{\frac{1-\delta}{4}} \left(-E_{\rho}\right)^{(1+\delta\theta)/2} (\Lambda + E)^{(1-\delta\theta)/2} \\ &\le C \left(\bar{b} + \overline{E} + \overline{\Lambda}\right)^{(1-\delta+2(1-\delta\theta))/4} \left(-E_{\rho}\right)^{(1+\delta\theta)/2} \end{aligned}$$

or

$$|I| \le \varepsilon\left(\overline{b}+\overline{E}+\overline{\Lambda}\right)+C_\varepsilon\left(-E_\rho\right)^{1/\nu}, \qquad \nu=\frac{1+\delta(1+2\theta)}{2(1+\delta\theta)}<1 \tag{4.59}$$

with $\varepsilon \in (0,1)$. It follows now from (4.58), (4.59) that the energy function satisfies the already-known inequality

$$C\overline{E}^{\nu}+\overline{E}'\le 0,$$

whence the assertion of the theorem. □

5 Equations with nonhomogeneous absorption terms

The results of Sections 2 and 3 can be extended to equations with nonhomogeneous absorption terms. Let us consider the problem

$$\left(|u|^{\gamma-1}u\right)_t-\operatorname{div}\left(\mathbf{A}(x,t,u,\nabla u)\right)+C(x,t,u)=f(x,t) \quad \text{in } Q=\Omega\times(0,T), \tag{5.1}$$

$$u(x,0)=u_0(x) \quad \text{in } \Omega. \tag{5.2}$$

Equation (5.1) follows from (2.1) if $\psi(\tau)=\tau|\tau|^{\gamma-1}$ and $B(x,t,u,Du)\equiv 0$.

Let us subject the functions $\mathbf{A}$ and C to the following structural conditions: there exist constants $\sigma>0$ and $p>1$ such that

$$\begin{gathered}\forall\,(s,\boldsymbol{\rho})\in\mathbb{R}\times\mathbb{R}^N \quad \text{and a.e. } x\in\Omega,\ t\in\mathbb{R}^+\\ C_1|\boldsymbol{\rho}|^p\le\mathbf{A}(x,t,s,\boldsymbol{\rho})\cdot\boldsymbol{\rho}\le C_2|\boldsymbol{\rho}|^p,\end{gathered} \tag{5.3}$$

$$\begin{gathered}\forall\, s\in\mathbb{R} \quad \text{and a.e. } x\in\Omega,\ t\in\mathbb{R}^+\\ s\,C(x,t,s)\ge C_3\,a(x,t)\,|s|^{\widetilde{\sigma}+1},\end{gathered} \tag{5.4}$$

with is a given measurable bounded function $a(x,t)\ge 0$ satisfying the condition

$$a^{-1}\in L^{(1+\sigma)/(\widetilde{\sigma}-\sigma)}(Q), \qquad 0<\sigma<\widetilde{\sigma}. \tag{5.5}$$

In (5.1)–(5.4), C_i $(i=1,2,3)$ are positive constants. The additional (and crucial for all further considerations) assumption is

$$\sigma<\gamma. \tag{5.6}$$

The right-hand side $f(x,t)$ of equation (5.1) and the initial data $u_0(x)$ are assumed to satisfy the conditions

$$u_0\in L^{\gamma+1}(\Omega), \qquad f\in L^{(1+\sigma)/\sigma}(Q), \qquad f\,a^{-1/(1+\widetilde{\sigma})}\in L^{(1+\widetilde{\sigma})/\widetilde{\sigma}}(Q). \tag{5.7}$$

We are interested in the qualitative properties of solutions to problem (5.1), (5.2) understood in the following sense.

Definition 5.1. A measurable-in-Q function $u(x,t)$ is said to be a weak solution of problem (5.1), (5.2) if

(i) $u \in L^p(0,T; W^{1,p}(\Omega)) \cap L^\infty(0,T; L^{\gamma+1}(\Omega))$;

(ii) $\lim_{t\to 0} \|u(x,t) - u_0(x)\|_{L^{\gamma+1}(\Omega)} = 0$;

(iii) for any test function $\zeta \in W^{1,\infty}(0,T : W_0^{1,p}(\Omega))$ vanishing at $t = T$, the integral identity

$$\int_Q \left\{ |u|^{\gamma-1} u\zeta_t - \mathbf{A} \cdot \nabla\zeta - C\zeta + f\zeta \right\} dxdt + \int_\Omega |u_0|^{\gamma-1} u_0 \zeta(x,0)dx = 0. \tag{5.8}$$

holds.

Let us introduce the local energy functions (cf. with (4.3))

$$\begin{aligned} E(P) &:= \int_{P(t,\rho)} |\nabla u|^p dxd\tau, \\ C(P) &:= \int_{P(t,\rho)} |u|^{\sigma+1} dxd\tau, \\ C_a(P) &:= \int_{P(t,\rho)} a(x,\tau)\, |u(x,\tau)|^{\widetilde{\sigma}+1} dxd\tau, \\ b(T) &:= \operatorname*{ess\,sup}_{s\in(t,T)} \int_{|x-x_0|<\rho+\vartheta(s-t)^\upsilon} |u|^{\gamma+1} dx, \end{aligned} \tag{5.9}$$

associated with a local weak solution of problem (5.1). Assume that

$$\|1/a\|_{L^{(1+\sigma)/(\widetilde{\sigma}-\sigma)}(Q)} \le K, \quad K = \text{const}, \tag{5.10}$$

whence

$$C^{(1+\widetilde{\sigma})/(1+\sigma)} \le C_a\, \|1/a\|_{L^{(1+\sigma)/(\widetilde{\sigma}-\sigma)}(Q)} \le K\, C_a. \tag{5.11}$$

We are now in position to give the precise statement of our results. The only global information we need will be formulated in terms of the *global energy function*

$$D(u(\cdot,\cdot)) := b(T,\Omega) + \int_Q \left(|\nabla u|^p + a\, |u|^{\widetilde{\sigma}+1} \right) dxdt,$$

where

$$b(T,\Omega) := \operatorname*{ess\,sup}_{t\in(0,T)} \int_\Omega |u(x,t)|^{\gamma+1} dx.$$

Our first result concerns the waiting time property of a local weak solution $u(x,t)$ of equation (5.1). Let us assume that

$$u_0 \equiv 0 \quad \text{in } B_{\rho_0}(x_0) \quad \text{for some } x_0 \in \Omega \text{ and } \rho_0 > 0, \tag{5.12}$$

$$f \equiv 0 \quad \text{in the cylinder } P = P(0,\rho_0) \tag{5.13}$$

and claim convergence (near $\rho = \rho_0$) of the auxiliary integral

$$I := \int_{\rho_0+0} (\rho - \rho_0)^{\beta} \left[\|u_0\|^{\gamma+1}_{L^{\gamma+1}(B_\rho(x_0))} + \left\| f\, a^{-1/(1+\widetilde{\sigma})} \right\|^{(1+\widetilde{\sigma})/\widetilde{\sigma}}_{L^{(1+\widetilde{\sigma})/\widetilde{\sigma}}(P(0,\rho))} \right]^{p/(p-1)} d\rho, \tag{5.14}$$

with the exponents

$$\beta = (1 - \delta\widetilde{\theta})(1+\kappa), \quad \delta = -\left(1 + \frac{p-1-\sigma}{p(1+\sigma)} N\right),$$
$$\widetilde{\theta} = \frac{pN - (1+\sigma)(N-1)}{(N+1)p - N(1+\sigma)}, \tag{5.15}$$

and

$$\kappa \in \left(0, \frac{p(1+\gamma)}{(p-1-\sigma)(1-\widetilde{\theta})}\right), \tag{5.16}$$

Condition (5.14) imposes a restriction on the vanishing rates as $\rho \to \rho_0$ of the functions

$$\|u_0\|_{L^{\gamma+1}(B_\rho(x_0))} \quad \text{and} \quad \|f(\cdot, t)\, a^{-1/(1+\widetilde{\sigma})}\|_{L^{(\sigma+1)/\sigma}(B_\rho(x_0))}.$$

Theorem 5.1. *Assume that conditions* (5.3), (5.4), (5.10) *are fulfilled and that*

$$\sigma < \gamma \leq p - 1. \tag{5.17}$$

Let u_0 and f satisfy conditions (5.12), (5.13), *and* (5.14). *Then there exists a positive constant M (depending only on the constants in* (5.3), (5.4), ρ_0, *the difference* $\widetilde{\sigma} - \sigma$, *and* $\operatorname{dist}(x_0, \partial\Omega)$*) such that every weak solution of* (5.1) *with bounded global energy, $D(u) \leq M$, possesses the property*

$$u(x,t) \equiv 0 \quad \textit{in } B_{\rho_0}(x_0) \times (0, T).$$

Under some additional assumptions on the structural exponents γ, σ, p and the function f we may prove a stronger result asserting that the support of $u(\cdot, t)$ strictly shrinks with respect to the initial support.

Theorem 5.2. *Assume the fulfillment of conditions* (5.3)–(5.6), (5.10), (5.17) *and let*

$$1 + \sigma \leq \gamma \frac{p}{p-1}. \tag{5.18}$$

Let u_0 satisfy (5.12). *Assume that*

$$f \equiv 0 \quad \textit{in the truncated cone } P \equiv P(0, \rho_0 : \vartheta, 1) \tag{5.19}$$

for some $\vartheta > 0$, and let (5.14) *be true. Then there exist positive constants M, $\widetilde{\sigma} > \sigma$ and t^* such that every weak solution of problem* (5.1) *with global energy satisfying the inequality $D(u) \leq M$, possesses the property*

$$u(x,t) \equiv 0 \quad \textit{in } P(0, \rho_0 : \vartheta, 1) \cap \{t \leq t^*\}.$$

The next result refers to the case when the initial data need not vanish, that is, parameter ρ_0 in the conditions of Theorems 5.1 and 5.2 is equal to zero.

Theorem 5.3. *Let conditions* (5.3)–(5.6), (5.17)–(5.18), (5.10) *be true. Assume that* $f \equiv 0$. *Then there exist positive constants* M, t^*, *and* $\upsilon \in (0,1)$ *such that every weak solution of problem* (5.1) *satisfying the inequality* $D(u) \le M$ *possesses the property*

$$u(x,t) \equiv 0 \quad in\ P(t^*, 0:1, \upsilon).$$

The proofs of Theorems 5.1–5.3 are in essence analogous to those of Theorems 3.1 and 4.2. From (5.8), we derive the energy relation

$$\begin{aligned} i_1 + i_2 + i_3 &:= \frac{\gamma}{\gamma+1}\int_{P\cap\{t=T\}} |u|^{\gamma+1} dx + \int_P \mathbf{A}\cdot\nabla u\, dxd\theta + \int_P u\, Cdxd\theta \\ &= \int_{\partial_l P} \mathbf{n}_x \cdot \mathbf{A} u\, d\Gamma d\theta + \frac{\gamma}{\gamma+1}\int_{\partial_l P} n_\tau |u|^{\gamma+1} d\Gamma d\theta \\ &\quad + \frac{\gamma}{\gamma+1}\int_{P\cap\{t=0\}} |u_0|^{\gamma+1} dx + \int_P uf\, dxd\theta \\ &:= j_1 + j_2 + j_3 + j_4. \end{aligned} \tag{5.20}$$

The terms j_i on the right-hand side of (5.20) are estimated in the standard way. We now turn to estimating the left-hand side of (5.20). By (5.3), (5.4), (5.9), we have at once that

$$i_1 + i_2 + i_3 \ge i_1 + C_1\, E + C_3\, C_a \ge C_4\,(E + C + i_1)^m\,. \tag{5.21}$$

Here

$$C_4 = C_4(C_1, C_3, m)2^{-m}, \qquad m = \frac{1+\widetilde{\sigma}}{1+\sigma} > 1.$$

Without loss of generality, we may assume that the global energy satisfies $D(u) \le 1$. From (5.20), (5.21), we obtain the inequality

$$C_4\,(E + C + i_1)^m \le |j_1| + |j_2| + |j_3| + |j_4|. \tag{5.22}$$

Since the right-hand side of (5.22) is an increasing function of T, we may always replace i_1 by $b(T)$.

Let us assume that the conditions of Theorem 5.3 are fulfilled. If this is the case, P is a paraboloid, and $j_3 = j_4 = 0$. The following estimates hold:

$$|j_1| \le L_1\, \Lambda(t)\, D(u)^{(q-1)(1-\widetilde{\theta})/qr}\, (E+C)^{1-\gamma} \left(-\frac{d(E+C)}{dt}\right)^{(p-1)/p}, \tag{5.23}$$

$$|j_2| \le L\,(E + C + b(T,\Omega))\,(b(T,\Omega))^{\kappa-1} \left(\int_t^T \left(K^{s(\gamma+1)/\widetilde{\theta}p}\right)^{\varepsilon} d\tau\right)^{1/\varepsilon}. \tag{5.24}$$

Let us claim that $\widetilde{\sigma}$ satisfies the inequality

$$1 < m = \frac{1+\widetilde{\sigma}}{1+\sigma} \le \kappa(\gamma, \sigma, p), \tag{5.25}$$

and assume that $T - t$ and $D(u)$ are so small that

$$L\,(b(T,\Omega))^{\kappa - m} \left(\int_t^T \left(K^{s(\gamma+1)/\widetilde{\theta}p} \right)^{\varepsilon} d\tau \right)^{1/\varepsilon} < \frac{C_4}{2}.$$

Then

$$\begin{aligned} &(E+C)^{m-1+\gamma} \\ &\le L_2\, \Lambda(t)\, D(u)^{(m+1)(q-1)(1-\widetilde{\theta})/qr}\, (E+C)^{1-\gamma} \left(-\frac{d(E+C)}{dt} \right)^{(p-1)/p}, \end{aligned} \tag{5.26}$$

whence the differential inequality for the energy function $Y(t) := E + C$:

$$Y^{\nu}(t) \le c(t)\,(-Y(t))', \quad \nu = \frac{(m-1+\gamma)}{p-1}, \tag{5.27}$$

with

$$c(t) = \left(L_1\, (M_*)^{(q-1)(1-\widetilde{\theta})/qr}\, \Lambda(t) \right)^{p/(p-1)}, \quad L_1 = \text{const} > 0.$$

The proof of Theorem 5.3 is thus reduced to the routine study of the properties of functions satisfying the differential inequality (5.27).

The proofs of Theorems 5.1, 5.2 can be done in a similar way which just requires some obvious changes in the derivation of the corresponding a priori estimates.

6 Equations with anisotropic nonlinearities

Let us consider now a class of parabolic equations which degenerate at the null level of both the solution and its gradient. We will study equations of the form

$$\frac{\partial}{\partial t}\left(u|u|^{\gamma-1} \right) = \sum_{i=1}^{N} \frac{\partial}{\partial x_i} \left(\left| \frac{\partial u}{\partial x_i} \right|^{(p_i-1)} \frac{\partial u}{\partial x_i} \right) + f(x,t) \tag{6.1}$$

with the exponents of nonlinearity

$$1 < p_i < \infty, \qquad 0 < \gamma < \infty.$$

Analogous elliptic equations were considered in Chapter 1. Parabolic equations with nonlinearities of this type were studied in Chapter 2, where we also indicated several classes of equations that can be transformed into (6.1) by a change of the sought function.

Equation (6.1) will be considered in the domain $Q = \Omega \times (0, T)$. It is assumed that

$$u(x,0) = u_0(x) \quad x \in \Omega. \tag{6.2}$$

Definition 6.1. A measurable-in-Q function $u(x,t)$, bounded along with its generalized derivatives $\frac{\partial u}{\partial x_i}$, $i = 1, \dots, N$, is said to be a weak solution of problem (6.1), (6.2) if

(i) $\lim_{t\to 0} \|u(x,t) - u_0(x)\|_{L^{\gamma+1}(\Omega)} = 0$;

(ii) for any test function $\zeta \in W^{1,\infty}(0,T; W_0^{1,\infty}(\Omega))$, vanishing at $t = T$, the integral identity holds

$$\int_Q \left\{ |u|^{\gamma-1} u\zeta_t - \sum_{i=1}^N \left|\frac{\partial u}{\partial x_i}\right|^{(p_i-2)} \frac{\partial u}{\partial x_i}\frac{\partial \zeta}{\partial x_i} + f\zeta \right\} dxdt + \int_\Omega u_0\zeta(x,0)dx = 0. \tag{6.3}$$

Our aim is to prove the following assertion.

Theorem 6.1. *Let $u(x,t)$ be a weak solution of problem* (6.1), (6.2), *with the right-hand side $f(x,t) \equiv 0$ and*

$$u(x,0) = u_0(x) \quad x \in \Omega), \tag{6.4}$$

such that

$$|u(x,t)| + \sum_{i=1}^N \left|\frac{\partial u}{\partial x_i}\right| \le M. \tag{6.5}$$

Assume that the structural constants γ, p_i, $i = 1, \dots, N$ satisfy the conditions

$$(\beta - \alpha)\frac{\theta(1+\gamma)}{\beta(1-\theta)} + 1 + \gamma < \alpha, \quad \alpha = \min_i p_i, \quad \beta = \max_i p_i, \tag{6.6}$$

and

$$\theta = \left(\frac{1}{1+\gamma} - \frac{N-1}{N\beta}\right) / \left(\frac{1}{N} - \frac{p^*}{N} + \frac{1}{1+\gamma}\right), \quad 0 < \theta < 1, \quad p_* = \sum_i^N \frac{1}{p_i}. \tag{6.7}$$

Then $u(x,t)$ possesses the property of finite speed of propagation in the sense of Definition 1.1.

If, instead of (6.4), *the data of the problem are assumed to satisfy the condition*

$$\left(\int_{B_\rho} |u_0(x)|^{1+\gamma} dx + \int_0^T \|f(\cdot,t)\|^{(1+\gamma)/\gamma}_{L^{(1+\gamma)/\gamma}(B_\rho)} dt\right) \le \varepsilon(\rho_0 - \rho)_+^{\frac{1}{1-\nu}}, \quad \rho_0 \le \rho, \tag{6.8}$$

$$\nu = \left(1 - \frac{1-\theta}{1+\gamma} - \frac{\theta}{\beta}\right)\frac{\alpha}{\alpha - 1}, \tag{6.9}$$

then $u(x,t)$ possesses the property of finite speed of propagation in the sense of Definition 1.1.

Remark 6.1. If $\alpha = \beta = p_i = p$, $i = 1, \dots, N$, then conditions (6.6), (6.7) coincide with the already known condition $p > 1 + \gamma$ which we imposed in the study of the case of nonisotropic degeneracy (Theorem 2.1). It is easy to see that condition (6.6) is surely fulfilled if the difference $\beta - \alpha$ is sufficiently small.

The proof is based on standard arguments. We introduce the energy functions

$$\begin{aligned} b(\rho, t) &= \int_{B_\rho} |u|^{1+\gamma} dx, \\ \overline{b}(\rho, t) &= \operatorname*{ess\,sup}_{0 \le \tau \le t} b(\rho, \tau), \\ E(\rho, t) &= \int_0^t \int_{B_\rho} \sum_{i=1}^N \left| \frac{\partial u}{\partial x_i} \right|^{p_i} dx dt \equiv \sum_{i=1}^N E_i, \end{aligned}$$

and then check that every weak solution $u(x,t)$ satisfies the energy relation (the integration-by-parts formula)

$$\frac{\gamma}{1+\gamma} b(\rho, t) + E(\rho, t) = J_1 + J_2. \tag{6.10}$$

In this equality,

$$\begin{aligned} J_1 &= \int_0^t \int_{S_\rho} u \sum_{i=1}^N \left| \frac{\partial u}{\partial x_i} \right|^{p_i - 2} \frac{\partial u}{\partial x_i} \mathbf{n}_i dS d\tau, \\ J_2 &= \frac{\gamma}{1+\gamma} b(\rho, 0) + \int_0^t \int_{B_\rho} u f dx d\tau, \end{aligned}$$

and $\mathbf{n} = (\mathbf{n}_1, \dots, \mathbf{n}_N)$ is the outward normal vector to $S_\rho = \partial B_\rho$.

We begin with the proof of the first assertion of the theorem. In this case, $\rho \le \rho_0$ in (6.10) and, correspondingly, $J_2 \equiv 0$. Using Hölder's inequality and (6.5) we derive the following estimate for J_1:

$$\begin{aligned} |J_1| &\le \sum_{i=1}^N \left(\int_0^t \int_{S_\rho} |u|^{p_i} dS d\tau \right)^{1/p_i} \left(\int_0^t \int_{S_\rho} |u_{x_i}|^{p_i} dS d\tau \right)^{(p_i - 1)/p_i} \\ &\le \left(\int_0^t \int_{S_\rho} |u|^{\beta} dS d\tau \right)^{1/\beta} \sum_{i=1}^N (\text{meas } S_\rho)^{(\beta - p_i)/\beta} \left(\frac{\partial E_i}{\partial \rho} \right)^{(p_i - 1)/p_i} \\ &\le K \left(\int_0^t \|u\|^\beta_{L^\beta(S_\rho)} d\tau \right)^{1/\beta} \left(\frac{\partial E}{\partial \rho} \right)^{(\alpha - 1)/\alpha}, \end{aligned} \tag{6.11}$$

with $K = K(\rho_0, T, M)$. Next, let us make use of the interpolation-trace inequality (Lemma 3.7 of the appendix). We have

$$\|u\|^\beta_{L^\beta(S_\rho)} \le C b^{\beta(1-\theta)/(1+\gamma)} \left(\sum_{i=1}^N \rho^{\delta_i} \| \frac{\partial u}{\partial x_i} \|_{L^{p_i}(B_\rho)} + \rho^{-\delta} b^{1/(1+\gamma)} \right)^{\beta\theta}, \tag{6.12}$$

with the exponents

$$\delta = \frac{N(\beta - 1 - \gamma) + \beta(1+\gamma)}{\beta(1+\gamma)}, \quad \delta_i = N\left(\frac{1}{\beta} - \frac{1}{p_i}\right).$$

The parameter θ is already defined in the conditions of Theorem 6.1. Integrating (6.12) in t and using Hölder's inequality and (6.5), we can write the following chain of inequalities

$$\begin{aligned}\left(\int_0^t \|u\|^{\beta}_{L^{\beta}(S_\rho)} d\tau\right)^{1/\beta} &\leq K_1 \rho^{-\delta\theta} \left(\int_0^t \left(E_t + b^{\beta/(1+\gamma)}\right)^{\theta} b^{(1-\theta)\beta/(1+\gamma)}\right)^{1/\beta} \\ &\leq K_2 \rho^{-\delta\theta} \overline{b}^{(1-\theta)/(1+\gamma)} t^{(1-\theta)(\beta} \left(E + \overline{b}\right)^{\theta/\beta}\end{aligned} \tag{6.13}$$

with $K_i = K_i(\rho_0, T, M, \beta, \delta, \theta)$. In so doing we have to use the condition $\beta \geq 1 + \gamma$. Gathering (6.10) with (6.13), we arrive at the final estimate

$$|J_1| \leq K_3 \rho^{-\delta\theta} t^{(1-\theta)/\beta} \left(\frac{\partial E}{\partial \rho}\right)^{(\alpha-1)/\alpha} \overline{b}^{(1-\theta)/(1+\gamma)} \left(E + \overline{b}\right)^{\theta/\beta}. \tag{6.14}$$

Now substituting b by $\overline{b}$ and making use of (6.14), we arrive at the standard differential inequality

$$E^{\nu} \leq \left(\overline{b} + E\right)^{\nu} \leq K_4 t^{\alpha(1-\theta)/\beta(\alpha-1)} \rho^{-\delta\theta\alpha/(\alpha-1)} \frac{\partial E}{\partial \rho}, \tag{6.15}$$

where

$$\nu = \left(1 - \frac{1-\theta}{1+\gamma} - \frac{\theta}{\beta}\right) \frac{\alpha}{\alpha - 1}, \qquad K_4 = K_4(\rho_0, T, M).$$

It is easy to check by a straightforward calculation that $\nu < 1$ if condition (6.6) is fulfilled.

Under the assumptions of the second assertion of Theorem 6.1, $J_2 \not\equiv 0$ in (6.10). In this case we use (6.8) and rely on the usual estimate

$$|J_2| \leq K_5 \varepsilon \left(\rho - \rho_0\right)_+^{\nu(\alpha-1)/\alpha(1-\nu)} + t\overline{b}, \quad K_5 = K_5(\gamma). \tag{6.16}$$

Gathering (6.10), (6.14), and (6.16) with $2t < \gamma/(1+\gamma)$, we obtain the well-studied inequality

$$\begin{aligned} E^{\nu} &\leq \left(\overline{b} + E\right)^{\nu} \\ &\leq K_6 t^{\alpha(1-\theta)/\beta(\alpha-1)} \rho^{-\delta\theta\alpha/(\alpha-1)} \frac{\partial E}{\partial \rho} + K_7 \varepsilon^{\alpha/(\alpha-1)} \left(\rho - \rho_0\right)_+^{\nu/(1-\nu)}, \end{aligned} \tag{6.17}$$

which can be given the standard analysis.

7 Systems of parabolic equations

The above arguments extend to the study of localization properties of solutions to systems of parabolic equations

$$\frac{\partial \boldsymbol{\psi}(x,\mathbf{u})}{\partial t} - \operatorname{div}(\mathbf{A}(x,t,\mathbf{u},\, D\mathbf{u})) + \mathbf{B}(x,t,\mathbf{u},\, D\mathbf{u}) + \mathbf{C}(x,t,\mathbf{u}) = \mathbf{f}(x,t), \tag{7.1}$$

where $\mathbf{u} = (\mathbf{u}_1(x,t),\ldots,\mathbf{u}_m(x,t))$ is the sought vector-solution, $\boldsymbol{\psi}(x,t)$ is a given vector-valued function. We shall assume that operators $\mathbf{A}$, $\mathbf{B}$, $\mathbf{C}$ for almost all $t \in (0,T)$ satisfy conditions (5.3)–(5.6) of Chapter 1.

All the operations we use in the further proceeding are justified in Section 3 of Chapter 1. We consider the weak solutions $\mathbf{u} \in V(Q)$,

$$V(Q) \equiv L^{\infty}(0,T;\, L^{1+\gamma}_{\text{loc}}(\Omega)) \cap L^{p}(0,T;\, W^{1,p}_{\text{loc}}(\Omega)) \cap L^{1+\sigma}_{\text{loc}}(Q),$$

which satisfy the initial condition

$$\mathbf{u}(x,0) = \mathbf{u}_0(x) \quad x \in \Omega, \qquad u_0(x) = 0 \quad \text{in } x \in B_\rho(x_0),\, \overline{B}_\rho(x_0) \Subset \Omega. \tag{7.2}$$

Operator $\mathbf{M} = \partial\boldsymbol{\psi}(x,t)/\partial t$ is subject to the conditions

$$\int_0^t \int_{B_\rho} \frac{\partial \boldsymbol{\psi}(x,\mathbf{u})}{\partial t} \cdot \mathbf{u}\, dx dt \geq b(\rho,\tau)|_{\tau=0}^{\tau=t}, \tag{7.3}$$

$$\frac{1}{K}\|\mathbf{u}(\cdot,t)\|^{1+\gamma}_{L^{1+\gamma}(B_\rho)} \leq b(\rho,t) \leq K\|\mathbf{u}(\cdot,t)\|^{1+\gamma}_{L^{1+\gamma}(B_\rho)}, \tag{7.4}$$

analogous to (2.2a)–(2.2e), for some constant K, and for every $\mathbf{u}(x,t) \in V(Q)$ and a.a. $\rho > 0$ such that $(\overline{B}_\rho \Subset \Omega)$.

Let a weak solution $\mathbf{u}(x,t) \in V(Q)$ of problem (7.1), (7.2) satisfy the standard energy relation (the formula of integration by parts)

$$\begin{aligned} b(\rho,\tau)|_{\tau=0}^{\tau=t} &+ \int_0^t \int_{B_\rho} (\mathbf{A}\cdot D\mathbf{u} + (\mathbf{B}+\mathbf{f})\cdot \mathbf{u})\, dx d\tau \\ &\leq \int_0^t \int_{S_\rho} \mathbf{u}\mathbf{A}\cdot \mathbf{n}\, dS d\tau \equiv I \end{aligned} \tag{7.5}$$

for almost all ρ. We introduce the energy functions

$$\begin{aligned} E(\rho,t) &= \int_0^t \int_{B_\rho} |D\mathbf{u}|^p dx d\tau, \\ C(\rho,t) &= \int_0^t \int_{B_\rho} |\mathbf{u}|^{1+\sigma} dx d\tau. \end{aligned} \tag{7.6}$$

Then with the use of assumptions (5.3)–(5.6) of Section 3 of Chapter 1 and of a counterpart of estimate (2.27), the energy relation (7.5) can be written in the form

$$b(\rho,\tau)|_{\tau=0}^{\tau=t} + C_2 E(\rho,t) + C_4 C(\rho,t) \leq I_1 + I_2 + I_3, \tag{7.7}$$

where

$$I_1 = C_3 \int_0^t (|\mathbf{u}|^{1+\alpha}, |D\mathbf{u}|^\beta)_{B_\rho} d\tau,$$

$$I_2 = \int_0^t \int_{B_\rho} |\mathbf{u}||\mathbf{f}| dx d\tau,$$

$$|I| \le I_3 = C_1 C_2^{-(p-1)/p} \left(\frac{dE}{d\rho}\right)^{(p-1)/p} \left(\int_0^t \int_{S_\rho} (|\mathbf{u}|^p dS d\tau\right)^{1/p}.$$

The estimates for each term on the right-hand side of (7.7) follow by analogy with the estimates of Sections 2 and 2.2 of this chapter, which allows for literal repetition of the proofs of Theorems 2.1 and 3.1.

8 Higher-order parabolic equations

This section is devoted to extensions of the energy method to higher-order equations. The typical difficulty is the same as that of Section 6 in Chapter 1: in such equations the boundary integrals which appear after multiplication of the equation by the solution under study and integration by parts cannot be interpreted as the derivative of the "energy" function associated with the solution. A way to circumvent this difficulty was proposed by F. Bernis. It consists of using special *weighted energy functions*. The weights are chosen so that the boundary integrals vanish, so that instead of embedding theorems one has to apply the *weighted interpolation inequalities*. Unlike the case of a second-order equation where the energy function satisfied a first-order nonlinear ordinary differential inequality, the weighted energy function satisfies a *fractional differential inequality*.

The presentation of this section follows the papers [77, 75] by F. Bernis.

Let us consider the initial and boundary-value problem

$$\begin{cases} \dfrac{\partial}{\partial t}\left(|u|^{q-1}\operatorname{sign} u\right) + \mathbf{A}u + |u|^{r-1}\operatorname{sign} u = f & \text{on } Q = \Omega \times (0,\infty), \\ u = 0 & \text{on } \partial\Omega \times ((0,\infty), \\ u(x,t) = u_0(x) & \text{in } \Omega \end{cases} \tag{8.1}$$

under the notation

$$\mathbf{A}u = (-1)^m \sum_{|\alpha = m|} D^\alpha \left(\left|D^\alpha u\right|^{p-1} \operatorname{sign} u\right).$$

Here $p, q, r > 1$ are real numbers, $m \ge 1$ is an integer. $\alpha = (\alpha_1, \ldots, \alpha_N)$ is a multiindex and $|\alpha| = \sum \alpha_i$. For $p = 2$, we obviously have $\mathbf{A} = (-\Delta)^m$, and if $m = 1$, then $\mathbf{A} = -\Delta_p$. Ω is an arbitrary open set of $\mathbb{R}^N$.

We always assume that

$$u_0 \in L^q(\Omega), \qquad f \in L^{p'}(0,T;\, W') + L^{r'}(\Omega \times (0,T)) \quad \forall T < \infty \tag{8.2}$$

with the conjugate exponents p', r' such that

$$\frac{1}{p}+\frac{1}{p'}=1, \qquad \frac{1}{r}+\frac{1}{r'}=1.$$

Denote by W' the dual space to

$$W=W_0^{m,p}(\Omega)\cap L^q(\Omega) \text{ if } 1<q\le p \text{ or } 1<r\le p \text{ or } \Omega \text{ is bounded.}$$

Definition 8.1. Let (8.2) be fulfilled. We say that $u(x,t)$ is an *energy solution* to (8.1) if

$$u\in C\left([0,\infty);\ L^q(\Omega)\right)\cap L^r(\Omega\times(0,T)) \quad \forall T<\infty, \tag{8.3}$$

$u(x,t)$ satisfies the equation in (8.1) in the sense of distributions in $\mathcal{D}'(Q)$ and $u(x,0)=u_0$.

The existence of such solutions is proved in [76]. Moreover, the integration-by-parts formula

$$\begin{aligned}\frac{1}{q'}\int_\Omega |u(x,T)|^q dx &+\int_0^T\int_\Omega \left(\left|D^m u\right|^p+|u|^r\right)dxdt\\ &=\frac{1}{q'}\int_\Omega |u_0|^q dx+\int_0^T (f(t),u(t))\,dx,\end{aligned}\tag{8.4}$$

holds, where $(\cdot,\cdot)$ denotes the duality between $W'+L^{r'}(\Omega)$ and $W\cap L^r(\Omega)$.

Adopt the following notation: $x=(x_1,\dots,x_N)$, $z=x_N$,

$$\zeta_N(T)=\sup\{z: x\in \operatorname{supp} u(\cdot,T)\},$$
$$S_{Nf}(T)=\sup\{z:(x,t)\in\operatorname{supp} f,\ 0\le t\le T\}, \quad S_N(T)=\sup\{S_{Nf}(T),\ \zeta_N(0)\}.$$

8.1 Localization in space. Let the hypotheses

$$\begin{cases} p,q,r\in\mathbb{R}, \quad p,q,r>1, \quad m,N\in\mathbb{N}, \quad m,N\ge 1,\\ u(x,t) \text{ is an energy solution of (8.1)},\end{cases}\tag{8.5}$$

$$S_N(T) \text{ is finite}\tag{8.6}$$

hold.

Theorem 8.1 (Finite speed of propagation). *Under hypotheses* (8.5), (8.6),

$$\zeta_N(T)-S_N(T)\le C_{mpqN}T^{\beta_0}E^{\lambda_0}(T) \quad \text{if } q<p\ (\forall r),$$

where

$$\begin{aligned}\lambda_0&=\frac{p-q}{q}\beta_0,\\ \beta_0&=\frac{1}{mp+N(p-q)/q},\end{aligned}\tag{8.7}$$

and the "energy" $E(T)$ is defined by

$$E(T) = E(0,T), \quad E(T_1,T_2) = \int_{T_1}^{T_2} \int_\Omega |D^m u|^p \, dxdt. \tag{8.8}$$

Theorem 8.2 (Space localization). *Under hypotheses* (8.5), (8.6),

$$\zeta_N(T) - S_N(T) \le C_{mpqN} E^\mu(T) \quad \text{if } r < p \ (\forall q), \qquad \mu = \frac{p-r}{mpq + N(p-r)}. \tag{8.9}$$

Theorem 8.3. *Under hypotheses* (8.5), (8.6),

$$\zeta_N(T) - S_N(T) \le C_{mpqN} \ln\left(1 + \widehat{C}_{mpqN} T^{\beta_0} E^{\lambda_0}(T)\right) \quad \text{if } q < p = r.$$

Set

$$\begin{cases} S_f(T) = \sup\{|x| : (x,t) \in \operatorname{supp} f, 0 \le t \le T\}, \\ S(T) \ \ = \sup\{S_f(T), \zeta(0)\}. \end{cases} \tag{8.10}$$

Theorem 8.4. *Let us assume that* (8.5) *holds and $S(T)$ is finite. Then*

$$\zeta(T) \le C_{mpqrN}\left(S(T) + T^\delta\right) \quad \text{if } q < p < r$$

with

$$\delta = \frac{r-p}{mp(r-q)}.$$

Proof of Theorem 8.1. Take an arbitrary fixed T. By translation, we can assume $D_N(T) = 0$. Another assumption is that the half-space $\{z > 0\}$ intersects Ω, otherwise we immediately have $\zeta_N(T) \le S_N(T)$.

Lemma 8.1. *Let $\rho \in W^{m,\infty}(\Omega)$ depend only on $z = x_N$ and*

$$\rho(z) \ge \begin{cases} 0 & \text{for } z \ge 0, \\ 0 & \text{for } z \le 0. \end{cases}$$

Set $\rho^{(i)} \equiv d^i\rho/dz^i$. For all $T \ge 0$,

$$\begin{aligned} \sup_{0\le t\le T} \frac{1}{q'} \int_\Omega \rho(z)|u(x,t)|^q dx + \int_0^T \int_\Omega \rho\,|D^m u|^p \, dxdt + \int_0^T \int_\Omega \rho |u|^r dxdt \\ \le C_m \sum_{i=1}^m \int_0^T \int_\Omega \left|\rho^{(i)}\right| |D^m u|^{p-1} \left|D^{m-i} u\right| dxdt. \end{aligned}$$

Proof. Taking the duality product of equation (8.1) and ρu and using the assumptions made on the supports of u_0, ρ, and f, we obtain that for all $t \geq 0$

$$\frac{1}{q'}\int_\Omega \rho(z)|u(x,t|^q dx + \int_0^t (\mathbf{A}u, \rho(\tau))dxdt + \int_0^t \int_\Omega \rho|u|^r dxdt = 0. \quad (8.11)$$

Now we compute $(\mathbf{A}u, \rho u)$. We have

$$\begin{aligned}
(\mathbf{A}u, \rho u) &= \sum_{|\alpha|=m} \int_\Omega \left|D^\alpha u\right|^{p-1} \operatorname{sign}\left(D^\alpha u \cdot D^\alpha(\rho u)\right) dx, \\
D^\alpha &= D_z^j D_y^\beta, \quad z = x_N, \quad y = (x_1, \dots, x_{N-1}), \\
\sum_{|\alpha|=m} &= \sum_{j=0}^m \sum_{|\beta|=m-j}, \quad |\beta| = |\alpha| - j;
\end{aligned}$$

by virtue of Leibniz's formula

$$D^\alpha(\rho u) = \rho D^\alpha u + \sum_{i=1}^j \binom{j}{i} \rho^{(i)} D_z^{j-i} D_y^\beta u \quad \text{for almost all } \tau > 0$$

and

$$\begin{aligned}
(\mathbf{A}u, \rho u) = &\int_\Omega \rho \left|D^m u\right|^p dx \\
&+ \sum_{j=1}^m \sum_{|\beta|=m-j} \sum_{i=1}^j \binom{j}{i} \int_\Omega \rho^{(i)} \left|D^\alpha u\right|^{p-1} \operatorname{sign}\left(D^\alpha u\right) \cdot D_z^{j-i} D_y^\beta u dx.
\end{aligned}$$

Plugging this formula in (8.11) and taking into account that $\left|D_z^{j-i} D_y^\beta u\right| \leq \left|D^{m-i} u\right|$ since $|\beta| + j = m$, we get the result. The constant C_m can be chosen to depend only on m because $i \leq j \leq m$. □

Adopt the notation

$$\begin{aligned}
E_s(z_0) &= \int_0^T \int_{\Omega \cap \{z > z_0\}} (z - z_0)^s \left|D^m u\right|^p dxdt, \\
F_s(z_0) &= \sup_{0 \leq t \leq T} \frac{1}{q'} \int_{\Omega \cap \{z > z_0\}}^s |u(s,t)|^q dx.
\end{aligned}$$

Recall that T is fixed.

Lemma 8.2. *Let $s \geq m$ be a real number. Then for all $z_0 \geq 0$,*

$$\begin{aligned}
&F_s(z_0) + E_s(z_0) \\
&\quad \leq C_{ms} \sum_{i=1}^m \int_0^T \int_{\Omega \cap \{z > z_0\}} (z - z_0)^{s-i} \left|D^m u\right|^{p-1} \left|D^{m-i} u\right| dxdt,
\end{aligned} \quad (8.12)$$

where all integrals are finite.

Proof. For Ω bounded and $s \geq m$, the function

$$\rho(x) = \begin{cases} 0 & \text{if } z \leq z_0, \\ (z - z_0)^s & \text{if } z > z_0 \end{cases}$$

belongs to $W^{m,\infty}(\Omega)$. The assertion follows now from the previous lemma. □

Lemma 8.3. *If $s \geq mp$, then for all $z_0 \geq 0$,*

$$F_s(z_0) \leq C_{mps} E_s(z_0), \tag{8.13}$$

$$E_s(z_0) \leq C_{mps} \int_0^T \int_{\Omega \cap \{z > z_0\}} (z - z_0)^{s-p} \left| D^{m-1} u \right|^p \text{`} dx dt < \infty. \tag{8.14}$$

Proof. We write the integrand of the right-hand side of (8.12) in the form

$$(z - z_0)^{s/p'} \left| D^m u \right|^{p-1} \cdot (z - z_0)^{s/p - i} \left| D^{m-i} u \right|$$

and apply Hölder's inequality with the exponents p, p' to bound (8.12) by

$$C_{ms} E_s^{1/p'}(z_0) \sum_{i=1}^{m} \left(\int_0^T \int_{\Omega \cap \{z > z_0\}} (z - z_0)^{s-pi} \left| D^{m-i} u \right|^p dx dt \right)^{1/p}.$$

Note that $s - pi \geq 0$ because $s \geq mp$. We obtain (8.14) repeatedly applying the following inequality.

Lemma 8.4 (Hardy inequality). *Let $p > 1$ and $\lambda > -1$. Set $H = \{x \in \mathbb{R}^N : x_N > 0\}$, $z = x_N$, and $D_z = \partial / \partial z$. Then*

$$\int_H z^\lambda |u|^p dx \leq \left(\frac{p}{\lambda + 1} \right)^p \int_H z^{\lambda + p} |D_z u|^p dx.$$

(This follows from the one-dimensional Hardy inequality [191, Theorem 330].) Then we obtain (8.13), applying the Hardy inequality once more. □

Lemma 8.5. *If $s - p$ is an integer and $s - p \geq mp$, then for all $z_0 \geq 0$,*

$$E_s(z_0) \leq C_{mpqNs} T^{1-a} E_{s-p}^{a + (p/q)(1-a)}(z_0), \tag{8.15}$$

where a is given by

$$\frac{1}{p} = \frac{m-1}{N+s-p} + a \left(\frac{1}{p} - \frac{m}{N+s-p} \right) + \frac{1-a}{q}. \tag{8.16}$$

Proof. We will need the following interpolation inequality.

Lemma 8.6 ([73, Lemma 8]). *Let $m, j, k \in \mathbb{N}$, $m \geq 1$, $0 \leq j < m$, $k \geq 0$. Let $p, q \in \mathbb{R}$ be such that $1 \leq q \leq p$. Then*

$$\left(\int_H z^k \left|D^j u\right|^p dx\right)^{1/p} \leq C\left(\int_H \left|D^p u\right|^p dx\right)^{a/p}\left(\int_H z^k |u|^q dx\right)^{(1-a)/q}$$

if the Lebesgue integrals on the right-hand side exist, where a satisfies the inequality $j/m \leq a < 1$ and is given by (8.16). The constant C depends only on N, m, j, p, q, k.

We apply this lemma to (8.14) with $k = s - p$, $j = m - 1$, then raise the result to the power p, integrate in t between 0 and T, and apply Hölder's inequality in the variable t. In this way we obtain that

$$E_s(z_0) \leq C_{mpqNs} T^{1-a} E^a_{s-p}(z_0) F^{(1-a)p/q}_{s-p}(z_0)$$

with a defined in (8.16). To obtain (8.15) we apply (8.13) with s replaced by $s - p$. □

We are now in position to finish the proof of Theorem 8.1. We are going to infer from (8.15) that $D^m u = 0$ for a.e. $z > b(T)$ which implies that $u = 0$ a.e. on the same set.

Inequality (8.15) is a fractional differential inequality in the variable z_0. To deal with such inequalities, we need the following result.

Lemma 8.7 ([77, Lemma A.4]). *Let $g \in L^1(\mathbb{R}_+)$, $g \geq 0$ a.e. in $\mathbb{R}_+$, and*

$$f_s(z) = \int_z^\infty (x - z)^s g(x)dx \quad \text{(Weyl's fractional integral of order } s+1 \text{ of } g).$$

Let $K, \alpha, \theta, s, h \in \mathbb{R}$ be such that $K > 0$, $\alpha > 0$, $\theta > 1$, $s \geq 1$, $0 < h \leq s < \omega$, where

$$\omega = \frac{\theta h}{\theta - 1}.$$

Assume that $f_{s-h}(0)$ is finite and

$$f_s(z) \leq K^\alpha f^\theta_{s-h}(z) \quad \text{for all } z \geq 0.$$

Then supp f *is a bounded interval $[0, b]$ with*

$$b \leq (\omega - s + 1)K^{\alpha/(\theta-1)(\omega-s)} f_0^{1/(\omega-s)}(0).$$

(Notice that in the particular case where l and s are integers and $s \geq l$, we have $E_l^{(s)}(z_0) = (-1)^s (l!/s!) E_{l-s}(z_0)$).

The assertion of Theorem 8.1 immediately follows now from Lemma 8.7 if we choose $K = \text{const} \cdot T$, $\alpha = 1 - a$, $\theta = a + \frac{p}{q}(1 - a)$, $h = p$. □

The proofs of Theorems 8.2–8.4 follow the same steps. The key point is the derivation of the fractional differential inequality for the weighted energy function $E_s(z_0)$. Under the conditions of Theorem 8.2 this inequality has the form

$$E_s(z_0) \le C_{mpqrsN} E_{s-p}^{1+p/(1/\mu+s-p)}(z_0)$$

with the exponent μ defined in (8.9).

The proof of Theorem 8.3 is a particular case of the proof of Theorem 8.4 which we give in full detail.

Proof of Theorem 8.4. Since $S(T)$ is finite and $q < p$, by virtue of Theorem 8.1, there exists a finite R such that

$$\operatorname{supp} u(\cdot, t) \subset B_R(0) \quad \text{for all } t \in [0, T]. \tag{8.17}$$

We will rely on the following weighted interpolation inequalities. Let

$$\begin{cases} H = \{x \in \mathbb{R}^N : x_N > 0\}, \quad x_N = z, \\ p, q, r \in \mathbb{R}, \quad p, q, r \ge 1, \\ m, j, k \in \mathbb{N}, \quad m \ge 1, \quad 0 \le j < m, \quad k \ge 0, \\ u \in L^p\left(0, T;\, W^{m,p}_{\text{loc}}(H)\right), \quad T > 0. \end{cases} \tag{8.18}$$

Lemma 8.8 ([75, Lemma II.1]). *Under conditions* (8.17), (8.18),

$$\int_0^T \int_H z^k \left|D^j u\right|^p dxdt \le C(TR^{N+k})^{(1-p/r)(1-j/m)} \times \left(\int_0^T \int_H z^k \left|D^m u\right|^p dxdt\right)^{j/m} \left(\int_0^T \int_H z^k |u|^r dxdt\right)^{(1-j/m)p/r}$$

where the constant C depends only on N, m, j, p, r, k. If $p = r$, condition (8.17) *can be dropped.*

This assertion implies that if s is the smallest real number such that $s - p$ is an integer and $s - p \ge mp$, then for all $z_0 \ge 0$,

$$E_s(z_0) \le C_{mprN} \left(TR^{N+s-p}\right)^{(r-p)/mr} E_{s-p}^{1-(r-p)/mr}(z_0). \tag{8.19}$$

Next, it follows from Lemma 8.7 and (8.19) that for such s,

$$E_1(z_0) \le C_{mprN} \left(TR^{N+s-p}\right)^{1/(s-\omega)} E_0^{1-1/(s-\omega)}(z_0) \tag{8.20}$$

with $\omega = p - mpr/(r-p) < 0$.

Lemma 8.9. *Let s be as in* (8.19). *Then*

$$\zeta_N(T) - S_N(T) \le C_{mpqrN} R^d T^e$$

with the exponents

$$1-d=\frac{e}{\delta}, \quad e=\frac{p}{(p-q)(1/\lambda_0+s-\omega)},$$

where δ, λ_0 *and* ω *are defined as in Theorems* 8.1 *and* 8.4 *and* (8.20).

Proof of Lemma 8.9. It follows from (8.20) that

$$\frac{1}{a}E_1^b(z_0)+\frac{1}{A}E_1^B(z_0)\le K|E_1'(z_0)| \quad \text{a.e. on } \mathbb{R}_+$$

with the constants

$$b=\frac{1}{1+\lambda_0}, \qquad a=T^{\beta_0/(1+\lambda_0)},$$
$$B=\frac{s-\omega}{s-\omega-1}, \qquad A=\left(TR^{N+s-p}\right)^{1/(s-\omega-1)}.$$

The assertion of the lemma immediately follows because the integral

$$\int_0^t \frac{ds}{s^b/a+s^B/A}$$

is convergent for all $t\ge 0$. □

Since $S(T)$ is bounded, the bound of Lemma 8.9 is applicable in every direction x_i, which yields the estimate

$$\zeta(T)\le \widehat{C}_{mpqrN}\left(S(T)+R^dT^e\right)\equiv h(T).$$

Since h is an increasing function, we can take $h(T)=R$. On the other hand, by Young's inequality

$$R^dT^e\le \epsilon R+C_\epsilon T^\delta \quad \text{with } 1-d=\frac{e}{\delta}.$$

Choosing ϵ appropriately small, we get the desired assertion. □

Remark 8.1. The results stated in Theorems 8.1–8.4 extend to the case where the domain Ω is unbounded. Let the set supp u_0 and the projection of supp f on Ω be bounded. Then $\zeta(T)<\infty$ for all $T>0$ and, as $T\to\infty$,

$$\begin{cases}\zeta(T)=\mathcal{O}(1) & \text{if } 1<r<p, (\forall q>1),\\ \zeta(T)=\mathcal{O}(\ln T) & \text{if } 1<q<p=r,\\ \zeta(T)=\mathcal{O}\left(T^\delta\right) & \text{if } 1<q<p<r\le p+mpq/N,\\ \zeta(T)=\mathcal{O}\left(T^{\beta_0}\right) & \text{if } 1<q<r \text{ and } r\ge p+mpq/N.\end{cases}$$

The exponents β_0 and δ are defined in the conditions of Theorems 8.1 and 8.4.

Remark 8.2. In these arguments, the plain energy sets can be substituted by the radial energy sets; cf. Section 6 of Chapter 1.

8.2 Decay rates as $t \to \infty$ and extinction in finite time. If $f = 0$ and Ω is bounded, the integration-by-parts formula gives sharp decay rates of $\|u(\cdot,t)\|_q$ as $t \to \infty$. Let $\Omega \subset B_R$ and

$$g(t) = \|u(\cdot,t)\|_q^q = \int_\Omega |u(x,t)|^q dx.$$

Theorem 8.5. *Under condition* (8.5), *for all* $r > 1$,

(I) *if* $q < p$, *then for all* $t > 0$,

$$g(t) \le \left(g^{-(p-q)/q}(0) + C_{mpqN} R^{-1/\beta_0} t\right)^{-q/(p-q)} \le \left(\frac{R^{1/\beta_0}}{C_{mpqrN} t}\right)^{q/(p-q)};$$

(II) $g(t) \le g(0)\exp(-C_{mpqN} t/R^{mp})$ *if* $q = p$,

(III) *if* $p < q$ *and* $\frac{1}{q} \ge \frac{1}{p} - \frac{m}{N}$, *then* $u(\cdot,t) = 0$ *for all* $t \ge T_0$ *with*

$$T_0 \le C_{mpqrN} R^{1/\beta_0} g^{(q-p)/q}(0).$$

We interpret $\frac{1}{\beta_0} = 0$ *if* $\frac{1}{q} = \frac{1}{p} - \frac{m}{N}$.

Theorem 8.6. *Under condition* (8.5) *for all* $t > 0$ *and all* $p > 1$,

(I) *if* $q < r$,

$$g(t) \le R^N \left(C_{qrN} t\right)^{-q/(r-q)};$$

(II) $g(t) \le g(o)\exp(-q't)$ *if* $q = r$.

Proof of Theorem 8.5. The function $g(t)$ is absolutely continuous, nonincreasing and

$$-\frac{1}{q'} g'(t) = \int_\Omega |D^m u|^p dx + \int_\Omega |u(x,t)|^r dx \quad \text{for a.a. } t > 0. \tag{8.21}$$

Extending $u(x,t)$ by zero to B_R, we estimate

$$\int_\Omega |u|^q dx \le C R^{q/(p\beta_0)} \left(\int_\Omega |D^m u|^p dx\right)^{q/p}. \tag{8.22}$$

It follows from (8.21), (8.22) that

$$g(t) \le C R^{q/(p\beta_0)} |g'(t)|^{q/p}.$$

The assertion follows by the straightforward integration of this inequality. □

Proof of Theorem 8.6. By Hölder's inequality

$$\int_\Omega |u|^q dx \le C R^{N(r-q)/r} \left(\int_\Omega |u|^r dx\right)^{q/r}.$$

Gathering this with (8.21), we arrive at the inequality

$$g(t) \le C R^{N(r-q)/r} |g'(t)|^{q/r}.$$ □

9 Bibliographical notes

The term "rate of propagation" was proposed in [208] in order to describe the situation when the zero cavern in a solution of a nonlinear parabolic equation with one space variable does not disappear instantly while the instant velocity of the interfaces need not be finite. A similar approach was proposed in [104] to describe the evolution of interfaces in solutions of the multidimensional porous medium equation (1.2). It was proven in [104] that given a point $x_0 \notin \operatorname{supp} u(x,t)$, the function $\rho(t) = \operatorname{dist}(x_0, \partial(\operatorname{supp} u(x,t)))$ (the optimal rate at the point x_0) satisfied a first-order differential inequality.

As was already mentioned in Section 1, in the case $N = 1$ the optimal rate at a point x_0 coincides with the traditionally accepted definition of the outer (or inner) interface in a nonnegative solution to a parabolic equation.

A review of results concerning the questions of existence, uniqueness, and the qualitative properties of solutions to nonlinear degenerate parabolic equations can be found in the survey paper [204] and [56, 304, 184]. The occurrence of interfaces (given by (1.1)) in solutions of nonlinear parabolic equations of the form

$$u_t = a_{xx}(u) + b_x(u) + c(u)$$

is analyzed in [184, 264]. The smoothness of interfaces in the solutions to equations of this class is studied in [171, 172, 173, 174] via the method of intersection comparison.

The proof of the main Lemma 2.1 follows paper [149]. The phenomenon described in Corollary 2.1 is already known for the simpler case $\mathbf{A}(x,t,u,Du) = |Du|^{p-2} \cdot Du$ and $\beta = 0$; see [57, 124, 138, 207, 223], and [148] for the first-order quasilinear equations. Theorem 2.2 establishes the property of stable localization of solutions of equation (2.39). Similar assertions are already known for specific first- and second-order quasilinear variational inequalities under some monotonicity assumptions—see [147, 148, 206]. For the numerical approach to the study of the waiting time property see Nakaki and Tomoeda [259] and their references.

The presentation of Section 3 follows papers [33, 34]. For the previous work on this subject see [18, 31, 52, 53, 54, 290, 291].

If $p < 1 + \gamma$, the speed of propagation of disturbances in the solutions of equation (2.39) is infinite—see [57, 138, 205, 297].

The class of equations (2.1) includes, in particular, the following model equation, by now well studied:

$$v_t - \Delta_p(|v|^{m-1}v) + |v|^{\lambda-1}v = f(x,t), \tag{9.1}$$

Equation (9.1) (with $p = 2$) is usually referred to as the nonlinear heat equation with absorption. If $v(x,t)$ is interpreted as the temperature of some continuum, the first and the second terms on the right-hand side of (9.1) represent the diffusion and the volume absorption of heat. The term $f(x,t)$ models an external source or sink of heat. The first of assumptions (4.1) holds if

$$m \geq 1, \qquad \lambda \in (0,1).$$

These relations mean that we study the processes of *linear* ($m = 1$) or *slow* ($m > 1$) diffusion under the influence of *strong absorption* ($\lambda \in (0, 1)$). In this choice of the exponents of nonlinearity the solutions display finite speed of propagation of disturbances from the data: see [184] and references therein. Moreover, it is known [284, 219, 218] that in this range of parameters the supports of nonnegative weak solutions to equation (9.1) may shrink with time t. It is also known [62, 135] that solutions of the Cauchy problem and the Cauchy–Dirichlet problem for equation (9.1) may even vanish on a subset of the problem domain Q despite of the fact that u_0 and the boundary data are strictly positive. These properties were discovered by means of comparison of solutions of (9.1) with suitable sub and supersolutions of these problems. For the application of the energy method to the study of equations with lower-order terms, see [90, 91].

The assertions of Theorems 3.1–4.1 are local in the sense that different parts of the boundary of $\operatorname{supp} u_0$ may originate pieces of the boundary of the null-set of $u(x,t)$, which display different shrinking properties. Having a possibility to control the rate of vanishing of u_0 and $f(x,t)$, one may design solutions of equation (2.1) which have prescribed shapes of supports. For the model equation (9.1) with $p = 2$ this phenomenon is known as "the heat crystal" [284, Chapter 3, Section 3]. In [18], this effect was studied via the local energy method.

Under additional structural assumptions on ψ, $\mathbf{A}$ and B, estimate (4.41) allows one to establish existence of a solution for problem (4.33)–(4.36). For instance, let us assume that $\psi(x,u) = u|u|^{\gamma-1}$ and $\mathbf{A}$ and C are monotone,

$$\forall s, s_1, s_2 \in \mathbb{R},\ \rho_1, \rho_2 \in \mathbb{R}^N \quad \begin{aligned} \big(\mathbf{A}(x,t,s,\rho_1) - \mathbf{A}(x,t,s,\rho_2)\big) \cdot (\rho_1 - \rho_2) \geq 0, \\ (C(x,t,s_1) - C(x,t,s_2))\,(s_1 - s_2) \geq 0, \end{aligned}$$

and B satisfies certain growth conditions. Under these assumptions the existence of a weak solution to problem (4.33)–(4.36) is proved in [203, 233, 243].

In the case $\Omega \equiv \mathbb{R}^N$, Theorem 4.1 and the estimate on the total energy allow us to establish the property of "instantaneous support shrinking" first found by Brézis and Friedman in [99] and then by Evans and Knerr in the paper [155] for a special case of equation (4.33). (See also [204, 1, 92] for some generalizations of this result).

It follows from Theorem 4.1 that solutions of problem (4.33)–(4.36) possess the property of a dead core formation. This property was earlier discovered in [62] and [135] for the solutions of special problems like (4.33)–(4.36): it was assumed that $\Gamma_N = 0$, $g > 0$ and $u_0 > 0$, which provided that the vanishing region was formed far from the initial plane $t = 0$ and the lateral boundary of Q.

The first assertion of Theorem 6.1 was announced without proof in the note [282].

It seems that Bamberger [60] was the first to study the free boundaries occurring in solutions of some doubly nonlinear parabolic equations with nonisotropic nonlinearities. The localization due to the one-dimensional phenomena seems to be new in the literature.

Systems of degenerate parabolic equations were studied by many authors. Many different variants of such systems arise from applications. For instance, the components of the reaction term $\mathbf{C}(x,t,\mathbf{u})$ may have different signs. This is what happens in the theory of combustion—see, e.g., Díaz and Hernández [136]. In other family of problems one of the components of the diffusion operator $\mathbf{A}(x, t, \mathbf{u}, D\mathbf{u})$ identically vanishes (Díaz and Stakgold [146], van Duijn and Knabner [303], Galiano and M. Peletier [180], van Duijn, Galiano, and M. Peletier [302]).

The research lines proposed in Chapters 1 and 2 are applicable to the study of the properties considered in the present chapter. Thus a list of some open problems could be as follows:

1. To introduce a version of the local energy method for fully nonlinear parabolic equations. We refer to the books by Krylov [230] and Lieberman [242] and the paper by Crandall et al. [117] for an application of the energy approach in a global framework and in special (but relevant) cases.

2. The cases where the solutions do not belong to the energy spaces require new ideas. See Blanchard and Murat [88] for the treatment by renormalizing of L^1 solutions. See also the results on existence and uniqueness of *entropy solutions* obtained in J. Carrillo [105].

3. The study of the free boundaries for discretized parabolic problems. See Pietra and Verdi [270] for the discrete free-boundary problems associated with the obstacle problem.

4. The local energy method for equations with nonpower nonisotropic nonlinearities (local versions of results from the paper by Andreucci and Tedeev [7]).

5. To apply energy methods to the study of multivalued or unilateral problems. For the results obtained with the methods based on comparison see Brezis and Friedman [99] and Evans and Knerr [155] for the obstacle problem and Yamada [310] for the one-phase Stefan problem.

4

Applications to Problems in Fluid Mechanics

1 Introduction

The results of formal considerations of the previous chapters are applied here to study the behavior of supports of solutions to some mathematical problems (models) of fluid mechanics. By *mathematical model* we mean a differential equation or a system of differential and functional equations completed, perhaps, by initial and boundary conditions, whose solutions describe certain parameters of a definite physical process.

Mathematical models appear as a byproduct of the main conservation and empirical laws of fluid mechanics. The description of a fluid medium with simple internal properties usually leads to a mathematical model composed of a single equation of elliptic, parabolic, or hyperbolic type. Most results of the previous chapters are directly applicable to models which involve only a single elliptic or parabolic equation. However, the mathematical modelling of processes in a medium with complicated internal properties or the character of motion may lead to models which involve systems of nonlinear differential equations. In such systems the components of the sought solutions—such as density, temperature, saturation, pressure, velocity—may satisfy equations of different types and even need not be defined on the same domain. These systems may degenerate with respect to the type or order at certain values of the solution or its derivatives, and the solutions may display all the variety of localization properties.

Some of the mathematical models discussed below admit immediate application of previous results, while others require certain preliminary reduction. Recall that we are interested in the vanishing (localization) properties of solutions from suitable function classes and that for most of the below-studied mathematical models the existence of solutions in appropriate function classes is already proved.

2 The balance laws of fluid mechanics

The reader is supposed to be acquainted with the mathematical principles of fluid mechanics. Nonetheless, for convenience we give here a brief description of several frequently used mathematical models of fluid mechanics. More detailed information about the mathematical principles of fluid mechanics, the balance laws, and related topics can be found in [152, 188, 248, 287].

The system of equations governing the motions of a fluid is constituted by the laws of conservation or, more generally, of balance of mass, momentum, and energy. These laws are traditionally formulated first in the integral form from which the equivalent differential formulations can be inferred. We first present the differential formulation of the balance laws.

Let $\rho(x,t)$, $\mathbf{v}(x,t)$, and $e(x,t)$ denote *density*, *velocity*, and *specific inner energy* of a fluid. The main balance laws are formulated as follows:

(i) the *mass balance* (the *continuity equation*)

$$\frac{\partial \rho}{\partial t} + \operatorname{div}(\rho \mathbf{v}) = h, \tag{2.1}$$

(ii) the *momentum balance* (*equations of motion*)

$$\rho D_t \mathbf{v} \equiv \rho \left(\frac{\partial \mathbf{v}}{\partial t} + (\mathbf{v} \cdot \nabla) \mathbf{v} \right) = \operatorname{div} \mathbf{S} + \rho \mathbf{f}, \tag{2.2}$$

where the symbol D_t denotes the *material derivative* calculated by the rule

$$D_t \phi = \frac{\partial \phi}{\partial t} + \mathbf{v} \cdot \nabla \phi,$$

(iii) the *energy balance*

$$\rho D_t e = \mathbf{S} : \mathbf{D} - \operatorname{div} \mathbf{q} + \rho Q. \tag{2.3}$$

Here and throughout the chapter, we use the following notation: h is the density of the external mass sources; $\mathbf{S}$ is the stress tensor; $\mathbf{f}$ is the density of the external mass forces; $\mathbf{D} = \frac{1}{2}(\nabla \mathbf{v} + \nabla^* \mathbf{v})$ is the rate of strain tensor (the symmetric part of $[\nabla \mathbf{v}]_{ij} = \partial v_j / \partial x_i$); $\nabla^* \mathbf{v}$ is the matrix transposed to $\nabla \mathbf{v}$; $\mathbf{S} : \mathbf{D}$ is the contraction of tensors $\mathbf{S}$ and $\mathbf{D}$ calculated by the formula

$$\mathbf{S} : \mathbf{D} = \sum_{i,j} S^{ij} D^{ij};$$

$\mathbf{q}$ is the rate of heat transport by conduction; and Q is the external heat supply per unit volume.

The system of balance laws (2.1), (2.2), (2.3) is incomplete because it includes five equations for 11 unknown functions (density ρ, specific inner energy e, three components of the velocity vector $\mathbf{v}$, and six components of the symmetric stress tensor S). It is completed by accepting the *constitutive laws* that express experimental information about the properties of the medium. Let us show how different choices of the constitutive laws may affect the form of the main balance laws.

Let us assume that

1. the medium is *homogeneous* and *incompressible*, i.e., $\rho(x, t)$ is constant and $h = 0$;

2. the *shear stresses are absent*, which means that tensor $\mathbf{S}$ is represented by a diagonal matrix, $\mathbf{S} = -p(x, t)\mathbf{I}$, where $p(x, t)$ is *pressure*, and $\mathbf{I}$ is the unit matrix.

Under these assumptions, system (2.1), (2.2) simplifies and transforms into the complete system of four scalar equations for four unknowns (pressure p and the components of the velocity vector $\mathbf{v}$):

$$D_t\mathbf{v} = -\frac{1}{\rho}\nabla p + \mathbf{f}, \quad \operatorname{div}\mathbf{v} = 0.$$

This is the famous *Euler system*, which describes the motions of the *ideal fluids*. Moreover, accepting the hypothesis that the fluid is *barotropic*, i.e., pressure and density are connected by the relation $p = p(\rho)$, and assuming that the flow is potential and stationary, we can reduce the problem to a nonlinear elliptic equation for the potential of the velocity. This problem is studied in Section 3.

Accepting the Stokes postulate that the dependence of the stress tensor $\mathbf{S}$ on the matrix $\nabla\mathbf{v}$ is affine, and assuming that the fluid is homogeneous and isotropic, we obtain the expression

$$\mathbf{S} = -p\mathbf{I} + \lambda \operatorname{div}\mathbf{v} + 2\mu\mathbf{D},$$

where λ and μ are the coefficients of *bulk* and *shear viscosity*. The fluids satisfying this condition are called *Newtonian*. If the viscosity coefficients are constant, we obtain the *Navier–Stokes* equations for viscous incompressible fluids,

$$D_t\mathbf{v} = -\nabla p + \mu\Delta\mathbf{v} + (\lambda + \mu)\nabla(\operatorname{div}\mathbf{v}) + \rho\mathbf{f}, \quad \operatorname{div}\mathbf{v} = 0.$$

A more complicated model corresponds to the case where the coefficients of shear and bulk viscosity are functions of the pressure p. Under the assumption that the fluid is barotropic, such a problem can be reduced to solving a single nonlinear degenerate parabolic equation for the density ρ. Results on the localization properties of solutions to this equation are given in Section 5.

According to the Reiner–Rivlin principle of material objectivity (see Gurtin [188]), the stress tensor in its most general form is given by the expression

$$\mathbf{S} = \phi_1(I_2, I_3)\mathbf{D} + \phi_2(I_2, I_3)\mathbf{D}^2,$$

where I_i denote the principal invariants of the tensor $\mathbf{D}$, and ϕ_i are given scalar functions.

The special choice $\phi_2 = 0$ and $\phi_1 = 2\mu > 0$ ($\mu > 0$ is the shear viscosity) leads to the model of the *Newtonian incompressible fluid* with the coefficient of bulk viscosity $\lambda = 0$.

To describe *purely viscous non-Newtonian fluids*, it is usually assumed that $\phi_2 \equiv 0$ and $\phi_1 \neq$ const. A subclass of such fluids is constituted by the Ostwald–de Waele (or the power-type) fluids, for which

$$\phi_1 = 2(\mu + \tau|\mathbf{D}|^{n-1}) \quad \text{for some } \tau > 0 \text{ and } n > 0$$

with $|\mathbf{D}|^2 = \mathbf{D} : \mathbf{D}$. The fluid is called *dilatant* if $n > 1$ and *pseudoplastic* if $0 < n < 1$, Showalter [292]. If $n = 1$, we revert into the class of Newtonian fluids.

A few models of dilatant fluids are studied in Sections 7.6 and 8. In Section 7.6, we consider an equation arising from the mathematical modelling of the flow of a dilatant fluid in a thin pipe when the motion is caused by a given drop of pressure.

The Navier–Stokes system associated with the two-dimensional stationary flow of an incompressible dilatant fluid has the form

$$\begin{aligned}
\mathbf{v} \cdot \nabla u &= -\frac{1}{\rho}\pi_x + \nu\left(|\mathbf{D}|^{p-2}u_x\right)_x + \frac{\nu}{2}\left(|\mathbf{D}|^{p-2}\left(u_y + v_x\right)\right)_y, \\
\mathbf{v} \cdot \nabla v &= -\frac{1}{\rho}\pi_y + \nu\left(|\mathbf{D}|^{p-2}v_y\right)_y + \frac{\nu}{2}\left(|\mathbf{D}|^{p-2}\left(u_y + v_x\right)\right)_x, \\
\operatorname{div} \mathbf{v} &= 0,
\end{aligned} \tag{2.4}$$

where $\mathbf{v} = (u, v)$ is the velocity, π is the pressure, and

$$\mathbf{D} = \begin{pmatrix} u_x & \frac{1}{2}\left(u_y + v_x\right) \\ \frac{1}{2}\left(u_y + v_x\right) & v_y \end{pmatrix}.$$

Under the assumption that the Reynolds number is small, the boundary layer approximation is valid, which allows one to reduce system (2.4) to a single degenerate parabolic equation posed in a half-strip in the plane of the von Mises variables. The localization properties of the solutions to this equation are studied in Section 8.

Several models of *non-Newtonian* fluids with the stress tensor

$$\mathbf{S} = -p\mathbf{I} + \mathbf{F}(\mathbf{D})$$

are considered in Section 7. Specifically, we study the localization effects for fluids with unbounded or vanishing density, in either case in the presence of outer forces nonlinearly depending on the velocity. The localization effects in such flows are due to the nonlinear structure of the momentum balance equation, which has the form

$$\rho D_t \mathbf{v} = -\nabla p + \operatorname{div} \mathbf{F}(\mathbf{D}) + \mathbf{f}(x, t, \mathbf{v}),$$

where $\mathbf{F}$ and $\mathbf{f}$ are given tensor and vector-valued functions correspondingly. The dependence of $\mathbf{f}$ on the velocity $\mathbf{v}$ may be caused by friction (see, e.g., [140]).

In the descriptions of *nonhomogeneous incompressible flows* that appear in various physical contexts (see Mosolov and Mjasnikov [256], Fernández-Cara, Guillén, and Ortega [156, 157], Antontsev, Kazhikhov, and Monakhov [45, 46], P. L. Lions [244, 245], and Simon [293]), the mass balance law (2.1) transforms into

$$D_t\rho = 0, \qquad \operatorname{div}\mathbf{v} = 0.$$

To describe *compressible flows*, the constitutive law $p = \Pi(\rho, \theta)$ is usually accepted. Since the temperature θ is closely connected with the inner energy e, such models also involve the energy balance equation.

The functions involved in the law of balance of the inner energy (2.3) are often subject to the constitutive law

$$\mathbf{q} = -k(x, \theta, |\nabla\theta|)\nabla\theta, \tag{2.5}$$

where k is a nonnegative function and $\theta(x,t)$ is the temperature of the medium. If the coefficient of heat conductivity k is a positive constant, relation (2.5) transforms into the classical Fourier law of heat transfer. In many physically important cases, k is not constant and, moreover, may vanish for certain values of θ and/or $|\nabla\theta|$ (see Kalashnikov [204] and references therein). Under the assumptions that $e = C\theta$ (C = const), the law (2.5) holds, the heat conduction by radiation is negligible, and the medium is immobile, the energy balance law leads to a separate nonlinear degenerate parabolic equation for the temperature

$$C\rho\frac{\partial\theta}{\partial t} = \operatorname{div}\left[k(\theta, |\nabla\theta|)\nabla\theta\right] + F.$$

In this special case, the resulting equation belongs to the class of equations studied in the previous chapters.

Some of the conservation laws (2.1)–(2.3) can sometimes be replaced by empirical laws. In fact, the mathematical theory of filtration in a porous medium is based upon the empirical Darcy law, which replaces the law of balance of momentum (2.2). Darcy's law in its simplest form is given by the relation

$$\mathbf{v} = -K\nabla p, \tag{2.6}$$

where K is the prescribed filtration coefficient. If the law (2.6) is accepted, the mathematical description of the process of filtration of an *ideal perfect* gas for which (2.6) reads as $\rho = Cp$, C = const, the continuity equation (2.1) can be written in the form

$$\frac{\partial\rho}{\partial t} - C\operatorname{div}(\rho\nabla\rho) = h.$$

For *politropic gases*, where $p \sim \rho^\gamma$ with $\gamma > 1$, the mass balance law transforms into

$$\frac{\partial\rho}{\partial t} - C\gamma\operatorname{div}(\rho^\gamma\nabla\rho) = h.$$

More sophisticated equations of the same class result from consideration of flows of the isothermal barotropic gases in a porous medium with the nonlinear Darcy law

$$\mathbf{v} = -K(x, p)|\nabla p|^{n-1}\nabla p, \quad n > 0,$$

with a nonconstant coefficient $K(x, p)$.

In Section 4, we consider the joint motion of two immiscible incompressible fluids through a porous medium. It is assumed that in each fluid,

1. the Darcy–Muskat law, which expresses the velocity through the gradient of the pressure, holds;

2. the mass balance law is true and the presence of the ith substance is given in terms of the saturation s_i; besides, $s_1 + s_2 = 1$;

3. the pressure is related by the empirical Laplace law, which reads $p_2 - p_1 = p_c(s_1, x)$; p_c is the prescribed capillary pressure defined from the experiment.

Under some physically reasonable assumptions this system is reduced to a system composed of a degenerate parabolic equation for the saturation s_1 and a degenerate elliptic equation for the reduced pressure p. The coefficients of the latter depend on s_1.

A similar system arises from modelling of the process of heat propagation in the electric conductor in the presence of Joule heating, a thermistor. Gathering the energy balance law and Ohm's law leads to a degenerate parabolic equation for the specific energy and a degenerate elliptic equation for the potential of the electric field. The blowup solutions of this problem are studied in Section 13.

3 Stationary problems of gas dynamics with free boundaries

3.1 Governing equations. Let us consider an ideal barotropic gas for which

$$\mathbf{P} = -p\mathbf{I}, \quad p = p(\rho) \quad \text{with } \frac{dp}{d\rho} > 0. \tag{3.1}$$

$\mathbf{I}$ is the unit matrix. The inner sources of mass and the outer mass forces are absent, $h \equiv 0$, $\mathbf{f} \equiv 0$, and the motion is potential and stationary:

$$\mathbf{v} = \nabla\varphi, \qquad \frac{\partial\rho}{\partial t} = 0, \qquad \frac{\partial\mathbf{v}}{\partial t} = 0,$$

where φ is the potential of the flow. Under these assumptions, the conservation laws (2.1), (2.2) simplify and yield the equations

$$\operatorname{div}(\rho\mathbf{v}) = \operatorname{div}(\rho\nabla\varphi) = 0, \tag{3.2}$$

$$\rho(\mathbf{v}\cdot\nabla)\,\mathbf{v} = -\nabla p. \tag{3.3}$$

Equation (3.3) admits Bernoulli's integral

$$\int_1^p \frac{dp}{\rho} + \frac{|\nabla \varphi|^2}{2} = \text{const}, \tag{3.4}$$

which holds along the stream lines. Equations (3.1) and (3.4) define a function $\rho = \rho(|\nabla \varphi|)$ such that $\frac{d\rho(\tau)}{d\tau} < 0$. By virtue of (3.2), the equation for the potential φ [10, 83, 87, 255, 262]:

$$\operatorname{div}(\rho(|\nabla \varphi|)\nabla \varphi) \equiv \sum_{i,j=1}^{3} a_{ij} \frac{\partial^2 \varphi}{\partial x_i \partial x_j} = 0, \tag{3.5}$$

holds, where $\rho = \rho(q)$, $q = |\nabla \varphi|$, and

$$a_{ij} = \left(\rho\delta_{ij} + \frac{1}{q}\frac{d\rho}{dq}\frac{\partial\varphi}{\partial x_i}\frac{\partial\varphi}{\partial x_j}\right), \quad \delta_{ij} = \begin{cases} 1 & \text{if } i = j, \\ 0 & \text{otherwise.} \end{cases}$$

The functions ρ, $\nabla\varphi$, abd q are the density of the gas, the velocity, and the module of the velocity. An equation analogous to (3.5) can be derived in the presence of the potential outer mass force $\mathbf{f} = -\nabla \Pi$.

The determinant of equation (3.5) is given by the formula

$$0 \le \Delta = \rho(1 - M^2) \equiv c^2(c^2 - q^2), \quad M^2(q) = \frac{q^2}{c^2} = -\frac{q}{\rho}\frac{d\rho}{dq} \le 1,$$

were M is the Mach number, q_s is the sonic velocity defined by the relation $M(q_s) = 1$, and c is local sonic speed. Equation (3.5) is thus elliptic for subsonic flows, where $|v|^2 < c^2$, $M^2 < 1$. It may degenerate into a parabolic equation if either $M^2(q_s) = 1$, or $\rho(q_{cr}) = 0$. The magnitude q_{cr}, $(q_s < q_{cr})$, is called the critical velocity. In what follows, we study only sonic and subsonic flows for which

$$q^2 = |v|^2 \le q_s^2 \quad \text{and} \quad M^2 \le 1. \tag{3.6}$$

We formulate below the conditions on the geometry of the prescribed parts of the boundaries and on the given module of the velocity vector $q_0\tau$ that provide the fulfillment of (3.6) for a physically reasonable equation of state $p = p(\rho)$.

Remark 3.1. Relations between pressure p, density ρ, the velocity module $|v| = q$, and Mach's number M can be readily calculated for the perfect politropic gases [10, 83, 87]. They have the form

$$p = p_0\rho^\gamma, \quad \rho(q) = \left(1 - \frac{\gamma - 1}{2}q^2\right)^{1/(\gamma-1)}, \quad \gamma > 1,$$

$$M^2(q) = q^2 / \left(1 - \frac{\gamma - 1}{2}q^2\right), \quad c^{-2} = \left(1 - \frac{\gamma - 1}{2}q^2\right),$$

$$q_s^2 = \frac{2}{\gamma + 1} < q_{cr}^2 = \frac{2}{\gamma - 1}, \quad M(q_s) = 1, \quad \rho(q_{cr}) = 0.$$

Further analysis is confined to the study of properties of the *free boundaries*. By a free boundary, we mean a line of *contact discontinuity* between two fluids with different physical parameters. The flow in one of these fluids is known and the free boundary is defined by the conditions

$$\mathbf{v} \cdot \mathbf{n} = \nabla \varphi \cdot \mathbf{n} = 0, \qquad |\mathbf{v}| = |\nabla \varphi| = q_0(\tau). \tag{3.7}$$

In these formulas $\mathbf{n}$ is the unit outer normal vector to the free boundary and $q_0(\tau)$ is a given function of the parameter τ. In particular, q_0 is constant if the given fluid is motionless in the plane orthogonal to the direction of the free-fall acceleration. For the parameter τ, we may take one of the independent variables x, the arc length s of the free boundary, or the potential φ. The free boundary is the boundary of the jet (finite or infinite). At the given impermeable parts of the boundary only the first of conditions (3.7) must be satisfied. We will consider the cases of subsonic or sonic plane or axis-symmetric flows.

Let the potential and the stream function of the flow be defined by the formulas

$$\varphi(x_1, x_2, x_3) = \varphi(x, r), \qquad \psi(x_1, x_2, x_3) = \psi(x, r),$$

where $x = x_1, r^2 = x_2^2 + x_3^2$. From (3.5), we infer that

$$r^\lambda \rho \frac{\partial \varphi}{\partial r} = -\frac{\partial \psi}{\partial x}, \qquad r^\lambda \rho \frac{\partial \varphi}{\partial x} = \frac{\partial \psi}{\partial r}. \tag{3.8}$$

The value $\lambda = 0$ corresponds to the plane case, $\lambda = 1$ if the flow is axis-symmetric. It is assumed that on the free boundary

$$\psi = 1 \quad \left(\text{or } \frac{\partial \varphi}{\partial n} = 0\right) \quad \text{and} \quad |\nabla \varphi|^2 = q_0(\tau), \tag{3.9}$$

while on the given boundary

$$\psi = 0 \quad \left(\text{or } \frac{\partial \varphi}{\partial n} = 0\right). \tag{3.10}$$

We will consider only a part of the domain occupied by the stream, under the assumption that (1) a solution of the complete problem does exist and (2) there exists a map of the stream domain in the plane of the physical variables (x, r) to the plane of complex potential (of the variables (φ, ψ)). Let us introduce the new unknown functions

$$U(\varphi, \psi) = \int_q^{q_s} \frac{\rho(\tau)}{\tau}\, d\tau, \qquad \theta(\varphi, \psi) = \arctan \frac{\varphi_r}{\varphi_x}. \tag{3.11}$$

According to (3.8), (3.11), the functions $x(\varphi, \psi)$, $r(\varphi, \psi)$, $U(\varphi, \psi)$, $\theta(\varphi, \psi)$ satisfy the system of equations

$$\begin{aligned} q\frac{\partial x}{\partial \varphi} &= \cos\theta, & r^\lambda \rho q \frac{\partial x}{\partial \psi} &= -\sin\theta, \\ q\frac{\partial r}{\partial \varphi} &= \sin\theta, & r^\lambda \rho q \frac{\partial r}{\partial \psi} &= \cos\theta. \end{aligned} \tag{3.12}$$

Making use of the relations

$$x_{\varphi\psi} = \left(\frac{\cos\theta}{q}\right)_\psi = x_{\psi\varphi} = -\left(\frac{\sin\theta}{r^\lambda q\rho}\right)_\varphi,$$
$$r_{\varphi\psi} = \left(\frac{\sin\theta}{q}\right)_\varphi = r_{\psi\varphi} = -\left(\frac{\cos\theta}{r^\lambda q\rho}\right)_\psi,$$

we obtain the system of equations for functions θ, q:

$$-\frac{\sin\theta}{q}\theta_\psi - \frac{\cos\theta}{q^2}q_\psi = -\frac{\cos\theta}{r^\lambda q\rho}\theta_\varphi + \frac{\sin\theta}{(r^\lambda q\rho)^2}(r^\lambda q\rho)_\varphi,$$
$$\frac{\cos\theta}{q}\theta_\varphi - \frac{\sin\theta}{q^2}q_\varphi = \frac{\sin\theta}{r^\lambda q\rho}\theta_\psi + \frac{\cos\theta}{(r^\lambda q\rho)^2}(r^\lambda q\rho)_\psi,$$

which can be written in the form

$$-\theta_\varphi = r^\lambda U_\psi, \quad r^\lambda\theta_\psi = KU_\varphi - \frac{\lambda\sin\theta}{r\rho q} \tag{3.13}$$

with $K = (1 - M^2)/\rho^2$. Differentiating the second equation in (3.13) in φ and eliminating θ, for the function $U(\varphi, \psi)$, we obtain the equation

$$\frac{\partial}{\partial\varphi}\left(K(U)\frac{\partial U}{\partial\varphi}\right) + \frac{\partial}{\partial\psi}\left(r^{2\lambda}\frac{\partial U}{\partial\psi}\right) - \frac{2\lambda\sin\theta}{rq}K(U)\frac{\partial U}{\partial\varphi} + \frac{2\lambda\sin^2\theta}{\rho r^2 q^2} = 0. \tag{3.14}$$

Given the functions $\theta(\varphi, \psi)$, $r(\varphi, \psi)$, $q(\varphi, \psi)$, $\rho(\varphi, \psi)$, equation (3.14) is an elliptic equation for the function $U(\varphi, \psi)$. However, it ceases to be elliptic and degenerates at the points where either $K(U) = 0$, or $r = 0$, $r \to \infty$.

As for the localization properties of solutions to this equation, let us note in advance that under the assumptions made on the data such effects are due to the anisotropic nature of degeneracy of equation (3.14) with respect to the independent variables φ, ψ.

According to (3.9), (3.10), (3.13), the functions $U(\varphi, \psi)$, $\theta(\varphi, \psi)$ satisfy the boundary conditions

$$U(\varphi, 1) = U(q_0(\tau)) = U_0(\varphi), \qquad r^\lambda\frac{\partial U(\varphi, 0)}{\partial\psi} = -\frac{\partial\theta_0(\varphi, 0)}{\partial\varphi}. \tag{3.15}$$

If $(\theta_0)_\varphi(\varphi, 0) \geq 0$, the prescribed part of the boundary is concave outward from the stream. It then follows from the maximum principle for solutions of elliptic equations that $0 \leq \min U_0 \leq U(\varphi, \psi)$, whence one infers that $q \leq q_s$, $M(q) \leq 1$. Conditions (3.6) are thus satisfied and the flow is either subsonic or sonic.

The study of system (3.12), (3.13) was performed in [9, 10, 295]. It was proved that the system admits solutions describing a large class of axis-symmetric (or plane) sonic flows with free boundaries such that $M(q) \leq 1$, $q \leq q_s$, $0 \leq U(\varphi, \psi)$). A few examples of such problems follow:

(1) collision of two axis-symmetric coaxial jets;

(2) discharge of gas from an axis-symmetric nozzle;

(3) discharge of gas from a plane nozzle.

Figures 3.1–3.3 illustrate the flows on the plane of the physical variables. We will consider only the flows in the shadowed zones. On the plane of the complex potential (ϕ, ψ) these parts of the flows correspond to the half-strip

$$Q = \left\{(\varphi, \psi) : (\varphi, \psi) \in \mathbb{R}^2, 0 < \varphi < \infty, 0 < \psi < 1\right\}.$$

Our considerations will be confined to the class of solutions to system (3.12), (3.13) that satisfy (3.15) and possess in Q the properties

$$0 \le U(\varphi, \psi) \le C_0, \qquad -\frac{\pi}{2} < -\frac{\pi}{2} + \delta \le \theta(\varphi, \psi) \le 0, \tag{3.16}$$

$$\left(|\ln \rho|, \, |\ln q|, \, \lambda\left(\left|\ln \frac{r^2}{\psi}\right| + \left|\sin \frac{\theta}{r}\right|\right)\right) \le C_0 < \infty, \tag{3.17}$$

$$E(t) \equiv \int_t^\infty \int_0^1 \left(KU_\varphi^2 + r^{2\lambda} U_\psi^2\right) d\varphi d\psi \le E(0) = E_0 \tag{3.18}$$

with some positive constants C_0, E_0, γ, and δ. Let us also assume that

$$0 \le K(U) \le C_0 U^\alpha, \quad 0 < \alpha < \infty. \tag{3.19}$$

Proposition 3.1 ([9, 10, 295]). *Under the assumptions*

$$0 \le K'(U), \quad K''(U) \le 0,$$

equation (3.14) *has* strong solutions *such that* $(U(\varphi, \theta), \theta(\varphi, \theta)) \in C^\alpha(Q)$ *and*

$$\int_\rho^\infty \int_0^1 \left(KU_{\varphi\varphi}^2 + r^{2\lambda} U_{\varphi\psi}^2 + |r^{2\lambda} U_\psi|^2 + |U_\psi|^\gamma\right) d\varphi d\psi \le C < \infty$$

with some constant $0 < \gamma < 1/2$.

3.2 Homogeneous boundary conditions: Axis-symmetric flows. We study here local properties of an axis-symmetric jet that moves along the symmetry axis with the sonic speed $q_0(\tau) \equiv q_s$, $U_0(q_s) \equiv 0$. Conditions (3.15) take on the form

$$-r^\lambda \frac{\partial U(\varphi, 0)}{\partial \psi} = \frac{\partial \theta(\varphi, 0)}{\partial \varphi} = 0 \quad (\theta(\varphi, 0) = 0), \quad U(\varphi, 1) = 0. \tag{3.20}$$

Theorem 3.1. *Let* $(U(\varphi, \psi), \theta(\varphi, \psi))$ $(U \not\equiv 0)$ *be a weak solution of system* (3.13) *satisfying* (3.20). *Let conditions* (3.16)–(3.19) *be fulfilled. Then there exists a constant* $T = T(E_0, C_0, \alpha, \delta)$, *such that*

$$U(\varphi, \psi) = \theta(\varphi, \psi) = 0 \quad \textit{for } 0 \le \psi \le 1, \, \varphi \ge T.$$

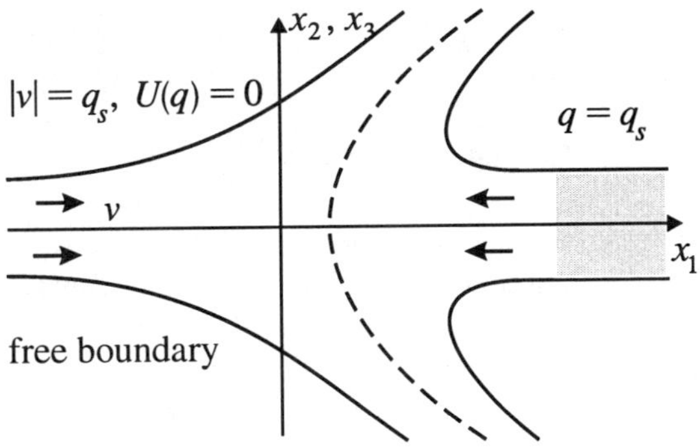

Figure 3.1: Collision of two axis-symmetric coaxial gas jets.

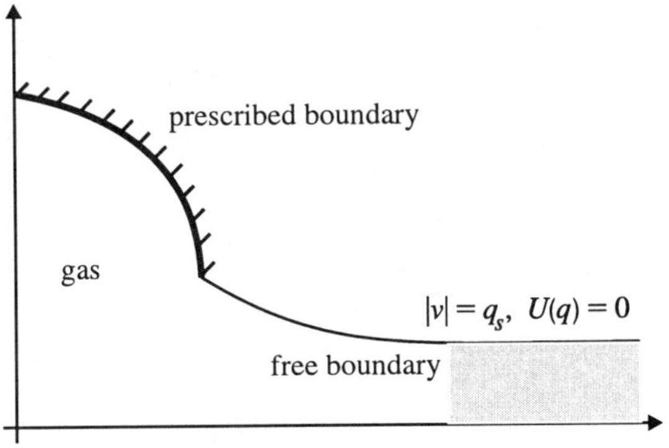

Figure 3.2: The velocity module along the free boundary is constant (sonic flow).

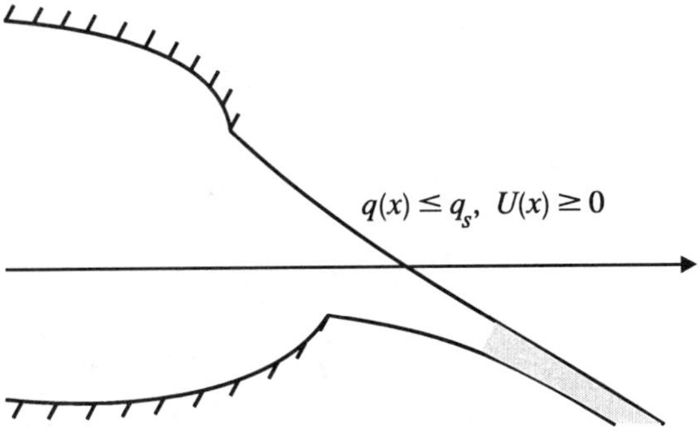

Figure 3.3: Along the free boundary the velocity module is given.

The physical interpretation of Theorem 3.1 is as follows. The axis-symmetric jet that moves along the axis of symmetry with sonic speed flattens in a finite distance. This means that the gas jet starts moving like a rigid body. Correspondingly, the free boundary becomes a straight line for $\varphi \geq T$. For plane sonic jets the fact of the speed flattening in a finite distance was first proved by L. V. Ovsyannikov in [266] via analysis of a linear degenerate elliptic equation for the function $\psi(U, \theta)$ on the plane of the godograph variables (U, θ).

Proof. We shall follow the scheme documented in Chapter 1, Section 2. Under conditions (3.20), for the solutions of equation (3.14) satisfying (3.16)–(3.19), the energy relation

$$E(t) \equiv \int_t^\infty \int_0^1 \left(KU_\varphi^2 + r^{2\lambda}U_\psi^2\right) d\varphi d\psi = I_1 + I_2 + I_3. \tag{3.21}$$

holds. In this relation,

$$\begin{aligned} I_1(t) &= -\int_0^1 K(U(t,\psi)) \frac{\partial U(t,\psi)}{\partial \varphi} U(t,\psi) d\psi, \\ I_2 &= -\int_t^\infty \int_0^1 \frac{2\lambda \sin\theta}{rq} K(U) \frac{\partial U}{\partial \varphi} U(\varphi,\psi) d\varphi d\psi, \\ I_3 &= \int_t^\infty \int_0^1 \frac{2\lambda \sin^2\theta}{\rho r^2 q^2} U(\varphi,\psi) d\varphi d\psi. \end{aligned}$$

It is easy to verify that

$$\frac{dE}{dt} \equiv E_t = -\int_0^1 \left(KU_\varphi^2 + r^{2\lambda}U_\psi^2\right) d\psi < 0, \quad E(\infty) = 0,$$

i.e., $E(t)$ is a decreasing function that tends to zero when $t \to \infty$.

For the sake of simplicity we set $\lambda = 1$ and make the convention to denote by C all the constants which depend only on C_0, α, E_0, and δ. We will rely upon the relations

$$\begin{aligned} |U(t,\psi)| &\leq \left| \int_\psi^1 U_\psi d\psi \right| \leq C\left(\int_0^1 r^2 U_\psi^2 d\psi \right)^{1/2} \left(\int_\psi^1 \frac{ds}{s} \right)^{1/2} \\ &\leq C(-E_t)^{1/2} |\ln \psi|^{1/2}, \end{aligned} \tag{3.22}$$

$$\begin{aligned} \int_0^1 |U|^{\alpha+2} d\psi &\leq C(-E_t)^{(2+\alpha)/2} \left(\int_0^1 |\ln \psi|^{(\alpha+2)/2} d\psi \right) \\ &\leq C(-E_t)^{(\alpha+2)/2}. \end{aligned} \tag{3.23}$$

Integrating the second equation of (3.13) in ψ and taking into account (3.20), we have that

$$r\theta(t,\psi) + \int_0^\psi (\sin\theta - \theta\cos\theta) \frac{ds}{rq\rho} = \int_0^\psi KU_\varphi ds.$$

Recall that $\theta \le 0$ and $g(\theta) = \sin\theta - \theta\cos\theta \le 0$. It follows that

$$r|\theta(t,\psi)| \le C\psi^{1/2}(-E_t(t))^{1/2}, \quad |\theta(t,\psi)| \le C(-E_t(t))^{1/2}, \tag{3.24}$$

$$\int_t^\infty |\theta|^2\varphi \le C\int_t^\infty (-E_t)dt = CE(t). \tag{3.25}$$

Using (3.13) and (3.25), we can write

$$\begin{aligned} |\theta(t,\psi)|^2 &\le 2\int_t^\infty |\theta|\left|\frac{\partial\theta}{\partial\varphi}\right| d\varphi = 2\int_t^\infty |\theta| r\left|\frac{\partial U}{\partial\psi}\right| d\varphi \\ &\le 2\left(\int_t^\infty |\theta|^2\, d\varphi\right)^{1/2}\left(\int_t^\infty r^2\left|\frac{\partial U}{\partial\psi}\right|^2 d\varphi\right)^{1/2} \\ &\le CE^{1/2}(t)\left(\int_t^\infty r^2\left|\frac{\partial U}{\partial\psi}\right|^2 d\varphi\right)^{1/2}. \end{aligned}$$

Raising both sides of this inequality to the power $\kappa > 0$ and integrating the resulting inequality in ψ, we have

$$\int_0^1 |\theta(t,\psi)|^{2\kappa} d\psi \le CE^\kappa(t), \quad 0 < \kappa \le 2. \tag{3.26}$$

Now combining (3.16), (3.17), (3.19) with (3.22), (3.23), (3.24)–(3.26) and using Hölder's inequality, we arrive at the estimate for I_i

$$\begin{aligned} |I_1| &\le C\left(\int_0^1 KU_\varphi^2 d\psi\right)^{1/2}\left(\int_0^1 |U|^{2+\alpha} d\psi\right)^{1/2} \\ &\le C(-E_t)^{(4+\alpha)/4}, \end{aligned} \tag{3.27}$$

$$\begin{aligned} |I_2| &\le C\left(\int_t^\infty\int_0^1 KU_\varphi{}^2 d\varphi d\psi\right)^{1/2}\left(\int_t^\infty\int_0^1 \frac{\theta^2}{r^2}U^{2+\alpha}\, d\varphi d\psi\right)^{1/2} \\ &\le CE^{1/2}\left(\int_t^\infty (-E_t)\int_0^1 \frac{|\theta|}{\psi^{1/2}}|\ln\psi|\, d\varphi d\psi\right)^{1/2} = CE^{1/2}I. \end{aligned}$$

On the other hand,

$$I = \int_0^1 \frac{|\theta|}{\psi^{1/2}}|\ln\psi|\, d\psi \le \left(\int_0^1 \theta^4 d\psi\right)^{1/4}\left(\int_0^1 \frac{|\ln\psi|^{4/3}}{\psi^{2/3}} d\psi\right)^{3/4} \le CE^{1/2},$$

and finally

$$|I_2| \le CE^{1/2}\left(\int_t^\infty (-E_t)E^{1/2}dt\right)^{1/2} = CE^{1/2}E^{3/4}. \tag{3.28}$$

The term I_3 is estimated analogously:

$$\begin{aligned}|I_3| &\le C\int_t^\infty\int_0^1 \frac{\sin^2\theta}{r^2}|U|d\varphi d\psi\\ &\le C\int_t^\infty(-E_t)\left(\int_0^1\frac{|\sin\theta|}{r^2}|\ln\psi|^{1/2}d\psi\right)d\varphi,\end{aligned}$$

but since

$$\begin{aligned}I &= \int_0^1 \frac{|\sin\theta|}{r^2}|\ln\psi|^{1/2}d\psi \le C\int_0^1\left(\frac{|\sin\theta|}{r}\right)^\gamma|\theta|^{1-\gamma}\frac{|\ln\psi|^{1/2}}{r^{2-\gamma}}d\psi\\ &\le C\int_0^1 \gamma|\theta|^{1-\gamma}\frac{|\ln\psi|^{1/2}}{\psi^{1-\gamma/2}}d\psi\\ &\le C\left(\int_0^1|\theta|^{2\kappa}d\psi\right)^{(1-\gamma)/2\kappa}\left(\int_0^1\left(\frac{|\ln\psi|^{1/2}}{\psi^{1-\gamma/2}}\right)^{2\kappa/(2\kappa-1+\gamma)}d\psi\right)^{(1-\gamma)/2\kappa}\\ &\le \frac{C}{1+\kappa}E^{(1-\gamma)/2\kappa} < \gamma < 1 \quad \text{with } 0 < \frac{2\kappa}{2\kappa-1+\gamma} < 1.\end{aligned}$$

In this way, we obtain the estimate

$$|I_3| \le C\int_t^\infty(-E_t)E^{(1-\gamma)/2\kappa}d\varphi = CE\times E^{(1-\gamma)/2\kappa}. \tag{3.29}$$

Gathering (3.21), (3.27) and (3.28), (3.29) and then choosing ρ sufficiently large to provide the inequality

$$C(E^{3/4}+E^{(1-\gamma)/2\kappa}) \le 1/2,$$

we obtain the ordinary differential inequality

$$\frac{1}{2}E \le C(-E_t)^{(4+\alpha)/4},$$

which can be written as

$$\frac{dE}{dt} + CE^\nu \le 0 \quad \text{with } \nu = \frac{4}{4+\alpha}.$$

The last inequality leads to the estimate

$$E^{1-\nu}(\rho) \le E^{1-\nu}(0) - C(1-\nu)t,$$

which completes the proof of the theorem with $T = E_0{}^{1-\nu}/C(1-\nu)$. □

Remark 3.2. A large and quite complete class of the uniformly subsonic flows with free boundaries was considered in [83, 87, 255, 262]. The existence and uniqueness theorems were proved, and some asymptotic properties of solutions were studied. The fluids had fictitious (i.e., not physical) state equation $p = p(\rho)$ which provided fulfillment of the inequality

$$M(q) \le M_0 = \text{const} < 1.$$

3.3 Nonhomogeneous boundary conditions: Plane case. We consider here the local properties of the plane gas jet presented in Figure 3.2. It is assumed that the free boundary of the jet is defined by the prescribed module of velocity

$$|\mathbf{v}| = q_0(\varphi) \le q_s \quad \text{and, correspondingly,} \quad U(\varphi, 1) = U(q_0) \equiv U_0(\varphi)$$

with $q_0(\varphi) \not\equiv \text{const.}$

According to formula (3.14) with $\lambda = 0$ and (3.9), (3.15), for the function $U(\varphi, \psi)$, the boundary-value problem

$$\frac{\partial}{\partial \varphi}\left(K(U)\frac{\partial U}{\partial \varphi}\right) + \frac{\partial^2 U}{\partial \varphi^2} = 0, \qquad (\varphi, \psi) \in Q, \tag{3.30}$$

$$\frac{\partial U}{\partial \psi}(\varphi, 0) = 0, \quad U(\varphi, 1) = U_0(\varphi). \tag{3.31}$$

is posed. We consider the weak solutions $U(\varphi, \psi)$ of problem (3.30), (3.31), which are a priori known to satisfy the conditions

$$0 \le U(\varphi, \psi) \le C_0,$$

$$E(t) \equiv \int_t^\infty \int_0^1 \left(K(U)U_\varphi{}^2 + U_\psi{}^2\right) d\psi d\varphi \le E_0. \tag{3.32}$$

Existence of such solutions is established in [9, 10, 295].

Let the given function $K(U)$ and the velocity module $q_o(\varphi)$ (the latter coincides with $U_0(\varphi)$ in (3.31)) be subject to the conditions

$$0 \le K(U) \le C_0 U^\alpha \quad \text{with some } 0 < \alpha \le 1,$$

$$H(U_0(\varphi)) \equiv U_0{}^2 + \left(\int_\varphi^\infty \left(U_0{}^2 + |U_0'|^{4/(2-\alpha)}\right) d\varphi\right)^{4/(4+\alpha)} \le \varepsilon\left(1 - \frac{\varphi}{T}\right)_+^{2/\alpha} \tag{3.33}$$

with some positive constants ε, T, C_0. This condition is obviously fulfilled for the function

$$U_0(\varphi) = \left(1 - \frac{\varphi}{T}\right)_+^{2/\alpha}.$$

Theorem 3.2. *Let $U(\varphi, \psi)$ be a weak solution of equation* (3.30) *satisfying* (3.31), *and let conditions* (3.32), (3.33) *be fulfilled. Then there exist constants ε_*, E_*, and C such that if $\varepsilon \le \varepsilon_*$, $E_0 \le E_*$, then*

$$0 \le U(\varphi, \psi) \le C\left(1 - \frac{\varphi}{T}\right)_+^{2/\alpha} \quad \textit{for } (\varphi, \psi) \in Q \tag{3.34}$$

and, in particular,

$$U(\varphi, \psi) \equiv \theta(\varphi, \psi) \equiv 0 \quad \textit{for } 0 \le \psi \le 1,\ T \le \varphi.$$

The physical meaning of the conclusion is that beginning with $\varphi = T$, the motion of the whole of the jet is sonic. Correspondingly, if $q_0(T) = q_s$ on the boundary of the jet, the free boundary is a straight line. One can also interpret Theorem 3.2 in the following way: for any constants E_0 and sufficiently small ε, there exists a finite T such that (3.34) is true.

Proof. Every weak solution of problem (3.30), (3.31) satisfies the energy relation

$$E(t) \equiv \int_t^\infty \int_0^1 \left(K(U)U_\varphi{}^2 + U_\psi{}^2\right) d\varphi d\psi = I_1 + I_2, \tag{3.35}$$

where

$$I_1 = -\int_0^1 K(U)U_\varphi(t,\psi)(U - U_0)d\psi,$$
$$I_2 = \int_t^\infty \int_0^1 K(U)U_\varphi(\varphi,\psi)U'_0 d\varphi d\psi,$$

and

$$\frac{dE}{dt} \equiv E_t = -\int_0^1 \left(K(U)U_\varphi{}^2 + U_\psi{}^2\right) d\psi \le 0.$$

Notice that

$$(U(\varphi,\psi) - U_0(\varphi))^2 \le \int_0^1 U_\psi{}^2 d\psi \le -E_t, \qquad U^2 \le 2\left(U_0{}^2 - E_t\right). \tag{3.36}$$

Using (3.18), (3.36), we estimate the terms I_1, I_2 on the right-hand side of (3.35) as

$$\begin{aligned} |I_1| &\le \left(\int_0^1 KU_\varphi^2 d\psi\right)^{1/2} \left(\int_0^1 U^\alpha (U - U_0)^2 d\psi\right)^{1/2} \\ &\le C(-E_t)\left(U_0^2 - E_t\right)^{\alpha/4} \le c\left(U_0^2 - E_t\right)^{(4+\alpha)/4}, \end{aligned} \tag{3.37}$$

$$\begin{aligned} |I_2| &\le C\left(\int_t^\infty \int_0^1 KU_\varphi^2 d\varphi d\psi\right)^{1/2} \left(\int_t^\infty \int_0^1 U^\alpha |U'_0|^2 d\varphi d\psi\right)^{1/2} \\ &\le CE^{1/2}\left(\int_t^\infty \int_0^1 (U_0^2 - E_t)^{\alpha/4} |U'_0|^2 d\varphi d\psi\right)^{1/2} \\ &\le CE^{1/2}\left(E + \int_t^\infty |U'_0|^2 d\varphi\right)^{\alpha/4} \left(\int_t^\infty |U'_0|^{4/(2-\alpha)} d\varphi\right)^{(2-\alpha)/4} \\ &\le E/2 + C\int_t^\infty \left(U_0^2 + |U'_0|^{4/(2-\alpha)}\right) d\varphi \end{aligned} \tag{3.38}$$

with some constant $C(\alpha, C_0)$. Gathering (3.35), (3.37), (3.38), we obtain the inequality

$$E(t) \le C\left[\left(U_0^2 - E_t\right)^{(4+\alpha)/4} + \int_t^\infty \left(U_0^2 + |U'_0|^{4/(2-\alpha)}\right) d\varphi\right].$$

Raising both sides to the power $4/(\alpha+4)$, we get the standard ordinary differential inequality

$$\frac{dE}{dt} + C_1 E^\nu \le C_2 H(U_0(t)) \le \varepsilon C_2 \left(1 - \frac{t}{T}\right)_+^{\nu/(1-\nu)}$$

with $\nu = 4/(4+\alpha)$ and some constants $C_i = C_i(\alpha, C_0)$.

The proof is completed by application of results from Section 2 of Chapter 1. □

Remark 3.3. The use of hodograph variables in the study of steady subsonic flows of a nonviscous fluid, past a given symmetric convex profile on a plane, leads to obstacle-type problems on unbounded domains which, as was mentioned in Section 7 of Chapter 1, motivated the study of solutions with compact support (see Brezis and Stampaccia [102, 103], Brezis [97], Brezis and Duvaut [98], Díaz and Dou [130], Díaz [127], and Santos [286]).

4 Two-phase filtration of immiscible incompressible fluids

In this section, we study the behaviour of supports of solutions to systems of equations of combined type that describe simultaneous filtration motions of two immiscible incompressible fluids in nonhomogeneous anisotropic porous media. Systems of this type arise, for example, from the mathematical modelling of the process of *secondary recovery of the oil reservoir*. Let us consider a reservoir in a porous medium (in our notation the set $\Omega \subset \mathbb{R}^3$) whose pores contain a hydrocarbon component (the "oil"). The process of secondary recovery consists in injecting a cheap liquid, for example water, into the porous medium in order to push the oil towards the production wells. During the time when the pressure exceeds the bubble pressure of the oil, the flow in the reservoir belongs to the type we are interested in.

4.1 Mathematical model. The mathematical model of such a motion is based on the following equations [36, 46, 110, 168, 226]:

1. the Darcy–Muskat law:

$$\mathbf{v}_i = -K_0(x) k_i(s_i) \nabla(p_i + \rho_i g h); \tag{4.1}$$

2. the mass balance law:

$$\frac{\partial (m s_i \rho_i)}{\partial t} + \operatorname{div}(\rho_i \mathbf{v}_i) = \rho_i q_i \quad (i = 1, 2), \quad s_1 + s_2 = 1, \tag{4.2}$$

3. Laplace's law:

$$p_2 - p_1 = p_c(s_1, x). \tag{4.3}$$

In these relations, $\mathbf{v}_i$, ρ_i, p_i, and s_i are velocity, density, pressure, and saturation of the ith phase. $m(x)$ is the porosity of the medium, $K_0(x)$ is the tensor of absolute permeability.

$$k_i(s) = k_{0i}/\mu_i, \quad i = 1, 2,$$

is the relative permeability, with $k_{0i}(s_i) \geq 0$, $k_{0i}(0) = 0$, and $k_{0i}(s_i)$ being an increasing function of s_i. μ_i is the viscosity, g is the free-fall acceleration, and h is the distance to a fixed horizontal reference plane. $q_i(x, t)$ $(q = q_1 = -q_2)$ is the injected mass of the ith phase, p_c is the capillary pressure (an increasing function of s_1 so that $\partial p_c/\partial s \leq 0$ and $p(x, s_1) = 0$).

Following [36, 46], let us reduce equations (4.1)–(4.3) to some elliptic–parabolic system for the pair of functions: the saturation $-s(x, t) = s_1(x, t)$ and the reduced pressure $p(x, t)$. The latter is defined as

$$p = p_1 - \int_0^s \frac{k_2}{k}\frac{\partial p_c}{\partial s} d\xi + \rho_1 g h, \quad k(s) = k_1 + k_2.$$

Dividing the ith equation of (4.2) by ρ_i = const and summing the results, we obtain the equation

$$\operatorname{div} \mathbf{v} = q + q_2 = 0, \qquad \mathbf{v} = \mathbf{v}_1 + \mathbf{v}_2.$$

Using (4.1), we get the relations

$$-\mathbf{v} = (K\nabla p + \mathbf{f}), \quad K(x, s) = k(s)K_0(x), \tag{4.4}$$

$$-\mathbf{v}_1 = K_0 a \nabla s + k_1 \nabla p + \mathbf{f}_0 \equiv K_0 a \nabla s - b\mathbf{v} + \mathbf{f}, \tag{4.5}$$

where

$$\mathbf{f} = \mathbf{f}_0 + K_2 \nabla p_c + K(\rho_1 - \rho_2) g \nabla h,$$

$$K_i = k_i K_0(x),$$

$$\mathbf{f}_0 = K \int_s^1 \frac{k_2}{k} \nabla (\partial p_c/\partial s)\, d\xi, \quad \mathbf{f} = \mathbf{f}_0 - b\mathbf{f}, \quad K = kK_0, \tag{4.6}$$

$$a(x, s) = \left|\frac{\partial p_c}{\partial s}\right| \frac{k_1 k_2}{k_1 + k_2}, \qquad b(s) = K_1 K^{-1} = \frac{k_1}{k}. \tag{4.7}$$

Substituting (4.5) into (4.2), we come to the system of equations for the functions (s, p),

$$m(x)\frac{\partial s}{\partial t} = \operatorname{div}(K_0(x)a(x, s)\nabla s + K_1(x, s)\nabla p + \mathbf{f}_0) + q, \tag{4.8}$$

$$0 = \operatorname{div}(K(x, s)\nabla p + \mathbf{f}(x, s)). \tag{4.9}$$

System (4.8), (4.9) can be written in an equivalent form as a system of equations for the functions $(s, p, \mathbf{v})$,

$$m\frac{\partial s}{\partial t} = \operatorname{div}(K_0 a \nabla s - b\mathbf{v} + \mathbf{f}) + q,$$
$$0 = \operatorname{div}(K\nabla p + \mathbf{f}), \quad -\mathbf{v} = K\nabla p + \mathbf{f}.$$

According to (4.4)–(4.7), the coefficients of system (4.8), (4.9) are defined by the functional parameters of the original model. These are the relative phase permeability $k_{0i} = \mu_i K_i$, the capillary pressure p_c, and the tensor of absolute permeability $K_0(x)$. The latter characterizes the porous medium and is assumed symmetric: $K_{0,ij} = K_{0,ji}$. It is also assumed that K_0 is positive definite,

$$\forall x, \xi \in \mathbb{R}^3 \setminus \{0\}, \quad \frac{1}{C}|\xi|^2 \leq (K_0(x)\xi, \xi) \leq C|\xi|^2. \tag{4.10}$$

Another assumption is that

$$\frac{1}{C} \leq m(x) \leq C \tag{4.11}$$

for some positive constant C. The behavior of the functional parameters k_1, k_2, p_c is defined from the physical experiments and is described by the relations

$$k_1(s) = k_1^0(s)s^{\lambda_1}, \quad k_2(s) = k_2^0(s)(1-s)^{\lambda_2}, \quad \lambda_i > 0, \tag{4.12}$$
$$-\frac{\partial p_c(x,s)}{\partial s} = p_0(x)p_c'(s)s^{\lambda_3}(1-s)^{\lambda_4}, \quad \lambda_i \geq 0,$$
$$\frac{1}{C} \leq \left(k_i^0, p_0, p_c'\right) \leq C, \quad i = 1, 2. \tag{4.13}$$

Assumptions (4.10)–(4.13) were proposed and justified in [46].

Conditions (4.10), (4.12) assure the uniform ellipticity of equation (4.9) since

$$\frac{1}{C} \leq k_1 + k_2 \leq C$$

and

$$\forall (\xi, r) \in \mathbb{R}^3 \times \mathbb{R}, \quad \frac{1}{C}|\xi|^2 \leq (K(x,r)\xi, \xi) \leq C|\xi|^2.$$

Under assumptions (4.12)–(4.13), the coefficient $a(x, s)$ in equation (4.8) admits, according to (4.7), the representation

$$a = a_0 s^{\alpha}(1-s)^{\beta} \tag{4.14}$$

with

$$\alpha = \lambda_1 + \lambda_3, \qquad \beta = \lambda_2 + \lambda_4, \qquad \frac{1}{C} \leq a_0 \leq C.$$

Hence the fulfillment of the inequalities

$$-1 < \alpha = \lambda_1 + \lambda_3 < 0 \quad (\text{or } -1 < \beta = \lambda_2 + \lambda_4 < 0)$$

implies that $a \to \infty$ when $s \to 0$ (or $s \to 1$) thus initiating the fast diffusion process. As opposed to this situation, the inequalities

$$0 < \alpha = \lambda_1 + \lambda_3 \quad (\text{or } 0 < \beta = \lambda_2 + \lambda_4)$$

provide that $a \to 0$ when $s \to 0$ (or $s \to 1$), and the process of slow diffusion takes place.

System (4.8), (4.9) is a complete quasilinear system involving a uniformly elliptic equation for $p(x, t)$ and a parabolic equation for $s(x, t)$ that degenerates at the levels $s = 0, 1$. It does not have a definite type and, moreover, does not fall into the scope of the classical Cauchy–Kovalevsky theorem.

Note that if the capillary pressure does not depend on x, $p_c(x, s) = p_c(s)$, then according to (4.6), we have $\mathbf{f}_0 = 0$, and system (4.8), (4.9) takes on the form

$$m\frac{\partial s}{\partial t} = \operatorname{div}(K_0 a \nabla s + K_1 \nabla p) + q, \tag{4.15}$$

$$0 = \operatorname{div}(K \nabla p + \mathbf{f}). \tag{4.16}$$

The simplest version of system (4.8), (4.9) is attained under the following assumptions:

1. The porous medium is homogeneous and isotropic, that is,

   $$m(x) = 1, \qquad p_c(x, s) = p_c(s), \qquad K_0(x) = I.$$

   There are no mass sources; $q = q_2 = 0$,

2. Either the motion is two dimensional in the plane $x_3 = \text{const}$ ($\mathbf{g} = (0, 0, g)$) or the densities of the fluids coincide.

If this is the case, taking into account formulas (4.5)–(4.7), one can rewrite system (4.8), (4.9) in the form

$$\frac{\partial s}{\partial t} = \operatorname{div}(a(s)\nabla s + k_1(s)\nabla p) \equiv -\operatorname{div} \mathbf{v}_1, \tag{4.17}$$

$$0 = \operatorname{div}(k(s)\nabla p) \equiv -\operatorname{div} \mathbf{v}. \tag{4.18}$$

Let us pose the following initial condition for the saturation s:

$$s(x, 0) = s_0(x), \quad x \in \Omega. \tag{4.19}$$

It is assumed that the boundary $\Gamma = \partial\Omega$ can be split into three parts: the impermeable part Γ_1, the boundary of the injection wells Γ_e, where the water is pumped in, and the boundary Γ_s of the production wells, where the oil is pumped out, so that $\Gamma = \Gamma_1 \cup \Gamma_e \cup \Gamma_s$. The natural boundary conditions for system (4.8), (4.9) are

$$\begin{aligned} &\Gamma_1 : -\mathbf{v}_1 \cdot \mathbf{n} = 0, \qquad -\mathbf{v} \cdot \mathbf{n} = 0,\\ &\Gamma_e : s = 0, \qquad -\mathbf{v} \cdot \mathbf{n} = g(x,t), \quad \text{or } s = 0,\ p = p_0(x,t),\\ &\Gamma_s : s \geq 0, \quad -\mathbf{v}_1 \cdot \mathbf{n} \geq 0, \quad s\mathbf{v} \cdot \mathbf{n} = 0, \end{aligned} \tag{4.20}$$

$$\Gamma_s : -\mathbf{v} \cdot \mathbf{n} = \mu(p - h(x,t)) \quad \text{on } \Gamma_s, \tag{4.21}$$

where g, h, and μ are given functions and $\mathbf{n}$ is the unit outer normal vector to $\partial\Omega$.

The unilateral boundary condition (4.20) reflects the fact that the water flux can only be directed outward Ω on Γ_1. The reader is referred to [110] and [167, 168] for a discussion of this condition.

The questions of existence, uniqueness, and stability of weak solutions to the main initial-boundary value problems for system (4.8), (4.9) are studied in [28, 29, 36, 46, 110, 168].

Works [36, 46] contain the proof of existence of the weak solutions s, p to system (4.8), (4.9) that satisfy the equations in the sense of the integral identities and are such that $s \in [0, 1]$,

$$\begin{aligned} &v(x,t) \equiv v[s(x,t)] = \int_0^{s(x,t)} \sqrt{a(\xi)}\, d\xi \in L^2\left(0, T;\ W^{1,2}(\Omega)\right),\\ &p(x,t) \in L^\infty(Q_T) \cap L^\infty\left(0, T;\ W^{1,q}(\Omega)\right), \quad q \in [2, \infty]. \end{aligned} \tag{4.22}$$

4.2 Stabilization to a constant profile in a finite time. First of all, note that the change of the dependent variable

$$u(s) = \int_0^s a(\xi) d\xi \equiv A(s), \quad s = \Psi(u), \quad u \in (0, A(1)),$$

transforms equation (4.15) into

$$\frac{\partial \Psi(u)}{\partial t} = \operatorname{div}\left(K_0 \nabla u + K_1(x,u) \nabla p\right) + f. \tag{4.23}$$

Equations of this type are studied in Chapter 2.

Let us first consider the behaviour of solutions $\mathbf{U} = (u, p)$ near the level $u = 0$ assuming that $u \in [0, \mathrm{A}(1)/2]$. Condition (4.14) yields the restrictions

$$\frac{1}{C}|r|^\gamma \leq \Psi(r) \leq C|r|^\gamma \quad \text{with } \gamma = \frac{1}{1-\alpha} > 1, \tag{4.24}$$

$$|k_1[s(u)]| \leq C|r|^\omega \quad \text{with } \omega = \frac{\lambda_1}{1+\alpha} \tag{4.25}$$

for every $r \in [0, A(1)]$. By (4.19)–(4.21), the solution $\mathbf{U} = (u, p)$ satisfies the initial and boundary conditions

$$u(x,0) = u_0(x) = A(s_0(x)) \quad \text{in } \Omega, \tag{4.26}$$

$$\begin{aligned} &u\left(K_0 \nabla u + K_1(x,u) \nabla p\right) \cdot \mathbf{n} = 0 \quad \text{on } \Gamma = \partial\Omega\\ &(u = 0 \quad \text{on } \Gamma_1 \subset \Gamma,\ \operatorname{meas} \Gamma_1 > 0). \end{aligned} \tag{4.27}$$

Theorem 4.1. *Let conditions* (4.10), (4.11), (4.14), (4.24), (4.27) *be fulfilled, and let* $\mathbf{U} = (u, p)$ *be a weak solution to system* (4.16), (4.23) *with* $f = 0$. *Let the solution be such that*

$$\forall t \geq 0, \quad \|\nabla p(\cdot, t)\|_{L^q(\Omega)} \leq M \quad \textit{for some } q > 2. \tag{4.28}$$

Assume that the inequalities

$$q \geq \frac{2(\gamma + k)}{\gamma - 1} \quad \textit{and} \quad \omega \geq \frac{\gamma + k}{2}\frac{q - 2}{q} - \frac{k - 1}{2} \tag{4.29}$$

hold with

$$k = \begin{cases} \max(1, N(\gamma - 1/2 - 1)) & \textit{if } \gamma + 1 \geq \dfrac{2N}{N - 2}, \\ 1 & \textit{otherwise}. \end{cases}$$

Then there exists $t^* = t^*(M, q, \alpha, \omega, \Omega, A(1)) < \infty$ *such that*

$$u(x, t) \equiv 0 \quad \textit{for } x \in \Omega,\ t \geq t^*$$

provided M *is sufficiently small.*

Let $f \not\equiv 0$, $f \in L^2(0, \infty; L^{(\gamma+k)/k}(\Omega))$, *and*

$$\|f(\cdot, t)\|^{(1+k)/k}_{L^{(\gamma+1)/\gamma}(\Omega)} \leq C_1 \left(T_f - t\right)_+^{\nu/(1-\nu)} \tag{4.30}$$

with $\nu = \frac{k+1}{\gamma+k}$, $T_q \geq t^*$, *and an appropriately small constant* C_1. *Then there exists a constant* C_2 *depending on* C_1, Ω, γ, k, *such that*

$$\|u(\cdot, t)\|^{\gamma+k}_{L^{(\gamma+k)}(\Omega)} \leq C_2 \left(T_q - t\right)_+^{\nu}$$

and $u(\cdot, t) = 0$ *in* Ω *for all* $t \geq T_f$.

The assertions of Theorem 4.1 are interpreted as follows:

1. if the domain Ω is initially occupied by the mixture "oil–water" and on the boundary of Ω conditions (4.26), (4.27) hold, then the "oil" component is completely forced out by "water" at the time t^*;
2. if Ω contains a source of the "oil," $f \not\equiv 0$, then the "oil" is completely forced out of Ω at the moment $T_f \geq t^*$ when the source vanishes.

Proof. We follow the general method explained in Chapter 2. Let us introduce the energy functions

$$y(t) := \int_\Omega m(x) G(k, u)\, dx,$$

$$E(t) := \int_\Omega u^{k-1} K_0(x) \nabla u \cdot \nabla u\, dx,$$

$$G(k, u) := \Psi(r)|r|^{k-1} r - k \int_0^r \Psi(\tau)|\tau|^{k-1}\, d\tau.$$

It follows from the assumptions (4.10), (4.11), (4.14), (4.24), (4.25) that

$$\frac{1}{C}\|u(\cdot,t)\|^{\gamma+k}_{L^{\gamma+k}(\Omega)} \le y(t) \le C\|u(\cdot,t)\|^{\gamma+k}_{L^{\gamma+k}(\Omega)}, \tag{4.31}$$

$$\frac{1}{C}\int_\Omega u^{k-1}|\nabla u|^2\,dx \le E(t) \le C\int_\Omega u^{k-1}|\nabla u|^2\,dx. \tag{4.32}$$

Multiplying equation (4.23) by u^k, integrating the result by parts, and taking into account the boundary condition (4.27), we obtain the energy relation (the integration-by-parts formula)

$$\frac{dy(t)}{dt} + (k-1)E(t) = I_1 + I_2, \tag{4.33}$$

where

$$I_1 = -(k-1)\int_\Omega K_1\nabla p u^{k-1}\nabla u dx, \qquad I_2 = \int_\Omega f u^k dx.$$

Using (4.25), (4.31), (4.32) and Hölder's inequalities, we estimate the term I_1 as

$$\begin{aligned}|I_1| &\le CE^{1/2}\|\nabla p\|_{L^q(\Omega)}\|u\|^{\lambda}_{L^{2\lambda q/(q-2)}(\Omega)} \le CE^{1/2}My^{\nu/2}\\ &\le \varepsilon E(t) + C^2/\varepsilon M^2 y^\nu, \quad \varepsilon\in(0,1).\end{aligned} \tag{4.34}$$

Here $C = C(\omega,\Omega,k,\gamma,A(1))$, $\lambda = \omega + (k-1)/2$, and $\nu = \frac{k+1}{\gamma+k}$. To derive this inequality, we relied on the following byproduct of conditions (4.29):

$$2\lambda q \ge (q-2)(\gamma+k) \quad \text{and} \quad (q-2)(\gamma+k) \ge q(1+k).$$

Gathering (4.33) and (4.34) with $q = 0$ and $2\varepsilon = k-1$, we arrive at the inequality

$$\frac{dy}{dt} + \frac{k-1}{2}E \le \frac{2C^2}{k-1}M^2y^\nu.$$

Now applying the inequality (see (2.44) in Chapter 2)

$$y^\nu \le CE, \quad \nu = \frac{k+1}{\gamma+k},$$

and choosing the constant M small enough, we obtain the ordinary differential inequality

$$y' + Cy^\nu \le 0, \quad \nu < 1.$$

The proof of the first assertion of the theorem is completed in a routine way by application of Lemma 2.2.

To prove the second assertion of the theorem with $f \not\equiv 0$, we apply Hölder's and Young's inequalities and (4.30) to estimate I_2 in the following way:

$$\begin{aligned}|I_2| &\le \|u\|^k_{L^{k+\gamma}(\Omega)}\|q\|_{L^{(k+\gamma)}(\Omega)} \le Cy^{k/(k+\gamma)}\|q\|_{L^{(k+\gamma)/\gamma}} \\ &\le \varepsilon y^{\nu} + C(\varepsilon)\|q\|^{(k+1)/k}_{L^{(k+\gamma)/\gamma}} \le \varepsilon y^{\nu} + C(\varepsilon)C_1\left(T_f - t\right)_+^{\nu/(1-\nu)}, \quad \varepsilon \in (0,1).\end{aligned} \tag{4.35}$$

Gathering (4.33), (4.34), (4.35) and choosing ε appropriately small, we come to the inequality

$$y' + Cy^{\nu} \le CC_1\left(T_f - t\right)_+^{\nu/(1-\nu)} \le 0$$

and apply Lemma 2.3. □

Remark 4.1. The same effect of stabilization in a finite time to a constant profile can be established in a much simpler way, and under the sole condition (4.24) on the coefficients, the solution to system (4.17), (4.18) is assumed to satisfy the boundary conditions

$$\mathbf{v}_1 \cdot \mathbf{n} = 0 \quad \text{and} \quad s = 0 \quad (\text{or } \mathbf{v} \cdot \mathbf{n} = 0),$$

If this is the case, the energy relation takes on the form

$$\frac{dy(t)}{dt} + (k-1)\int_{\Omega} u^{k-1}|\nabla u|^2\,dx = (k-1)I_1,$$

and one can readily verify that

$$\begin{aligned}I_1 &= \int_{\Omega} k_1 u^{(k-1)}\nabla u \cdot \nabla p\,dx = \int_{\Omega} k\nabla F(u)\cdot\nabla p\,dx \\ &= \int_{\partial\Omega} F\nabla p \cdot \mathbf{n}\,d\Gamma = 0,\end{aligned}$$

where

$$F(u) = \int_0^u \frac{\tau^{k-1}k_1(\tau)}{k(\tau)}\,d\tau.$$

Remark 4.2. An assertion similar to Theorem 4.1 is true for the value $u = A(1)$, respectively, $s = 1$, provided u is substituted by $\overline{u} = u - A(1/2)$ in conditions (4.26), (4.27). It is sufficient to claim that inequalities (4.24), (4.25) hold only near the level $r = 0$, say, for $r \in (0, A(1)/2)$. Equation (4.22) is multiplied by the function $v = \min(u^k, A(1)/2)$, which gives the energy relation.

4.3 Finite speed of propagation and the waiting time effect. Let the pair $\mathbf{U} = (u, p)$ be a weak solution of equations (4.16), (4.23). In the study of the character of propagation of disturbances, we rely only on local properties of u and p and confine all our considerations to a cylinder $B_{\rho_1}(x_0) \times (0, t_1) \subset \Omega \times (0, t_1)$.

Theorem 4.2. *Let conditions* (4.10), (4.11), (4.14), (4.24), (4.25), (4.28) *be fulfilled. Let* $\mathbf{U} = (u, p)$ *be a weak solution of system* (4.16)–(4.22) *in the cylinder* $B_{\rho_1}(x_0) \times (0, t_1)$. *It is assumed that* Ψ *satisfies condition* (4.24) *with* $\gamma < 1$ *and*

$$\left|\frac{db(u)}{du}\right| \le C|u|^{\mu} \quad \textit{for any } u \in (0, A(1)) \tag{4.36}$$

with $b = k_1/k$ *and* $\mu > 0$ *such that* $\mu \le (\gamma(q-2) - q - 2)/2q$. *Then there exist* $t^* \in (0, T)$ *and a function* $\rho(t)$, $0 \le \rho(t) \le \rho_1$, *such that*

$$u(x,t) = 0 \quad \textit{for all } t \in (0, t^*) \textit{ in } B_{\rho(t)}(x_0)$$

if $u_0 = u(x, 0) = 0$ *in* $B_{\rho_1}(x_0)$ *and* $q(x,t) = 0$ *in* $B_{\rho_1}(x_0) \times (0, T)$.

This assertion is interpreted as follows. Let the domain $B_{\rho_0} \subset \Omega$ be initially occupied by only one of the liquids. Say, by the oil, i.e., the water saturation $s(x, 0) = 0$, and, correspondingly, $u(x, 0) = 0$. Then regardless of the boundary conditions on $\partial\Omega$, under condition (4.40) the oil component is forced out from B_{ρ_0} only after a nonzero time t^*.

Proof. The method we follow is the same as that of Section 2 in Chapter 3. We introduce the energy functions

$$b(\rho, t) := \operatorname*{ess\,sup}_{0\le\tau\le t} \int_{B_\rho(x_0)} |u(x,\tau)|^{\gamma+1}\, dx,$$

$$E(\rho, t) := \int_0^t \int_{B_\rho(x_0)} K_0 \nabla u \cdot \nabla u \, dx d\tau,$$

multiply equation (4.23) by u, and integrate the result by parts in $B_\rho(x_0) \times (0, t)$. Using inequality (4.24), we conclude that

$$Cb(\rho, t) + E(\rho, t) \le Cb(x, 0) + I_1 + I_2 + I_3, \tag{4.37}$$

where

$$I_1 = \int_0^t \int_{S_\rho} u K_0 \nabla u \cdot \mathbf{n} dS\, d\tau,$$

$$I_2 = \int_0^t \int_{B_\rho} u \operatorname{div}(K_0 b(u) k \cdot \nabla p)\, dx d\tau,$$

$$I_3 = \int_0^t \int_{B_\rho} uq \, dx d\tau.$$

Note that in the case under consideration $b(\rho, 0) = 0$ if $\rho \le \rho_1$ and $I_3 = 0$. Applying Hölder's inequality and using (4.10), we obtain

$$\begin{aligned}|I_1| &\le C \int_0^t \int_{S_\rho} |u||\nabla u| dS d\tau \\ &\le C \left(\int_0^t \int_{S_\rho} |u|^2 dS d\tau\right)^{1/2} \left(\frac{\partial E}{\partial \rho}\right)^{1/2}.\end{aligned}$$

Now we use the interpolation-trace inequality (2.31) of Chapter 3,

$$\left(\int_0^t \int_{S_\rho} |u|^2 dx d\tau\right)^{1/2} \leq C\left(E) + tb^{2/(\gamma+1)}\right)^{\theta/2} \left(tb^{2/(\gamma+1)}\right)^{(1-\theta)/2}$$
$$\leq t^{(1-\theta)/2} K(T)\,(E+b)^{\theta/2+(1-\theta)/(\gamma+1)},$$

where

$$K(T) = \max\left(1, T^{\theta/2}\right) \max\left(1, b^{\theta(1-\gamma)/2(\gamma+1)}(\rho_1, T)\right)$$

and

$$\theta = \frac{N(1-\gamma+\gamma+1)}{N(1-\gamma+2\gamma+2)}.$$

Using Young's inequality, we then deduce that

$$|I_1| \leq \varepsilon\,(E+b) + C_\varepsilon (CK(T)t^\kappa)^{1/(1-\beta)} \left(\frac{\partial E}{\partial \rho}\right)^{1/\nu}, \tag{4.38}$$

where $\kappa = 1-\theta/2$, $\beta = \theta + \frac{1-\theta}{\gamma+1}$, and $\nu = 2(1-\beta)$. Using equation (4.16) and (4.36), we have for I_2

$$\begin{aligned}
|I_2| &= \left|\int_0^t \int_{B_\rho} u \operatorname{div}(bK_0 k\cdot\nabla p)\, dxdt\right| \\
&= \left|\int_0^t \int_{B_\rho} \left(ub'(u)\nabla u K_0 \cdot \nabla p\right) dxdt\right| \\
&\leq C \int_0^t \int_{B_\rho} |u|^{1+\mu} |\nabla p||\nabla u|\, dxdt \qquad (4.39)\\
&\leq CMt^{(q-2)/q} E^{1/2}(\rho,t) \left(\int_0^t \int_{B_\rho} |u|^{2q(\mu+1)/(q-2)}\, dxdt\right)^{(q-2)/2q} \\
&\leq CMt^{3(q-2)/2q} E^{1/2}(\rho,t) b^{(1+\mu)/(1+\gamma)} \\
&\leq CMt^{3(q-2)/2q}\,(E(\rho,t)+b(\rho,t)).
\end{aligned}$$

Gathering (4.37), (4.38), (4.39) with suitable constants $\varepsilon > 0$ and $t^* \geq t$, we come to the differential inequality

$$E \leq b + E \leq Ct^{\kappa/(1-\beta)} \left(\frac{\partial E}{\partial \rho}\right)^{1/\nu}$$

and, correspondingly, obtain the estimate

$$E^{1-\nu}(\rho,t) \leq E^{1-\nu}(\rho_1,t) - C^{-\nu} t^{-\kappa\nu/(1-\beta)}(\rho_1 - \rho),$$

whence the assertion of the theorem with

$$\rho(t) = \rho_1 - E^{1-\nu}(\rho_1,t) C^\nu t^{\nu\kappa/(1-\beta)}.$$

□

Theorem 4.3. *Let us assume that the conditions of Theorem* 4.2 *hold and there exist* $t_q > 0$, $\rho_0 \in (0, \rho_1)$, *and* $\varepsilon > 0$ *(small) such that*

$$b(\rho, 0) + \int_0^{t_q} \int_{B_\rho} |q(x,t)|^{(1+\gamma)/\gamma} \, dxdt \le \varepsilon \, (\rho - \rho_0)_+^{1/(1-\nu)} \tag{4.40}$$

with $\rho_0 \in (0, \rho_1)$ *and* $\nu \in (0, 1)$. *Then there exists* $t^* > 0$ $(t^* \le t_q)$ *such that*

$$u(x, t) = 0 \quad for\ |x| \le \rho_0,\ 0 \le t \le t^*.$$

Proof. We rely on the energy inequality (4.37). By virtue of (4.40), the additional term I_3 admits the estimate

$$|I_3| \le \delta t_q b(\rho, t) + C_\delta \int_0^{t_q} \int_{B_\rho 0} |q|^{(1+\gamma)\gamma} \, dx \, dt \le \varepsilon \, (\rho - \rho_0)_+^{1/(1-\nu)} \tag{4.41}$$

with $\delta \in (0, 1)$. Gathering (4.37), (4.40), (4.41), (4.40), we obtain the standard nonhomogeneous differential inequality of the type (1.20),

$$E \le b + E \le C \left(\frac{\partial E}{\partial \rho} \right)^{1/\nu} + C\varepsilon \, (\rho - \rho_0)_+^{1/(1-\nu)}$$

with $\rho \in (\rho_0, \rho_1)$. The statement of the theorem now immediately follows from results of Section 2. □

5 Flows of gas with density-dependent viscosity

5.1 Main equations. In this section we study the dynamics of supports of solutions to the system of equations of mixed type which describes a motion of a viscous not heat-conductive gas. The laws of balance of mass and momentum can be written in the form

$$\begin{gathered} \frac{\partial \rho}{\partial t} + \operatorname{div}(\rho \mathbf{v}) = 0, \\ D_t \mathbf{v} = \operatorname{div} \mathbf{S} + \rho \mathbf{f}. \end{gathered} \tag{5.1}$$

We assume that the connection between the stress tensor $\mathbf{S}$ and the rate of deformation tensor $\mathbf{D}$ is given by the Stokes law

$$\mathbf{S} = (-p + \lambda \operatorname{div} \mathbf{v}) \, I + 2\mu \mathbf{D}. \tag{5.2}$$

Here I is the unit matrix, p is the pressure, λ and μ are the coefficients of the bulk and shear viscosity. The second law of thermodynamics states that the total entropy of the heat-isolated volume does not increase. This means, in particular, that

$$\mu \ge 0, \quad 3\lambda + 2\mu \ge 0.$$

In the generic case, μ and λ may be functions of the independent thermodynamic parameters density ρ, pressure p, specific inner energy U, temperature θ, entropy

s. In our special case, $U \equiv$ const, $\theta \equiv$ const, and $s \equiv$ const so that the independent thermodynamic parameters are p and ρ. In addition, we accept the state equation

$$p = p(\rho) = \rho^{\gamma}, \quad \gamma \geq 1. \tag{5.3}$$

The gases with the state equation (5.3) are called *perfect* and *barotropic*.

We will assume that μ and λ are given functions of density ρ. It is easy to see that equations (5.1)–(5.3) constitute a complete system of equations for defining the functions $(\mathbf{v}, p, \rho)$. For the one-dimensional motions where $\mathbf{x} = (x, 0, 0)$, $\mathbf{v} = (u(x,t), 0, 0)$, $\mathbf{f} = (f, 0, 0)$, this system can be written in the form

$$\begin{cases} \dfrac{\partial \rho}{\partial t} + \dfrac{\partial(\rho u)}{\partial x} = 0, \\ \rho D_t u \equiv \rho\left(\dfrac{\partial u}{\partial t} + u\dfrac{\partial u}{\partial x}\right) = -\dfrac{\partial p}{\partial x} + \dfrac{\partial}{\partial x}\left(a\dfrac{\partial u}{\partial x}\right) + \rho f(x,t), \quad a = \lambda + \mu. \end{cases} \tag{5.4}$$

At the initial moment $t = 0$, the velocity $u(x, 0)$ and density ρ are given as

$$u(x,0) = u_0(x), \quad \rho(x,0) = \rho_0(x), \quad x \in \Omega. \tag{5.5}$$

Let us introduce the new independent variables (q, t), the *mass Lagrangian coordinates*, by the formulas

$$q = q(\xi) = \int_0^{\xi} \rho_0(\eta)d\eta, \qquad q = q(\xi(x,t)) = q(x,t),$$

$$x = x(\xi,t) = \xi(q) + \int_0^t u(\xi(q),\tau)d\tau, \qquad x(\xi,0) = \xi,$$

$$\frac{dx(\xi,t)}{dt} = u(x,t), \qquad x(\xi,0) = \xi.$$

In the new variables (q, t), problem (5.4), (5.5) becomes

$$\frac{\partial}{\partial t}\left(\frac{1}{\rho}\right) = \frac{\partial u}{\partial q}, \tag{5.6}$$

$$\frac{\partial u}{\partial t} = -\frac{\partial p}{\partial q} + \frac{\partial}{\partial q}\left(\rho a\frac{\partial u}{\partial q}\right) + \rho\tilde{f}(q,t), \tag{5.7}$$

$$\rho(q,0) = \rho_0(q).$$

Note that if the right-hand part $\tilde{f}(q,t) = f(x(q,t),t)$ of (5.7) is given in the original Euler variables (x, t), then in the new Lagrange variables it is an operator over $u(q, t)$.

Let us assume that the pressure and viscosity satisfy the relations

$$a(\rho) = \lambda(\rho) + \mu(\rho) = \rho^{\alpha}, \qquad p(\rho) = \rho^{\gamma}. \tag{5.8}$$

System (5.6), (5.7) can be reduced to a single equation for the density ρ [227]. Indeed, differentiating (5.7) in q and using the formulas

$$\rho a \frac{\partial u}{\partial q} = \rho a \frac{\partial}{\partial t}\left(\frac{1}{\rho}\right) = -\frac{\partial F}{\partial t}, \quad F(\rho) = \int_0^{\rho} \frac{a(\xi)}{\xi}\, d\xi, \quad \frac{\partial^2 u}{\partial q \partial t} = \frac{\partial^2}{\partial^2 t}\left(\frac{1}{\rho}\right),$$

$$u(q,t) + \frac{\partial F}{\partial q} + \int_0^t \left(\frac{\partial p}{\partial q} - \rho \tilde{f}\right) d\tau = u_0(q) + \frac{\partial F(\rho_0(q))}{\partial q} \equiv G(q), \tag{5.9}$$

we come to the equation

$$\frac{\partial^2}{\partial^2 t}\left(\frac{1}{\rho}\right) = \frac{\partial^2}{\partial^2 q}\left(-\frac{\partial F(\rho)}{\partial t} - p\right) + \frac{\partial}{\partial q}\left(\rho \tilde{f}\right),$$

whence we infer, integrating in time,

$$\frac{\partial}{\partial t}\left(\frac{1}{\rho}\right) = \frac{\partial^2}{\partial^2 q}\left(-F(\rho) - \int_0^t p(q,\tau) d\tau\right) + \int_0^t \frac{\partial}{\partial q}\left(\rho \tilde{f}\right) d\tau + g(q). \tag{5.10}$$

In this relation, $g(q) = G_q(q)$.

It is convenient to deal with pressure $p = \rho^{\gamma}$ instead of density ρ. Equation (5.10) rewrites for p as

$$\begin{aligned} &\frac{\partial}{\partial t}\left(\frac{1}{1-\gamma} p^{(1-\gamma)/\gamma} + \frac{1}{2}\left(\int_0^t \frac{\partial p(x,\tau)}{\partial q} d\tau\right)^2\right) + \frac{1}{\gamma} p^{(\alpha-\gamma)/\gamma}\left(\frac{\partial p}{\partial q}\right)^2 \\ &\quad = \frac{\partial}{\partial q}\left(p \int_0^t \frac{\partial p(x,\tau)}{\partial q} d\tau + \frac{1}{\gamma} p^{\alpha/\gamma} \frac{\partial p}{\partial q}\right) + \tilde{g}(q) p, \end{aligned} \tag{5.11}$$

where

$$\tilde{g}(q) = g(q) + \int_0^t \frac{\partial}{\partial q}\left(\rho \tilde{f}\right) d\tau \equiv \frac{\partial G(q)}{\partial q} + \int_0^t \frac{\partial}{\partial q}\left(\rho \tilde{f}\right) d\tau.$$

We will consider the weak nonnegative bounded solutions of (5.11) defined in $Q = \Omega \times (0,T)$, $\Omega = (-L, L)$, and satisfying the initial condition

$$p(q,0) = p_0(q) \equiv \rho_0^{\gamma}(q), \quad p_0 \begin{cases} \geq 0 & \text{in } \Omega, \\ \equiv 0 & \text{if } |q| \leq \rho_0 < L. \end{cases} \tag{5.12}$$

Definition 5.1. A nonnegative bounded function $p(x,t)$ ($0 \leq p \leq M$), defined in Q, is said to be a weak solution of (5.11), (5.12) if the following conditions hold:

$$\frac{\partial}{\partial q}\left(p^{(\alpha+\gamma)/2\gamma}\right) \in L^2\left((0,T), L^2(\Omega)\right),$$

$$\int_0^t \frac{\partial p}{\partial q} d\tau \in L^1\left((0,T), L^2(\Omega)\right),$$

$$\lim_{t\to 0} \|p(q,t) - p_0(q)\|_{L^{(\gamma-1)/\gamma}(\Omega)} = 0,$$

and for every test function $\varphi \in C^\infty((0,T), C_0^\infty(\Omega))$,

$$\begin{aligned}
&\int_Q -\left(\frac{1}{1-\gamma}p^{(1-\gamma)/\gamma} + \frac{1}{2}\left(\int_0^t \frac{\partial p(x,\tau)}{\partial q}d\tau\right)^2\right)\frac{\partial \varphi}{\partial t}dqd\tau \\
&\quad + \int_Q \left(\frac{1}{\gamma-1}p^{\alpha/\gamma}\frac{\partial p}{\partial q} + p\int_0^t \frac{\partial p(q,\tau)}{\partial q}d\tau\right)\frac{\partial \varphi}{\partial q}dqd\tau \\
&= \int_Q \left(\frac{1}{\gamma}p^{(\alpha-\gamma)/\gamma}\left(\frac{\partial p}{\partial q}\right)^2 - p\tilde{g}\right)\varphi dqd\tau \\
&\quad + \frac{1}{\gamma-1}\int_Q p_0^{(\gamma-1)/\gamma}\varphi(q,0)dq.
\end{aligned} \tag{5.13}$$

The existence of such solutions is proved in [211, 215, 216, 8, 307]. We will consider problem (5.11), (5.12) under the assumption that

$$f \equiv 0, \qquad |\tilde{g}| = |g| \le M_1. \tag{5.14}$$

5.2 Finite speed of propagation and the waiting time effect.

Theorem 5.1. *Let conditions* (5.8) *with* $1 < \gamma$, $0 < \alpha \le \gamma$ *and* (5.14) *be fulfilled. Then any weak solution* $p(q,t)$ *of problem* (5.11), (5.12) *possesses the property of finite speed of propagation of disturbances in the sense of Definition* 1.1. *That is, there exist an instant* $t^* > 0$ *and a function* $\rho(t)$ *such that* $0 < t^* < T$, $0 < \rho(t) < \rho_0$, $\rho(0) = \rho_0$, *and*

$$p(q,t) = 0 \quad \text{for } |q| < \rho(t),\ 0 \le t \le t^*.$$

The function $\rho(t)$ *is defined by the formula*

$$\rho^\nu(t) = \rho_0^\nu - Ct^\sigma,$$

where the positive constants ν, σ *depend only on* α, γ, *and* C *depend on* α, γ, M, M_1, *and* ρ_0.

Theorem 5.2. *Under the conditions of Theorem* 5.1, *let*

$$\int_{B_\rho} p_0^{(\gamma-1)/\gamma}(q)\,dq \le C\,(\rho-\rho_0)^\nu, \quad \rho \in (\rho_0, L), \quad \nu = \nu(\gamma,\alpha).$$

Then $p(q,t)$ *exhibits the waiting time effect, i.e., there exists an instant* $t^* > 0$ *such that*

$$p(q,t) = 0 \quad \text{if } |q| \le \rho_0,\ t \in [0,t^*].$$

The weak solution of problem (5.11), (5.12) describes the process of discharge of a gas from a domain with positive pressure ($p_0 > 0$, $|q| > L$) into vacuum ($p_0 = 0$, $|q| < L$). Theorems 5.1 and 5.2 describe the dynamics of the free boundary defined as the zero level curve of $p(q,t)$. Formula (5.9) gives the distribution of velocity $u(q,t)$ in the domain $B_{\rho(t)} \times (0,t)$.

The proofs presented below follow [50]. Let us introduce the energy functions

$$
\begin{aligned}
b(\rho,t) &= \int_{B_\rho} \left(\frac{1}{1-\gamma} p^{(1-\gamma)/\gamma} + \frac{1}{2} \left(\int_0^t \frac{\partial p(x,\tau)}{\partial q} d\tau \right)^2 \right) dq, \\
E(\rho,t) &= \int_0^t \int_{B_\rho} \frac{1}{\gamma} p^{(\alpha-\gamma)/\gamma} \left(\frac{\partial p}{\partial q} \right)^2 dq d\tau, \\
\overline{b}(\rho) &= \operatorname*{ess\,sup}_{\tau\in[0,t]} b(\rho,\tau), \qquad B_\rho = \{q \in \Omega : |q| < \rho\}.
\end{aligned}
$$

Let us set in (5.13) $\varphi(q,t) = \zeta_n(|q|)\chi_k(t)$, where the functions $\zeta_n(|q|)$, $\chi_k(t)$ are defined by formulas (2.23) and (2.10) of Chapter 3. Using the properties of these functions, let us pass in (5.13) to the limit as $n \to \infty$, $k \to \infty$. This procedure leads to the energy relation (the integration-by-parts formula)

$$
b(\rho,\tau)|_{\tau=0}^{\tau=t} + E(\rho,t) = I_1 + I_2 + I_3, \tag{5.15}
$$

where

$$
\begin{aligned}
I_1 &= \frac{1}{\gamma} \int_0^t p(\xi,\tau) \frac{\partial p(\xi,\tau)}{\partial q} \Big|_{\xi=-\rho}^{\xi=\rho} d\tau \\
I_2 &= \int_0^t p(\xi,\tau) \left(\int_0^\tau \frac{\partial p(\xi,s)}{\partial q} ds \right) \Big|_{\xi=-\rho}^{\xi=\rho} d\tau, \\
I_3 &= \int_0^t \int_{B_\rho} gp\, dq d\tau
\end{aligned}
$$

As compared with the previous considerations, this passage to the limit is equivalent to the formal integration of equation (5.11) over $B_\rho \times (0,t)$ without multiplication by $p(q,t)$. This difference is explained by the special structure of the terms on the left-hand part of(5.11). From now on we can argue in the standard way. The terms I_i, $i = 1,2,3$, are estimated in exactly the same way as in Theorems 2.1 and 2.2 of Chapter 3:

$$
\begin{aligned}
|I_1| &\le \left(\int_0^t \left[p^{\alpha/\gamma-1} p_q^2 \right] d\tau \right)^{1/2} \left(\int_0^t \left[p^{3\gamma-\alpha/\gamma-1} \right] d\tau \right)^{1/2} \\
&\le C \left(E_\rho\right)^{1/2} \rho^{-\delta_1} t^{\kappa_1} \left(E + \overline{b}\right)^{\lambda_1},
\end{aligned} \tag{5.16}
$$

$$
\begin{aligned}
|I_2| &\le \int_0^t [p] \left(\int_0^\tau \left[p^{\alpha/\gamma-1} p_q^2 \right] ds \times \int_0^\tau \left[p^{1-\alpha/\gamma} \right] ds \right)^{1/2} d\tau \\
&\le C \left(E_\rho\right)^{1/2} \rho^{-\delta_2} t^{\kappa_2} \left(E + \overline{b}\right)^{\lambda_2},
\end{aligned} \tag{5.17}
$$

$$
|I_3| \le M_1 M^{1/\gamma} t^* \overline{b}(\rho) \quad (0 \le t \le t^*) \tag{5.18}
$$

with $[u(\xi)] = u(-\xi) + u(\xi)$, and some positive constants $\delta_i = \delta_i(\gamma,\alpha)$, $\kappa_i = \kappa_i(\gamma,\alpha)$, $\lambda_i = \lambda_i(\gamma,\alpha)$, and $C = C(\rho_0, T, M, M_1, \gamma, \alpha)$.

Gathering (5.15), (5.16)–(5.18) (with t^* small enough), we come to the ordinary differential inequality of the form (1.17) of Chapter 3 in the case of Theorem 5.1 or (1.20) of Chapter 3 in the case of Theorem 5.2. The assertions follow from results of Sections 1.2 and 1.3 of Chapter 3.

The weak solution of problem (5.11), (5.12) describes the process of discharge of a gas from a domain with positive pressure ($p_0 > 0$, $|q| > L$) into vacuum ($p_0 = 0$, $|q| < L$). Theorems 5.1, 5.2 describe the dynamics of the free boundary defined as the zero level curve of $p(q, t)$. Formula (5.9) gives the distribution of velocity $u(q, t)$ in the domain $B_{\rho(t)} \times (0, t)$.

6 Viscous-elastic media

6.1 Nonlinear one-dimensional viscous-elasticity equation. Let us introduce the displacement vector

$$\mathbf{u} = \mathbf{x} - \xi, \qquad \mathbf{x}(0) = \xi$$

so that velocity $\mathbf{v}$ satisfies $\mathbf{v} = \frac{d\mathbf{u}}{dt}$. Calculating the material derivative by the rule

$$D_t \mathbf{v} = \frac{\partial \mathbf{v}}{\partial t} + \mathbf{v} \cdot \nabla \mathbf{v}$$

and assuming that the second term on the right-hand side is negligibly small ($\mathbf{v}$ and $|\nabla \mathbf{v}|$ are both small), we have

$$D_t \mathbf{v} = \frac{\partial^2 \mathbf{u}}{\partial t^2}.$$

For the one-dimensional motions with constant density (which may always be scaled to unit)

$$\mathbf{u}(\mathbf{x}, t) = (u(x, t), 0, 0), \qquad \mathbf{x} = (x, 0, 0), \qquad \mathbf{f} = (f(x, t), 0, 0),$$

and the components of the stress tensor $\mathbf{S}$ have the form

$$S^{11} = \sigma, \quad S^{ij} = 0 \quad \text{for } i = 1, 2, 3,\ j = 2, 3.$$

With this simplification, the momentum balance law (2.2) reads as

$$\frac{\partial^2 u}{\partial t^2} = \frac{\partial \sigma}{\partial x} + f(x, t). \tag{6.1}$$

In the models of viscous-elastic bodies, the stress tensor S is a function of the deformation gradient u_x and its time derivative u_{xt} (see [186, 119, 227]) so that accepting the state equation

$$\sigma = \sigma(u_x, u_{xt})$$

in (6.1), we obtain the equation

$$
\begin{aligned}
&\frac{\partial^2 u(x,t)}{\partial t^2} = \frac{\partial \sigma(u_x(x,t), u_{xt}(x,t))}{\partial x} + f(x,t), \\
&u(x,0) = u_0(x), \\
&u_t(x,0) = \varphi(x).
\end{aligned}
\tag{6.2}
$$

Equation (6.2) occurs in various problems concerning the motions of viscous-elastic bodies [186, 119, 227].

Let us assume that the function $\sigma(r,q) \in C^1(\mathbb{R} \times \mathbb{R})$ is subject to the conditions

$$
\begin{cases}
\sigma \equiv \sigma(r,q) = \sigma_0(r) + \dfrac{\partial \sigma_1(r)}{\partial q} q, \\
\forall r \in \mathbb{R}, \quad C_2|r|^p \le \sigma_1(r) r \le C_1 |r|^p, \quad 2 < p < \infty, \\
|\sigma_0| \le C_3 |r|^{p-1}.
\end{cases}
\tag{6.3}
$$

Then equation (6.1) can be rewritten in the form

$$
\frac{\partial^2 u}{\partial t^2} = \frac{\partial^2(\sigma_1(u_x))}{\partial t \partial x} + \frac{\partial \sigma_0(u_x)}{\partial x} + f(x,t), \tag{6.4}
$$

whence, after the integration in t,

$$
\frac{\partial u}{\partial t} = \frac{\partial(\sigma_1(u_x))}{\partial x} + \frac{\partial}{\partial x}\left(\int_0^t \sigma_0(u_x(x,s)) ds \right) + \tilde{f} \tag{6.5}
$$

with

$$
\tilde{f}(x,t) = \varphi - \frac{\partial \sigma_1(u_{0x}(x))}{\partial x} + \int_0^t f(x,s) ds \tag{6.6}
$$

and

$$
u(x,0) = u_0(x), \qquad u_t(x,0) = \varphi(x).
$$

6.2 Finite speed of propagation and the waiting time effect. We consider the weak solutions to equation (6.2) from the function class

$$
u(x,t) \in W \equiv L^\infty\left(0,T; L^2_{\mathrm{loc}}(\Omega)\right) \cap L^p\left(0,T; W^{1,p}_{\mathrm{loc}}(\Omega)\right)
$$

and satisfying the initial condition

$$
u(x,0) = u_0(x) \quad \text{in } \Omega = (-L,L), \qquad u_0(x) \equiv 0 \quad \text{for } |x| \le \rho_0 < L. \tag{6.7}
$$

Theorem 6.1. *Let $u(x,t)$ be a weak solution of problem* (6.5), (6.7) *under the assumptions $\tilde{f}(x,t) \equiv 0$, $|x| \le \rho_0$, and let conditions* (6.3) *be fulfilled. Then $u(x,t)$ possesses the finite speed of propagation property in the sense of Definition* 1.1

of Chapter 3, *i.e., there exist* $t^* > 0$ *and a function* $\rho(t)$ *such that* $0 < t^* < T$, $0 < \rho(t) < \rho_0$, $\rho(0) = \rho_0$, *and*

$$u(x,t) = 0 \quad \textit{for } |x| < \rho(t),\ 0 \le t \le t^*.$$

The function $\rho(t)$ *is defined by the formula*

$$\rho^\nu(t) = \rho_0^\nu - Ct^\sigma,$$

where positive constants ν, σ *depend only on* p, *and* $C \equiv C(p, T, \overline{b}(\rho_0))$. *Moreover, if*

$$\int_{B_\rho} |u_0|^2 dx + \int_0^T |\widetilde{f}(x,\tau)|^2 dx d\tau \le C\,(\rho - \rho_0)_+^{1/(1-\nu)} \tag{6.8}$$

with

$$\nu = \nu(p) = \frac{2p}{3p-2}, \qquad \rho \in (\rho_0, L),$$

$u(x,t)$ *possesses the waiting time property: there exists* $t^* > 0$ *such that*

$$u(x,t) = 0 \quad \textit{for } |x| \le \rho_0,\ t \in [0, t^*].$$

The proof of Theorem 6.1 follows Section 1 of Chapter 3 and Section 8 of this chapter. The energy relation (the integration-by-parts formula) has the form

$$b(\rho,\tau)|_{\tau=0}^{\tau=t} + E(\rho,t) = \sum_{i=1}^{4} \equiv I \tag{6.9}$$

under the notation

$$\begin{aligned}
b(\rho,t) &= \int_{B_\rho} |u(x,\tau)|^2 dx, \quad \overline{b}(\rho) = \operatorname*{ess\,sup}_{0\le\tau\le t^*} b(\rho,\tau),\\
E(\rho,t) &= \int_0^t \int_{B_\rho} \sigma_1(u_x(x,\tau))u_x(x,\tau)dxd\tau,\\
I_1 &= \int_0^t u(\xi,\tau)\sigma_1(u_x)|_{\xi=-\rho}^{\xi=\rho} d\tau,\\
I_2 &= -\int_0^t \int_{B_\rho} u_x(x,\tau) \int_0^\tau \sigma_0(u_x(x,s))dsdxd\tau,\\
I_3 &= \int_0^t u(x,\tau) \int_0^\tau \sigma_0(u_x(x,s))ds|_{\xi=-\rho}^{\xi=\rho} d\tau,\\
I_4 &= \int_0^t \int_{B_\rho} u\tilde{f} dxd\tau.
\end{aligned}$$

The terms I_1, I_4 have the standard form, while I_2, I_3 require additional integration in time. We make use of conditions (6.3) and follow the proofs of Propositions 1.1

and 1.2 of Chapter 3, which gives the estimate

$$|I| \le Ct^{\kappa}\left((E+\overline{b})+\rho^{-\delta}\left(\frac{\partial E}{\partial \rho}\right)^{1/\nu}\right)+C\left(\int_0^t\int_{B_\rho}|\tilde{f}|^2 dx d\tau\right)$$

with $C = C(\rho_0, T, \overline{b}(\rho_0))$, and some positive constants $\kappa(p)$, $\delta(p)$. Choosing t^* sufficiently small and taking into account (6.8), we obtain from (6.9) the ordinary differential inequality

$$E^{\nu} \le (E+\overline{b})^{\nu} \le C\left(\rho^{-\delta\nu} t^{\kappa\nu}\frac{\partial E}{\partial \rho} + (\rho-\rho_0)_+^{1/(1-\nu)}\right),$$

which is studied in the routine way.

Remark 6.1. The localization properties can also be studied if the function $\sigma = \sigma(r, q)$ is not subject to conditions (6.3). The study is performed in terms of the function

$$w(x,t) = u_t(x,t), \quad u(x,t) = \int_0^t w(x,\tau)d\tau + u_0(x),$$

which satisfies the equation

$$\frac{\partial w}{\partial t} = \frac{\partial}{\partial x}\left(\sigma\left(\int_0^t w(x,\tau)d\tau + u_0(x),\ u_x(x,t)\right)\right) + f(x,t).$$

7 Flows of nonhomogeneous non-Newtonian fluid

7.1 Governing equations. For nonlinear nonhomogeneous non-Newtonian fluids, the symmetric stress tensor **S** is a nonlinear isotropic function of the rate of deformation tensor **D** and some thermodynamic parameters. We assume that the tensor **S** has the form

$$\mathbf{S} = -p\mathbf{I} + \mathbf{F}(\mathbf{D}, \rho, \theta), \quad \mathbf{F}(0, \rho, \theta) = 0 \tag{7.1}$$

where $\theta(x, t)$ is the temperature of the medium. The symmetric tensor **F** is assumed to satisfy the conditions

$$\begin{cases} \forall \mathbf{D} \in \mathbb{R}^{N\times N} \text{such that } D^{ij} = D^{ji}, \\ \delta|\mathbf{D}|^q \le \mathbf{F}(\mathbf{D}) : \mathbf{D} = F^{ij}D^{ij}, \quad \text{where } |\mathbf{D}|^2 = D^{ij}D^{ij}, \\ 1 \le q < \infty, \quad q \ne 2, \quad 0 < \delta = \delta(\rho) < \infty. \end{cases} \tag{7.2}$$

(We use the summation convention $F^{ij}D^{ij} = \sum_{i,j=1}^{N} F^{ij}D^{ij}$.) The fluids satisfying condition (7.2) are called *dilatant* if $q > 2$ and *pseudoplastic* if $q < 2$.

For the classical Navier–Stokes incompressible homogeneous isothermal ($\rho \equiv$ const) viscous fluids, relation (7.1) has the form

$$\mathbf{S} = -pI + 2\mu\mathbf{D}, \quad \mu \equiv \text{const} \tag{7.3}$$

(see [234, 299]). For such fluids in condition (7.3), $q = 2$, and no localization effects can be established.

According to [89, 118], for viscous-plastic fluids,

$$\mathbf{S} = -pI + 2\left(\mu + \tau|\mathbf{D}|^{\sigma-1}\right)\mathbf{D}, \quad 0 \le \sigma < 1. \tag{7.4}$$

Here $\mu \equiv$ const is the shear viscosity, $\tau \equiv$ const is the limit shear stress. It follows from Young's inequality that in this case condition (7.2) holds with

$$q = \frac{2}{\theta} + \frac{(1+\sigma)(\theta-1)}{\theta} \in (\sigma+1, 2), \quad 1 < \theta < \infty. \tag{7.5}$$

For dilatant fluids, in (7.5) $\sigma < 1$. Correspondingly, in (7.2) $q < 2$, $\sigma = 0$, and $\mu = 0$. Such fluids are called *ideal-plastic*.

Let us assume that the fluid is incompressible and nonhomogeneous, there are no inner mass sources, and that the motion is isothermal, i.e., $\theta(x,t) \equiv$ const. Under these assumptions, the conservation laws (2.1), (2.2) yield the following complete system of equations posed in the cylinder $Q = \Omega \times (0,T) \subset \mathbb{R}^N \times \mathbb{R}^+$:

$$D_t\rho \equiv \frac{\partial \rho}{\partial t} + \mathbf{v}\cdot\nabla\rho = 0 \quad \text{(a hyperbolic equation)}, \tag{7.6}$$

$$\operatorname{div}\mathbf{v} = 0 \quad \text{(an elliptic equation)}, \tag{7.7}$$

$$\rho D_t\mathbf{v} \equiv \rho\left(\frac{\partial \mathbf{v}}{\partial t} + (\mathbf{v}\cdot\nabla)\mathbf{v}\right) = \operatorname{div}\mathbf{S} + \rho\mathbf{f} \quad \text{(a parabolic equation)}, \tag{7.8}$$

$$\mathbf{S} = -p\mathbf{I} + \mathbf{F}(\mathbf{D}), \quad D_{ij}(\mathbf{v}) = D^{ij}(\mathbf{v}) = \frac{1}{2}\left(\frac{\partial v_i}{\partial x_j} + \frac{\partial v_j}{\partial x_i}\right). \tag{7.9}$$

In these conditions, $\mathbf{v}(x,t)$, $\rho(x,t)$, and $p(x,t)$ are, respectively, velocity, density, and pressure in the fluid; $\mathbf{S}$ and $\mathbf{D}$ are the stress tensor and the tensor of rate of deformation; $\mathbf{I}$ is the unit tensor; and $\mathbf{f}(x,t)$ is the prescribed mass force.

System (7.6)–(7.9) is of combined type. If in (7.6) the velocity $\mathbf{v}(x,t)$ is given, then (7.6) is a hyperbolic equation for the density $\rho(x,t)$. If in (7.8) the density $\rho(x,t)$ is known, then (7.7)–(7.9) is a pseudoparabolic system of equations for the velocity $\mathbf{v}$ and pressure $p(x,t)$. The specific type of system (7.6)–(7.9) is not essential for our further arguments.

System (7.6)–(7.9) is endowed with the initial and boundary conditions

$$\mathbf{v}(x,0) = \mathbf{v}_0(x), \qquad \rho(x,0) = \rho_0(x) \quad \text{in } x \in \Omega, \tag{7.10}$$

$$\mathbf{v}(x,t) = 0 \quad \text{on } (x,t) \in \Gamma_T = \partial\Omega \times (0,T). \tag{7.11}$$

It is assumed that

$$E(0) = \frac{1}{2}\int_\Omega \rho(x,0)|\mathbf{v}(x,0)|^2 dx < \infty, \qquad \frac{1}{M} \le \rho_0 \le M \equiv \text{const}. \tag{7.12}$$

In [234, 299, 234], the question of existence of solutions to problem (7.6)–(7.11) was studied for the case of the classical Navier–Stokes incompressible homogeneous fluids, that is, under the assumptions that $\rho(x,t) \equiv$ const and (7.3) is fulfilled. Global in time existence of weak solutions to problem (7.6)–(7.11) for nonhomogeneous fluids where $\rho(x,t) \not\equiv \text{const}$, and with different relations (7.2), were proved in [89, 46, 156, 157, 293]. In particular, under the assumption that $\sigma = 1$ in (7.4), in [89] was proved the existence of a weak solution such that $\mathbf{v},\ \rho,\ p \in W_2$, where

$$W_q = \{\mathbf{v} \in L^\infty(0,T; L^2(\Omega) \cap L^2(0,T; W_0^{1,q}(\Omega)), \quad 1/M \le \rho \le M)\}.$$

In our further study, we assume the existence of a solution in an appropriate function class and analyze its localization properties via the energy method following the ideas of [19, 25, 27, 35, 51].

7.2 Pseudoplastic fluids: Localization in time. In our study we rely on Korn's inequality which reads as follows: for every $\mathbf{v} \in \mathring{W}^{1,p}(\Omega)$

$$\frac{1}{C}\|\nabla \mathbf{v}\|_{q,\Omega} \le \|\mathbf{D}(\mathbf{v})\|_{q,\Omega} \le C\|\nabla \mathbf{v}\|_{q,\Omega}, \qquad K\|\mathbf{v}\|_{r,\Omega} \le \|\mathbf{D}(\mathbf{v})\|_{q,\Omega}, \tag{7.13}$$

with $r \le \frac{qN}{N-q} < \infty$, $K = K(r,q,N,\Omega)$, $C = C(r,q,N,\Omega)$ (see [256, 265, 225]).

Theorem 7.1. *Let* $(\mathbf{v}(x,t), \rho(x,t), p(x,t)) \in W_q$ *be a weak solution of problem* (7.6)–(7.11), *condition* (7.12) *be satisfied and, additionally,*

$$(7.2) \textit{ holds with } \delta = \delta_0 = \text{const} \textit{ and } q \in \left(\frac{2N}{N+2}, 2\right). \tag{7.14}$$

(1) *If* $\mathbf{f} \equiv 0$, *then*

$$E(t) = \frac{1}{2}\int_\Omega \rho(x,t)|\mathbf{v}(x,t)|^2 dx = 0 \quad \textit{for } t \ge t^*$$

with t^* *defined by the formula*

$$t^* = \frac{2E^{(2-q)/2}(0)}{\delta_0(2-q)}\left(\frac{M}{2K^2}\right)^{q/2}.$$

Here K *is the constant from Korn's inequality* (7.13) *with* $r = 2$. *In particular,* $\mathbf{v}(x,t) \equiv 0$ *in* $Q \cap \{t \ge t^*\}$.

(2) *Let* $\mathbf{f} \not\equiv 0$ *satisfy the condition*

$$\|\, \mathbf{f}(\cdot,t)\,\|_{L^{r*}(\Omega)}^{q*} \le C_f\big(1 - t/t_f\big)_+^{q/(2-q)}, \tag{7.15}$$

with $C_f = \text{const}$, $u_+ = max\{0, u\}$, $t_f \in (0, T)$, $q^* = \frac{q}{q-1}$, $r \leq \frac{Nq}{N-q}$. *Assume the existence of* $\bar{\mu} \in (0,1)$ *and* $\epsilon \in (0, (q\delta_0)^{1/q})$ *such that*

$$\begin{aligned} C_f &\leq t_f^{q/(2-q)} \frac{q-1}{q} \left(\frac{q^*}{MK}\right)^{q^*} \left(\frac{2-q}{2}\right)^{q/(2-q)} (1-\bar{\mu}) \\ &\quad \times \bar{\mu}^{q/(2-q)} \epsilon^{q*} \left(\delta_0 - \frac{1}{q}\varepsilon^q\right)^{\frac{2}{2-q}} \\ &\equiv t_f^{q/(2-q)} a(q, M, K) b(\bar{\mu})\, d(\epsilon, q, \delta_0), \\ t_f - t_1 &\geq \frac{2}{2-q}\frac{1}{\bar{\mu}} M \left(2K^2\right)^{q/2} (E(t_1))^{(2-q)/2}. \end{aligned} \tag{7.16}$$

Then there exists a constant C such that

$$\frac{1}{2M}\|\mathbf{v}(\cdot, t)\|^2_{L^2(\Omega)} \leq E(t) \leq C\left(1 - \frac{t}{t_f}\right)_+^{q/(2-q)} \tag{7.17}$$

and $\mathbf{v}(x, t) \equiv 0$ *in* $Q \cap \{t \geq t_f\}$.

The mechanical sense of Theorem 7.1 is that if the flow of a non-Newtonian pseudoplastic fluid is generated by the initial data, then in a finite time the fluid becomes immobile. If the flow is stirred by the source term $\mathbf{f} \neq 0$, which vanishes at the instant t_f, then the fluid is still for all $t \geq t_f$ provided the intensity of the source is suitably small.

Proof. Every weak solution of problem (7.6)–(7.11) satisfies the energy relation

$$\frac{dE(t)}{dt} + (F : D, 1)_\Omega = (\rho\mathbf{f}, \mathbf{v})_\Omega \equiv I. \tag{7.18}$$

In this relation

$$E(t) = \frac{1}{2}(\rho\mathbf{v}, \mathbf{v})_\Omega, \quad (\mathbf{v}, \mathbf{U})_\Omega = \int_\Omega \mathbf{v}\mathbf{U}dx.$$

The derivation of (7.18) relies on (7.9), the symmetry of the tensor $\mathbf{P}$, and the integration-by-parts formula

$$\begin{aligned} \left(\rho\frac{d\mathbf{v}}{dt}, \mathbf{v}\right)_\Omega &= \frac{d}{2dt}(\rho\mathbf{v}, \mathbf{v})_\Omega \equiv \frac{dE(t)}{dt}, \\ (\operatorname{div}\mathbf{P}, \mathbf{v})_\Omega &= (-\nabla p + \operatorname{div} F, \mathbf{v})_\Omega = (p, \operatorname{div}\mathbf{v})_\Omega - (\mathbf{F} : \mathbf{D}, 1)_\Omega \\ &= -(\mathbf{F} : \mathbf{D}, 1)_\Omega. \end{aligned}$$

Relation (7.18) is obtained (formally) by multiplying (7.8) by $\mathbf{v}(x, t)$ and the consequent integration by parts. To be precise, (7.18) is established first for Galerkin's approximations to the weak solution of the problem and is shown then to be true for the limit function (see [89]).

Let $\mathbf{f} = 0$. Then (7.18) takes the form

$$\frac{dE(t)}{dt} + (\mathbf{F}(\mathbf{D}(\mathbf{v})) : \mathbf{D}(\mathbf{v}))_\Omega = 0. \tag{7.19}$$

By (7.14),

$$\frac{dE\,(t)}{dt} + \delta_0 \|\mathbf{D}(\mathbf{v})\|_{q,\Omega}^q \leq 0. \tag{7.20}$$

Using (7.13) with $r = 2$ and the inequalities

$$E(t) \leq \frac{M}{2}\|\mathbf{v}\|_{2,\Omega}^2 \leq \frac{M}{2K^2}\|\mathbf{D}(\mathbf{v})\|_{q,\Omega}^2,$$

we come to the ordinary differential inequality

$$\frac{dE(t)}{dt} + C_0 E^{q/2}(t) \leq 0, C_0 = \delta_0 K^q \left(\frac{2}{M}\right)^{q/2}.$$

Integrating it, we have

$$0 \leq E^{(2-q)/2}\,(t) \leq E^{(2-q)/2}\,(0) - C_0 \frac{2-q}{2} t.$$

Thus $E(t) \equiv 0$ for

$$t \geq t^* = \frac{2E^{(2-q)/2}\,(0)}{(2-q)C_0} = \frac{2E^{(2-q)/2}(0)}{\delta_0(2-q)} \left(\frac{M}{2K^2}\right)^{q/2}.$$

If $\mathbf{f} \neq 0$, we apply (7.14), Korn's inequality (7.13) with $r = Nq/(N-q)$, and Young's inequality to estimate the right-hand side of (7.18) as

$$\begin{aligned}
|I| &= |(\rho \mathbf{f}, \mathbf{v})_\Omega| \leq M\|\mathbf{f}\|_{L^{r^*}(\Omega)}\|\mathbf{v}\|_{L^r(\Omega)} \\
&\leq MK\|\mathbf{f}\|_{L^{r^*}(\Omega)}\mathbf{D}(\mathbf{v})\|_{L^q(\Omega)} \\
&\leq \frac{1}{q}\varepsilon^q \|\mathbf{D}(\mathbf{v})\|_{L^q(\Omega)}^q + \frac{1}{q^*}\varepsilon^{-q^*}(MK\|\mathbf{f}\|_{L^{r^*}(\Omega)})^{q^*}
\end{aligned}$$

with

$$r^* = \frac{r}{r-1}, \qquad q^* = \frac{q}{q-1}, \qquad \varepsilon \in (0,1).$$

Assuming $\frac{1}{q}\varepsilon^q < \delta_0$ and using (7.15), (7.20), we then arrive at the standard ordinary differential inequality for the energy function $E(t)$,

$$\frac{dE}{dt} + C_1 E^{q/2} \leq C_2 \left(1 - \frac{t}{t_f}\right)^{q/(2-q)}$$

with

$$C_1 = \left(\delta_0 - \frac{1}{q}\varepsilon^q\right)\left(\frac{2}{MK^2}\right)^{q/2}, \qquad C_2 = C_f\left(\frac{MK}{q*\varepsilon}\right)^{q^*}.$$

If $C_f = 0$, then the source is absent. Letting in (7.16) $\varepsilon = 0$ and $t_1 = 0$ we get (7.20) and, correspondingly, the equality $t_f = t^*$. If $C_f \neq 0$ and $\mathbf{f} \not\equiv 0$, then according to Lemma 2.3 and Remark 2.4 of Chapter 1, estimate (7.17) holds if

$$\begin{cases} C_2 \leq t_f^{q/(2-q)}\left(\dfrac{2-q}{2}\right)^{q/(2-q)}(1-\bar{\mu})\bar{\mu}^{q/(2-q)}C_1^{2/(2-q)}, \\ t_f - t_1 \geq \dfrac{2}{2-q}\dfrac{1}{\bar{\mu}C_1}(E(t_1))^{(2-q)/2}. \end{cases} \tag{7.21}$$

It is easy to verify that conditions (7.21) are equivalents to (7.16). □

Corollary 7.1. *The following are consequences of Theorem* (7.1):

1. *Given an arbitrary instant* $t_f \in (0, \infty)$, *there exist* $E(t_1)$ *and* C_f *such that* (7.17) *holds.*

2. *Given arbitrary constants* $E(t_1)$ *and* C_f (*sufficiently small*) *in* (7.12), (7.15), *one can find* $t_f \in (0, \infty)$ *such that* (7.17) *holds.*

Remarks.

1. The explicit value of t^* is sometimes easy to calculate. Let

$$\mathbf{P} = -p\mathbf{E} + 2\left(\mu + \tau|\mathbf{D}|^{\sigma-1}\right)\mathbf{D}, \quad 0 \leq \sigma < 1.$$

It follows from assertion 1 of Theorem 7.1 that the instant t^* when the fluid becomes immobile is defined by

$$t^* = \frac{\tau^{\theta*}}{\mu^{1/\theta}}\frac{2^{(2-q)/2}E^{(2-q)/q}(0)\left(K^2M\right)^{2/q}}{(2-q)\theta(\theta-1)^{(1-\theta)/\theta}},$$

where

$$\theta \in (1, \infty), \qquad q = \frac{2}{\theta} + \frac{(\sigma+1)(\theta-1)}{\theta},$$

and K is the constant in Korn's inequality. The values of the constants μ and τ depend on the properties of the medium.

2. If $q = \frac{2N}{N+2}$, the constant K in inequality (7.13) does not depend on Ω. In this exceptional case, Theorem 7.1 is applicable to the solutions of the Cauchy problem for system (7.6)–(7.9):

$$\mathbf{v}(x,0) = \mathbf{v}_0(x), \qquad \rho(x,0) = \rho_0(x) \quad \text{in } \Omega = \mathbb{R}^N.$$

The proofs do not need any essential change if

$$E(0) = \frac{1}{2}\int_{\Omega} \rho(x,0)|\mathbf{v}(x,0)|^2 dx < \infty, \qquad \frac{1}{M} \le \rho_0 \le M.$$

3. In [89], the existence of a weak solution to problem (7.6)–(7.9) was proved for nonisothermic flows. It was assumed that in the law (7.4) the coefficient of dynamic viscosity μ and the limiting shear stress τ depend on the temperature θ. The temperature θ solves the initial–boundary value problem

$$\rho\frac{d\theta}{dt} \equiv \rho\left(\frac{\partial\theta}{\partial t} + (\mathbf{v}\cdot\nabla)\,\theta\right) = \operatorname{div}\left(k(\theta)\nabla\theta\right) + L(x,t,\mathbf{v},\theta), \tag{7.22}$$

$$\theta = 0 \quad \text{on the lateral boundary of } Q, \qquad \theta(x,0) = \theta_0(x) \quad \text{in } \Omega. \tag{7.23}$$

It was assumed that the functions $\mu(\theta)$ and $\tau(\theta)$ and the coefficient of heat conductivity $k(\theta)$ satisfied the conditions

$$\frac{1}{M} \le (\mu, \tau, k) \le M < \infty. \tag{7.24}$$

The statement of Theorem 7.1 also remains true for solutions of problem (7.6)–(7.11), (7.22), (7.23). Moreover, if instead of (7.24) the function $k(\theta)$ is subject to the conditions

$$\frac{1}{k_0}|\theta|^{\alpha} \le k(\theta) \le k_0|\theta|^{\alpha}, \quad -1 < \alpha < 0, \quad k_0 = \text{const},$$
$$L(x,t,\mathbf{v},\theta) \equiv 0, \quad \theta_0 \in L^2(\Omega),$$

then according to Chapter 2 and Section 5 of this chapter, there exists a finite time t^* such that

$$\theta(x,t) \equiv 0 \quad \text{for } x \in \Omega,\ t \ge t^*.$$

7.3 Pseudoplastic fluid with vanishing or unbounded density.

7.3.1 *Formulation of the problem.* Let us study problem (7.6)–(7.9) with the initial and boundary conditions

$$\rho(x,0)\mathbf{v}(x,0) = \rho_0(x)\mathbf{v}_0(x), \quad \rho(x,0) = \rho_0(x) \quad \text{in } \Omega, \tag{7.25}$$

$$\mathbf{v}(x,t) = 0 \quad \text{on the lateral boundary } \Gamma_T \text{ of } \Omega\times(0,T), \tag{7.26}$$

under the assumptions

$$\left\|\frac{1}{\rho_0(x)}\right\|_{L^m(\Omega)} \le C_1, \qquad \|\rho_0(x)\|_{L^M(\Omega)} \le C_2, \quad \min\{m, M\} > 1, \tag{7.27}$$

$$\sqrt{\rho_0}\mathbf{v}_0 \in L^2(\Omega), \qquad E(0) = \frac{1}{2}\int_{\Omega} \rho_0\,|\mathbf{v}_0|^2\,dx < \infty. \tag{7.28}$$

$$\forall\,\mathbf{r} \in \mathbb{R}^N, \qquad \delta\,|\mathbf{D}(\mathbf{r})|^q \le \mathbf{F}(\mathbf{r}) : \mathbf{D}(\mathbf{r}) \tag{7.29}$$

with

$$\delta = \text{const} > 0, \quad q \in \left(\frac{2MN}{N(M-1)+2M}, 2 \right), \quad M > \frac{N}{2}.$$

Note that condition (7.27) allows the density ρ to become zero or infinite on any set of zero measure.

We are interested in the properties of those solutions to this problem that satisfy the conditions [51]

$$\begin{aligned}
&\mathbf{v} \in \mathbf{W}_{q,\sigma} \\
&\quad = \{\mathbf{v} \in L^{\infty}(0, T; L^2(\Omega) \cap L^{1+\sigma}(0, T; L^{1+\sigma}(\Omega) \cap L^2(0, T; W_0^{1,q}(\Omega))\}, \\
&\rho \geq 0, \quad \frac{1}{\rho} \in L^{\infty}\left(0, T; L^m(\Omega)\right), \quad \rho \in L^{\infty}\left(0, T; L^M(\Omega)\right), \\
&\sqrt{\rho}\mathbf{v} \in L^{\infty}\left(0, T; L^2(\Omega)\right), \quad \mathbf{v} \in L^q\left(0, T; W_0^{1,q}(\Omega)\right).
\end{aligned}$$

Additionally to these conditions, we assume that the solutions satisfy the energy relation (7.18).

7.3.2 *Estimates for the energy function.*

Lemma 7.1 (Conservation law for the density). *Let* $\rho(x,t) \geq 0$ *be a weak solution of equation* (7.6) *satisfying condition* (7.25), *and let* $\mathbf{v}$ *satisfy* (7.7), (7.26), *and* (7.25). *Then for every* $t \in (0, T)$,

$$\left\| \rho^{-1}(\cdot, t) \right\|_{L^m(\Omega)} = \left\| \rho_0^{-1}(x) \right\|_{L^m(\Omega)} \leq C_1, \tag{7.30}$$

$$\|\rho(\cdot, t)\|_{L^M(\Omega)} = \|\rho_0(x)\|_{L^M(\Omega)} \leq C_2. \tag{7.31}$$

The proof of this assertion can be found in [46, Lemma 2.1, Chapter 3], [293], and [89].

Lemma 7.2. *Under the conditions of Lemma* 7.1, *the energy function*

$$E(t) = \frac{1}{2} \int_{\Omega} \rho |\mathbf{v}|^2 \, dx$$

satisfies the estimate

$$\frac{1}{2C_1} \|\mathbf{v}\|^2_{L^{2m/(m+1)}}(\Omega)^2 \leq E(t) \leq \frac{C_2}{2} \|\mathbf{v}\|^2_{L^{2M/(M-1)}}(\Omega). \tag{7.32}$$

Proof. It follows from Hölder's inequality and (7.31) that

$$\begin{aligned}
E(t) &\equiv \frac{1}{2} \int_{\Omega} \rho |\mathbf{v}|^2 \, dx \leq \frac{1}{2} \left(\int_{\Omega} \rho^M \right)^{1/M} \left(\int_{\Omega} |\mathbf{v}|^{2M'} \right)^{2/2M'} \\
&\leq \frac{C_2}{2} \|\mathbf{v}\|^2_{L^{2M/(M-1)}(\Omega)}, \qquad M' = \frac{M}{M-1}.
\end{aligned}$$

On the other hand, applying the inverse Hölder inequality and (7.30), we obtain

$$E(t) = \frac{1}{2}\int_{\Omega} \rho\,|\mathbf{v}|^2\,dx \geq \frac{1}{2}\left(\int_{\Omega} \rho^{-m} dx\right)^{-1/m} \left(\int_{\Omega} |\mathbf{v}|^{2m'}\,dx\right)^{2/2m'}$$
$$\geq \frac{1}{2C_1}\,\|\mathbf{v}\|^2_{L^{2m/(m+1)}}(\Omega)\,. \qquad \square$$

Lemma 7.3. *Under the conditions* (7.30), (7.31), *the energy function* $E(t)$ *obeys the estimate*

$$C_3 E^{\frac{q}{2}} \leq ||\mathbf{D}(\mathbf{v})||^q_{L^q(\Omega)} = \int_{\Omega} |\mathbf{D}(\mathbf{v})|^q\,dx, \tag{7.33}$$

where

$$q \in \left(\frac{2MN}{N(M-1)+2M}, 2\right), \quad C_3 = \left(\frac{2K^2}{C_2}\right)^{q/2}, \quad M > \frac{N}{2}.$$

Proof. As in the proof of Lemma 7.1, we have

$$E(t) \leq \frac{1}{2}\,\|\rho\|_{L^M(\Omega)}\,\|\mathbf{v}\|^2_{L^{2M/(M-1)}(\Omega)} \leq \frac{C_2}{2}\,\|\mathbf{v}\|^2_{L^{2M/(M-1)}(\Omega)}\,.$$

Applying Korn's inequality (7.13), we obtain

$$\|\mathbf{v}\|^2_{L^{2M/(M-1)}(\Omega)} \leq K^{-2}\,\|\mathbf{D}(\mathbf{v})\|^2_{L^q(\Omega)} \quad \text{with } \frac{2M}{M-1} \leq \frac{Nq}{N-q}.$$

This inequality gives

$$2E(t) \leq C_2\,\|\mathbf{v}\|^2_{L^{2M/(M-1)}(\Omega)} \leq K^2 C_2\,\|\mathbf{D}(\mathbf{v})\|^2_{L^q(\Omega)}\,,$$

which is equivalent to (7.33). $\square$

7.3.3 *Localization in a finite time.*

Theorem 7.2. *Let* $(\mathbf{v}(x,t), \rho(x,t), p(x,t))$ *be a weak solution of problem I, and let* (7.29) *hold and the initial density satisfy conditions* (7.27), (7.28). *Then we have the following:*

1. *If* $\mathbf{f} \equiv 0$, *then* $E(t) \equiv 0$ *for all* $t \geq t^*$ *with*

$$t^* = \frac{2E^{(2-q)/2}(0)}{\delta(2-q)}\left(\frac{C_2}{2K^2}\right)^{q/2},$$

where K *is the constant from Korn's inequality* (7.13). *Moreover, by Lemma 7.2,*

$$\mathbf{v}(x,t) \equiv 0 \quad \textit{for } x \in \Omega \textit{ and } t \geq t^*.$$

2. *Let* $\mathbf{f} \neq 0$ *and the condition*

$$\|\mathbf{f}(\cdot,t)\|^{q*}_{L^{2M/(M-1)}(\Omega)} \le C_f \left(1-\frac{t}{t_f}\right)_+^{q/(2-q)}, \quad q^* = \frac{q}{q-1} \tag{7.34}$$

hold for some $C_f = \text{const}$ *and* $t_f \in (0,T)$. *Let there exist* $\bar{\mu} \in (0,1)$ *and* $\epsilon \in (0,1)$ *such that*

$$\begin{aligned} C_f &\le t_f^{q/(2-q)}(1-\bar{\mu})\bar{\mu}^{q/(2-q)}\frac{q-1}{q}\left(\frac{2-q}{2}\right)^{q/(2-q)} \\ &\quad\times \epsilon^{q/(q-1)}\left[\delta\left(\frac{2}{C_2K^2}\right)^{q/2} - \frac{1}{q}(\epsilon\sqrt{C_2})^q\right] \\ &\equiv t_f^{q/(2-q)}a(q)b(\bar{\mu})d(\epsilon,q), \\ t_f - t_1 &\ge \frac{2}{2-q}\frac{1}{\bar{\mu}C_5}(E(t_1))^{(2-q)/2}. \end{aligned}$$

Then one can find a constant C *such that*

$$\|\mathbf{v}(\cdot,t)\|^2_{L^{2m/(m+1)}(\Omega)} \le C\left(1-\frac{t}{t_f}\right)_+^{q/(2-q)},$$

whence $\mathbf{v}(x,t) \equiv 0$ *for* $x \in \Omega$ *and* $t_f \le t$.

Proof. Let $\mathbf{f} = 0$. Then the energy relation (4.50) becomes

$$\frac{dE(t)}{dt} + (\mathbf{F}(\mathbf{D}(\mathbf{v})) : \mathbf{D}(\mathbf{v}) = 0.$$

By (7.29),

$$\frac{dE\,(t)}{dt} + \delta|\mathbf{D}(\mathbf{v})|^q \le 0,$$

and by virtue of Lemma 7.3, we obtain the ordinary differential inequality

$$\frac{dE(t)}{dt} + C_4E^{q/2}(t) \le 0 \tag{7.35}$$

with the constants

$$C_4 = \delta C_3, \qquad q \in \left(\frac{2MN}{N\,(M-1)+2M}, 2\right), \qquad M > \frac{N}{2}.$$

If $\mathbf{f} \neq 0$, the energy relation has the form

$$\frac{dE\,(t)}{dt} + (F : \mathbf{D}, 1)_\Omega = (\rho\mathbf{f}, \mathbf{v})_\Omega\,.$$

Repeating the previous arguments, we can write

$$\frac{dE(t)}{dt} + C_4 E^{q/2} \leq I = \left|(\rho \mathbf{f}, \mathbf{v})_\Omega\right|. \tag{7.36}$$

Applying Lemma 7.3 and the inequalities of Hölder and Young, we have

$$\begin{aligned} I &= \left|\left(\frac{\rho^{1/2}}{\sqrt{2}}\mathbf{v}, \rho^{1/2}\mathbf{f}\sqrt{2}\right)\right|_\Omega \\ &\leq \frac{1}{q}\left(\epsilon\sqrt{C_2}\right)^q E^{q/2}(t) + \frac{1}{q^*}\epsilon^{q^*} \|\mathbf{f}\|^{q^*}_{L^{2M/(M-1)}(\Omega)}, \end{aligned} \tag{7.37}$$

where $\frac{1}{p} + \frac{1}{q} = 1$, $\epsilon > 0$. It follows from (7.36), (7.37) that

$$\frac{dE(t)}{dt} + C_3 E^{q/2}(t) \leq \frac{1}{q}\left(\epsilon\sqrt{C_2}\right)^q E^{q/2}(t) + \frac{1}{q^*}\epsilon^{q^*} \|\mathbf{f}\|^{q^*}_{L^{2M/(M-1)}(\Omega)},$$

whence

$$\frac{dE(t)}{dt} + C_5 E^{q/2}(t) \leq C_6 \|\mathbf{f}\|^{q^*}_{L^{2M/(M-1)}(\Omega)}$$

with the constants

$$C_5 = C_4 - \frac{1}{q}\left(\epsilon\sqrt{C_2}\right)^q = \delta\left(\frac{2K^2}{C_2}\right)^{q/2} - \frac{1}{q}\left(\epsilon\sqrt{C_2}\right)^q, \quad C_6 = q^*\epsilon^{-q^*}.$$

On the other hand, according to (7.34),

$$\|\mathbf{f}(\cdot, t)\|^{q^*}_{L^{2M/(M-1)}(\Omega)} \leq C_f \left(1 - \frac{t}{t_f}\right)_+^{q/(2-q)},$$

and we finally get

$$\frac{dE(t)}{dt} + C_5 E^{q/2}(t) \leq C_7 \left(1 - \frac{t}{t_f}\right)_+^{q/(2-q)}. \tag{7.38}$$

If we choose now $\epsilon > 0$ so that $C_5 > 0$, inequality (7.38) can be easily integrated, and the proof is completed as in Theorem 7.1. □

Remark 7.1. Inequality (7.38) yields the estimate for the energy function

$$\begin{aligned} E(t) &\leq C_7 \int_0^t \left(1 - \frac{s}{t_f}\right)^{q/(2-q)} ds + E(0) \\ &\leq C_7 \frac{t_f}{4-q}(2-q)\left[1 - \left(1 - \frac{t}{t_f}\right)_+^{(4-q)/(2-q)}\right] + E(0) \end{aligned} \tag{7.39}$$

with the constants $C_7 = C_f q^* \epsilon^{q^*}$.

7.4 Flows with nonlinear sources. Let us study problem (7.6)–(7.11) under the assumption that the forcing term in the momentum balance law (7.8) may depend on $\mathbf{v}$ and obeys the conditions

$$\mathbf{r} \in \mathbb{R}^N, \qquad \delta_1 |\mathbf{r}|^{1+\sigma} \le -\mathbf{f}(x,t,\mathbf{r}) \cdot \mathbf{r}, \qquad \sigma \in (0,1), \tag{7.40}$$

$$\forall \mathbf{r} \in \mathbb{R}^N, \quad \delta \ |\mathbf{D}(\mathbf{r})|^q \le \mathbf{F}(\mathbf{r}) : \mathbf{D}(\mathbf{r}) \equiv \sum_{i,j=1}^{n} F_{ij} D_{ij}, \quad q \in \left(\frac{2N}{N+2}, N\right) \tag{7.41}$$

with some constants δ, δ_1. We will consider the weak solutions

$$(\mathbf{v}, \rho, p) \in W_{q,\sigma}$$

where

$$W_{q,\sigma} = \{\mathbf{v} \in L^\infty(0,T; L^2(\Omega) \cap L^{1+\sigma}(0,T; L^{1+\sigma}(\Omega) \cap L^2(0,T; W_0^{1,q}(\Omega)), \\ 1/M \le \rho \le M)\}.$$

We begin by establishing the energy relation

$$\frac{dE(t)}{dt} + \int_\Omega \mathbf{F}(\mathbf{D}(\mathbf{v})) : \mathbf{D}(\mathbf{v}) dx = \int_\Omega \mathbf{f} \cdot \mathbf{v} dx. \tag{7.42}$$

Using conditions (7.40), (7.41), we infer the inequality

$$\frac{dE(t)}{dt} + \delta \int_\Omega |\mathbf{D}(\mathbf{v})|^q dx + \delta_1 \int_\Omega |\mathbf{v}|^{1+\sigma} dx \le 0$$

or

$$\frac{dE(t)}{dt} + C_1 Q(\mathbf{v}) \le 0, \tag{7.43}$$

where we use the notation

$$Q(\mathbf{v}) = \|\mathbf{D}(\mathbf{v})\|_q^q + \|\mathbf{v}\|_{1+\sigma}^{1+\sigma}, \qquad C_1 = \min\{\delta_1, \delta\} > 0.$$

Next, we apply the inequality

$$\|\mathbf{v}\|_2 \le \beta \|\mathbf{D}(\mathbf{v})\|_q^\theta \|\mathbf{v}\|_{1+\sigma}^{1-\theta}$$

with the exponent

$$\theta = \frac{Nq(1-\sigma)}{2\,(Nq - (1+\sigma)(N-q))}, \qquad q \ge \frac{2N}{(N+1)},$$

which is a byproduct of (7.13), Lemma 3.2, and Remark 3.5 of the appendix, and then make use of the inequalities

$$\begin{aligned} E(t) &\le \frac{M}{2} ||\mathbf{v}||_2^2 \le \frac{M}{2} \beta^2 \left(\|\mathbf{D}(\mathbf{v})\|_q^q\right)^{2\theta/q} \left(\|\mathbf{v}\|_{1+\sigma}^{1+\sigma}\right)^{2(1-\theta)/(1+\sigma)} \\ &\le C_2 \left(\|\mathbf{D}(\mathbf{v})\|_q^q + \|\mathbf{v}\|_{1+\sigma}^{1+\sigma}\right)^{1/\nu} = C_2 Q^{1/\nu}(\mathbf{v}), \quad C_2 = \frac{M}{2}\beta^2 \end{aligned} \tag{7.44}$$

with the exponent

$$\frac{1}{\nu} = \frac{2\theta}{q} + \frac{2(1-\theta)}{1+\sigma}$$
$$= 1 + \frac{q(1-\sigma)}{q(N+1+\sigma) - N(1+\sigma)} > 1 \quad \text{for some } \sigma \in (0, 1).$$

Plugging (7.43) into (7.44), we come to the ordinary differential inequality

$$\frac{dE(t)}{dt} + C_3 E^{\nu}(t) \leq 0, \quad \nu \in (0, 1), \quad C_3 = C_1 C_2^{-\nu},$$

which is studied in the standard way.

Theorem 7.3. *Let* $(\mathbf{v}, \rho, p) \in W_{q,\sigma}$ *be a weak solution of problem* (7.6)–(7.11), *and let conditions* (7.40), (7.41) *be satisfied. Then for any* $E(0) < \infty$, $0 < M < \infty$, *there exists* $t^* < \infty$, *depending on* $E(0)$, M, δ_1, δ, σ, q, *and* N, *such that*

$$\mathbf{v}(x, t) \equiv 0 \quad \textit{for } x \in \Omega,\ t \geq t_f.$$

Remark 7.2. Under condition (7.40) on the exterior mass forces, the effect of stabilization to zero in a finite time is also exhibited by dilatant fluids with the exponent $q \geq 2$. Specifically, for the classical Navier–Stokes equations for incompressible homogeneous fluids $\rho = \text{const}$ and

$$\mathbf{S} = -p\mathbf{I} + 2\mu\mathbf{D}, \quad \mu \equiv \text{const}, \quad q = 2.$$

7.5 Space localization in dilatant fluids. Let us examine local properties of special solutions of the form (7.49) to system (7.6)–(7.9) regardless of the boundary conditions.

Let $v(x, t)$ be a weak solution of the problem

$$\begin{cases} \dfrac{\partial v}{\partial t} = \operatorname{div} \mathbf{F}(\nabla v) + \dfrac{\partial p}{\partial x_3} + f(x, t), \\ v(x, 0) = v_0(x) \quad \text{in } B_{\rho_1}(x_0) \subset \Omega \subset \mathbb{R}^2. \end{cases} \tag{7.45}$$

Introduce the function

$$u(x, t) = \left(v(x, t) + \int_0^t a(\tau) d\tau \right), \quad a(t) = \frac{\partial p}{\partial x_3},$$

which solves the standard problem studied in Section 3:

$$\begin{cases} \dfrac{\partial u}{\partial t} = \operatorname{div} \mathbf{F}(\nabla u) + f(x, t), \\ u(x, 0) = u_0(x) \quad \text{in } B_{\rho_1}(x_0) \subset \Omega \subset \mathbb{R}^2. \end{cases} \tag{7.46}$$

The property of space localization of the solutions to this problem is formulated in terms of the energy functions

$$E(\rho, t) = \int_0^t \int_{B_\rho} \mathbf{F}(\nabla u) \cdot \nabla u dx, \qquad b(\rho, t) = \sup_{0 \leq \tau \leq t} \int_{B_\rho} |u(x, t)|^2 dx.$$

Let us assume fulfillment of the conditions

$$\begin{cases} \forall \mathbf{r} \in \mathbb{R}^2, \quad \delta |\mathbf{F}|^q \leq \mathbf{f}(\mathbf{r})\mathbf{r} \leq \dfrac{1}{\delta}|\mathbf{r}|^q \quad \text{for some } 2 < q < \infty, \\ |a(t)| + \| v_0(x) \|^2_{2,B_{\rho_1}} \leq M < \infty \quad \text{with } 0 < \rho_0 < \rho_1, \\ v(x,0) = u(x,0) = 0, \quad f(x,t) = 0 \quad \text{for } (x,t) \in B_{\rho_0}(x_0) \times (0,T), \end{cases} \tag{7.47}$$

$$\|u_0\|^2_{L^2(B_\rho)} + \int_0^T \| f(\cdot,t) \|^2_{L^2(B_\rho)} dt \leq C(\rho - \rho_0)_+^{1/(1-\alpha)}, \quad \alpha = \frac{3q-2}{4(q-1)}. \tag{7.48}$$

Results of Chapter 3 yield the following assertion.

Theorem 7.4. *Let $v(x,t)$ be a weak solution of problem* (7.45) *and let conditions* (7.47) *be fulfilled. Then $u(x,t) = 0$ a.e. in $B_{\rho(t)}(x_0)$ with*

$$\rho^\nu(t) = \rho_0^\nu - Ct^\lambda \min_{2/q \leq \tau \leq 1} \left(E^\gamma G(\rho_0)\right)$$

and

$$\begin{aligned} G &= \frac{1}{\tau q - 2} \max(1, {\rho_0}^{\nu-1}) \max(b^\mu(\rho_0,t), b^\eta(\rho_0,t)), \\ a(t) &= \frac{\partial p}{\partial x_3}, \qquad \nu = \frac{k}{2(q-1)}, \qquad \gamma = \frac{\tau q - 2}{4q}, \\ k &= \frac{4}{q-1}, \qquad \mu = \frac{q-2}{q-1} - \frac{2-\tau q}{k}, \qquad \lambda = \frac{2}{k}. \end{aligned}$$

If, additionally to (7.47), *condition* (7.48) *is fulfilled, then there exists $t^* = t^*(M, \delta, \rho_1, q) \in (0,T)$ such that $u(x,t) = 0$ in $B_{\rho_0}(x_0)$ for all $t \leq t^*$.*

The second assertion of the theorem means that if a non-Newtonian fluid satisfying (7.49), (7.48) is immobile in B_{ρ_0} at the instant $t = 0$, then regardless of the boundary conditions and the behaviour of the forcing terms in the equation beyond the ball B_{ρ_0}, the motion of the fluid is defined by the relation

$$v(x,t) = -\int_0^t a(\tau)d\tau = 0 \quad \text{for } 0 \leq t \leq t^*, x \in B_{\rho_0}.$$

Specifically, the fluid remains immobile in the absence of the pressure drop, $a(t) = 0$.

7.6 A flow of an ideal non-Newtonian fluid in a pipe. Let us consider a particular case of the law (7.1). Let us assume that $q = 1$ in (7.2), which corresponds to an ideal-plastic fluid (for example, $\mu = \sigma = 0$ in (7.4)). Under this choice of the parameters, Korn's inequality (7.13) ceases to be applicable.

We restrict ourselves to the study of a special class of solutions to system (7.6)–(7.9). Let us assume that

$$\begin{aligned} &\mathbf{v} = (0, 0, v(x,t)), \quad \rho(x,t) \equiv 1, \quad x = (x_1, x_2) \in \Omega \subset R^2, \\ &\mathbf{f} = (0, 0, f(x,t)), \quad p(x,t) = p(x_3, t), \quad \frac{\partial p}{\partial x_3} = a(t). \end{aligned} \tag{7.49}$$

It is assumed that the drop of pressure is given. The solutions of the form (7.49) correspond to flows in a pipe. Under the restrictions imposed, problem (7.6)–(7.11) takes the special form

$$\begin{cases} \dfrac{\partial v}{\partial t} = \operatorname{div} \mathbf{f}(\nabla v) + g, & g = -\dfrac{\partial p}{\partial x_3} + f(x,t), \\ v(x,0) = v_0(x), & x = (x_1, x_2) \in \Omega \subset \mathbb{R}^2. \end{cases}$$

Let us also suppose that the given vector $\mathbf{f}$ and that the functions $a(t)$ and $f(x,t)$ satisfy the conditions

$$\delta|\mathbf{r}| \le \mathbf{f}(\mathbf{r})\mathbf{r} \quad \forall \mathbf{r} \in \mathbb{R}^2, \tag{7.50}$$

$$v_0 \in L^2(\Omega), \qquad K\| \, g \, \|_{L^2(\Omega)} \le \delta\tau, \qquad \tau < 1. \tag{7.51}$$

Theorem 7.5. *Let* $v(x,t)$ *be a weak solution of problem* (7.6)–(7.11) *satisfying the inequality*

$$\frac{1}{2}\frac{dE}{dt} + (\mathbf{f}(\nabla v), \nabla v)_\Omega = (g, v)_\Omega, \quad E(t) = (v, v) = \int_\Omega |v|^2 d\Omega, \tag{7.52}$$

and let conditions (7.50)–(7.51) *hold.*

Then $v(x,t) \equiv 0$ *in* Ω *for all* $t \ge t^* = \dfrac{E(0)}{\delta(1-\tau)}$.

Proof. Using Korn's inequality (7.13),

$$\forall\, v \in W_0^{1,1}(\Omega) \|v\|_{L^2(\Omega)} \le K(\Omega)\|\nabla v\|_{L^1(\Omega)}, \quad \forall\, v \in W_0^{1,1}(\Omega),$$

we derive from (7.50), (7.52) the ordinary differential inequality

$$\frac{dE(t)}{dt} + 2\delta(1-\tau)E^{1/2}(t) \le 0.$$

Integrating this, we obtain the estimate

$$\int_\Omega |v(x,t)|^2 d\Omega = E(t) \le \left(E^{1/2}(0) - \delta(1-\tau)\right)^2,$$

which completes the proof. □

Remark 7.3. The assertion of Theorem 7.5 can be interpreted as follows. If the motion of an ideal plastic fluid in a pipe is caused by a (small) drop of pressure at the ends of the pipe, the fluid stops after a finite time.

8 Boundary layers in dilatant fluid

Let us demonstrate an application of the energy method to the investigation of the Prandtl equations for dilatant fluids. In our presentation, we follow [35].

First, we reformulate the Prandtl system of equations of the boundary layer in new unknowns introduced by von Mises.

We consider the problem

$$\frac{\nu}{2^{(n-1)}}\sqrt{w}\frac{\partial}{\partial\psi}\left(\left|\frac{\partial w}{\partial\psi}\right|^{(n-1)}\frac{\partial w}{\partial\psi}\right)-\frac{\partial w}{\partial x}-V_0(x)\frac{\partial w}{\partial x}+2UU_x=0, \tag{8.1}$$

$$\begin{gathered}1<n<\infty,\\ w(0,\psi)=w_0(\psi),\quad w(x,0)=0,\\ w(x,\psi)\to U^2(x)\quad \text{as } \psi\to\infty,\\ 0<x<X,\quad 0<\psi<\infty.\end{gathered} \tag{8.2}$$

We shall study the local properties of solutions of the problem (8.1), (8.2) near the point $(0,\psi_0)$, assuming that

$$w_0(\psi)\equiv U^2(0)\quad \text{for } \psi_0\le\psi<\infty. \tag{8.3}$$

It was proved in the paper [285] that a weak solution of problem (8.1), (8.2) exists, is unique and possesses the property of finite speed of propagation of disturbances from the initial data. The last property is understood in the sense that there exists a function $\psi_4\ge\psi_0$ such that

$$w(x,\psi)\equiv U^2(x)\quad \text{if } \psi_0\ge\psi_4. \tag{8.4}$$

Here we demonstrate the possibility of applying the energy method to estimate the size of the support of the function

$$z=U^2(x)-w(x,\psi) \tag{8.5}$$

and also to prove that this function demonstrates the waiting time phenomena.

For the function z defined by equality (8.5), equation (8.1) takes the form

$$\frac{\nu}{2^{(n-1)}}\sqrt{w}\frac{\partial}{\partial\psi}\left(\left|\frac{\partial z}{\partial\psi}\right|^{(n-1)}\frac{\partial z}{\partial\psi}\right)-\frac{\partial z}{\partial x}-V_0(x)\frac{\partial w}{\partial x}=0. \tag{8.6}$$

We assume the conditions (from [285])

$$w_0(\psi)\in C^{2+\alpha},\quad (U_x,v_0)\in C^1,\qquad v_0(x)\le 0,\quad 0\le U_x, \tag{8.7}$$

$$0<U^2(0)\equiv w_0(\psi),\qquad \psi\ge\psi_0. \tag{8.8}$$

We point out that our proof directly uses only a part of conditions (8.7), (8.8). The rest will only provide, according to [285], some properties of the solution.

Introduce the function

$$E(\rho, t) = \int_{\rho}^{\infty} \int_{0}^{t} \theta^2 \left| \frac{\partial z}{\partial \psi} \right|^{(n+1)} dx d\psi, \quad \rho > 0, \tag{8.9}$$

$$\theta = \frac{\nu}{2^{n-1}} \left(z^2 \sqrt{U^2 - z} \right)'_z = \frac{3w(x, \psi) - U^2(x)}{2\sqrt{w}}. \tag{8.10}$$

Due to [285], the following estimates hold for large ψ:

$$0 < C_0 \le (\theta, w) \le C_1 < \infty, \tag{8.11}$$

$$E(\rho_0, X) < \infty. \tag{8.12}$$

Multiplying equation (8.6) by $z(x, \psi)$ and integrating by parts (or, which is the same, using the definition of weak solution given in [285]), we obtain the energy relation

$$E(\rho, t) + 1/2 \int_{\rho}^{\infty} z^2(\rho, t) d\rho + \int_{0}^{t} |v_0(x)| \frac{z^2(x, 0)}{2} dx = I_1 + I_2, \tag{8.13}$$

where

$$I_1 := -\frac{\nu}{2^{n-1}} \int_{0}^{t} \sqrt{w} z |z_\psi|^{n-1} z_\psi(x, \rho) dx, \tag{8.14}$$

$$I_2 := \frac{1}{2} \int_{\rho}^{\infty} z^2(0, \rho) d\rho, \quad I_2 \equiv 0 \quad \text{for } \rho \ge \psi. \tag{8.15}$$

In order to establish the presence of waiting time phenomena, we shall assume additionally that

$$I_2 \equiv \frac{1}{2} \int_{\rho}^{\infty} \left(U^2(0) - w_0(\psi) \right)^2 d\psi \le C_2 (\rho_0 - \rho)_+^{\frac{3n+1}{n-1}}. \tag{8.16}$$

The energy equality (8.13) enables one to obtain the inequalities

$$\begin{aligned} |I_1| &\le C_3(\nu, n, C_0, C_1) \left(-\frac{\partial E}{\partial \rho} \right)^{\frac{n}{n+1}} \left(\int_{0}^{t} |z(x, \rho)|^n dx \right)^{1/n} \\ &\le C_4 \left(-\frac{\partial E}{\partial \rho} \right)^{\frac{n}{n+1}} t^{(2n+1)/n(3n+1)} E^{1/(3n+1)} \left(\sup_{\tau \le t} \int_{\rho}^{\infty} z(\tau, \psi)^2 d\psi \right)^{n/(n+1)} \\ &\le \varepsilon \left(E + \sup_{\tau \le t} \int_{\rho}^{\infty} z^2(\psi, \tau) d\psi \right) + C_5 t^{(2n+1)/2n^2} \left(-\frac{\partial E}{\partial \rho} \right)^{(3n+1)/2(n+1)}, \end{aligned} \tag{8.17}$$

$$\begin{aligned} &E + \sup_{\tau \le t} \int_{\rho}^{\infty} z^2(\tau, \rho) d\rho + \int_{0}^{t} |v_0(x)| \frac{z^2(0, x)}{2} dx \\ &\le C_6 \left(t^{(2n+1)/2n^2} \left(-\frac{\partial E}{\partial \rho} \right)^{(3n+1)/2(n+1)} + C_2 (\rho_0 - \rho)_+^{(3n+1)/(n-1)} \right). \end{aligned} \tag{8.18}$$

When we study the finite speed of propagation effect, we choose $\rho \geq \rho_0$, and it follows from (8.18) that

$$E \leq C_6 t^{(2n+1)/(2n^2)} \left(-\frac{\partial E}{\partial \rho}\right)^{(3n+1)/2(n+1)} \tag{8.19}$$

or, respectively,

$$E_\rho + C_7 t^{-\theta} E^\nu \leq 0, \quad \nu = \frac{2(n+1)}{3n+1} < 1 \quad \text{if } 1 < n, \tag{8.20}$$

$$C_7 = C_6^{2(n+1)/3n+1}, \quad \theta = \frac{2(2n+1)(n+1)}{2n^2(3n+1)},$$

$$E^{1-\nu}(\rho) \leq E^{1-\nu}(\rho_0) - C_7 t^{-\theta}(\rho - \rho_0). \tag{8.21}$$

Theorem 8.1 (Finite speed of propagation of disturbances). *Let assumptions* (8.7), (8.8) *hold. Then*

$$z(x, \psi) = U^2(x) - w(x, \psi) \equiv 0$$

for

$$\psi \geq \rho(t) = \rho_0 - t^\theta / C_7 E^{1-\nu}(\rho_0, t).$$

To prove the existence of the waiting time effect, we consider inequality (8.18) with $\rho \in (\rho_1, \rho_0)$, assuming that

$$E(\rho_1, X) = E_1 < \infty. \tag{8.22}$$

From (8.18), we have

$$E \leq C_6 \left(t^{(2n+1)/2n^2} \left(-E_\rho\right)^{(3n+1)/2(n+1)} + C_2(\rho_0 - \rho)^{(3n+1)/n-1}\right). \tag{8.23}$$

The solutions of inequality (8.23) admit the estimates

$$E(\rho, t) \leq E_1(\rho_0 - \rho_1)^{-\sigma} (\rho_0 - \rho)_+^\sigma, \quad \rho \in [\rho_1, \rho_0]$$

with $\sigma = (3n+1)/(n-1)$ if the values E_1, t, and C_2 from (8.15) satisfy the inequality

$$E_1 \geq C_6 \left(t^{(2n+1)/(2n^2)} (E_1 \sigma)^{(3n+1)/2(n+1)} + C_2\right). \tag{8.24}$$

Theorem 8.2 (The waiting time phenomenon). *Let conditions* (8.7), (8.8), (8.16) *hold and the given constants* E_1, C_2 *satisfy the condition*

$$E_1 > C_2 C_6.$$

Then there exists t_0, *defined by* (8.24),

$$t_0 = \left(\frac{E_1 - C_2 C_6}{C_6 (E_1 \sigma)^{1/\nu}}\right)^{2n^2/(2n+1)},$$

such that

$$E(\rho_0, t) \equiv 0, \qquad 0 \le t \le t_0,$$
$$z(x, \psi) \equiv U^2(x)w(x, \psi) = 0, \quad \psi_0 \le \psi < \infty, \quad 0 \le x \le t_0.$$

Remark 8.1. All the above energy estimates can be established directly for the original system of equations of boundary layer in terms of the physical variables (x, y).

Remark 8.2. The statement of Theorem 8.2 remains true if the difference $U^2(0) - w_0(\psi)$ (now vanishing like some power of $(\psi - \psi_0)$ near the point $\psi = \psi_0$) vanishes in an arbitrary way provided that

$$\int_{\rho_0}^{\rho} \frac{U^2(0) - w_0(\psi)}{(\psi - \psi_0)^{\mu}} d\psi < \infty$$

with some appropriate constant μ determined by the given value n.

9 Boussinesq system involving nonlinear thermal diffusion

9.1 The model. The Boussinesq system of hydrodynamics equations [93] arises from a zero-order approximation to the coupling between the Navier–Stokes equations and the thermodynamic equation [252]. The presence of density gradients in a fluid leads to the conversion of gravitational potential energy into motion through the action of buoyant forces. Density differences are induced, for instance, by gradients of temperature arising by nonuniform heating of the fluid. In the Boussinesq approximation of a large class of flow problems, thermodynamical coefficients such as viscosity, specific heat and thermal conductivity can be assumed constant leading to a coupled system with linear second-order operators in the Navier–Stokes and heat equations; see, e.g., [159, 164, 197, 298]. However, there are some fluids like lubricants or some plasma flow for which this is no longer an accurate assumption; see, e.g., [185, 278]. In this situation, the following system of quasilinear equations must be considered [252]:

$$\begin{cases} \mathbf{u}_t + (\mathbf{u} \cdot \nabla)\mathbf{u} - \operatorname{div}(\mu(\theta)\mathbf{D}(\mathbf{u})) + \nabla p = \mathbf{F}(\theta), \\ \operatorname{div} \mathbf{u} = 0, \\ \mathcal{C}(\theta)_t + \mathbf{u} \cdot \nabla \mathcal{C}(\theta) - \Delta\varphi(\theta) = 0, \end{cases} \tag{9.1}$$

with $\mathbf{u}$ the velocity field of the fluid, θ its temperature, p the pressure, $\mu(\theta)$ the viscosity of the fluid, $\mathbf{F}(\theta)$ the buoyancy force, $\mathbf{D}(\mathbf{u}) := \nabla\mathbf{u} + \nabla^*\mathbf{u}$, and

$$\mathcal{C}(\theta) := \int_{\theta_0}^{\theta} C(s)ds \quad \text{and} \quad \varphi(\theta) := \int_{\theta_0}^{\theta} \kappa(s)ds$$

with $C(\theta)$ and $\kappa(\theta)$ the specific heat and thermal conductivity of the fluid, respectively. Assuming, as usual, that $C > 0$, then $\mathcal{C}$ is inversible, and so $\theta = \mathcal{C}^{-1}(\bar{\theta})$

for some real argument $\bar{\theta}$. Then we can define the functions

$$\bar{\varphi}(\bar{\theta}) := \varphi \circ \mathcal{C}^{-1}(\bar{\theta}), \quad \bar{\mathbf{F}}(\bar{\theta}) := \mathbf{F} \circ \mathcal{C}^{-1}(\bar{\theta}), \quad \bar{\mu}(\bar{\theta}) := \mu \circ \mathcal{C}^{-1}(\bar{\theta}).$$

Substituting these expressions in (9.1), we obtain a formulation of the Boussinesq system,

$$\begin{cases} \mathbf{u}_t + (\mathbf{u} \cdot \nabla)\mathbf{u} - \operatorname{div}(\bar{\mu}(\theta)\mathbf{D}(\mathbf{u})) + \nabla p = \bar{\mathbf{F}}(\theta), \\ \operatorname{div} \mathbf{u} = 0, \\ \theta_t + \mathbf{u} \cdot \nabla\theta - \Delta\bar{\varphi}(\theta) = 0. \end{cases} \tag{9.2}$$

Notice that since C and κ are positive, their primitives $\mathcal{C}$ and φ are increasing functions. Suppose that a perturbation of a constant temperature θ_0 causes a small gradient of temperature between the boundary (higher temperature) and the interior (lower temperature) in a neighborhood, and assume that the behavior of $\mathcal{C}$ and $\bar{\varphi}$ near θ_0 can be approximated by

$$\mathcal{C}(s) \sim c_1 (s - \theta_0) + c_2 (s - \theta_0)^p, \qquad \bar{\varphi}(s) \sim k_1 (s - \theta_0) + k_2 (s - \theta_0)^q$$

for $s > \theta_0$ with $p, q > 0$. We have

$$\bar{\varphi}'(\mathcal{C}(s)) = \frac{\varphi'(s)}{\mathcal{C}'(s)} = \frac{k_1 + k_2 q (s - \theta_0)^{q-1}}{c_1 + c_2 p (s - \theta_0)^{p-1}}.$$

Therefore, when $s \to \theta_0$ (and then $\mathcal{C}(s) \to 0$), we get one of the following behaviors of $\bar{\varphi}'$ close to zero:

(i) if $p, q > 1$, then $\bar{\varphi}'(0) = k_1/c_1$;

(ii) if either $1 > q > p$ or $q > 1 > p$, then $\lim_{\mathcal{C}(s)\to 0} \bar{\varphi}'(\mathcal{C}(s)) = 0$;

(iii) if either $p > 1 > q$ or $1 > p > q$, then $\lim_{\mathcal{C}(s)\to 0} \bar{\varphi}'(\mathcal{C}(s)) = +\infty$.

In the first case, the linear parts of the decomposition of $\mathcal{C}$ and φ dominate; this case arises, for instance, when conductivity and specific heat are constant, leading to a convection-diffusion equation with a linear diffusion term. In the other two cases, the nonlinear parts dominate, leading to two qualitatively different situations:

1. If $p < q$, the specific heat dominates over the conductivity, i.e., when temperature approaches θ_0, the fluid stores more heat, and conductivity is sharply reduced. We shall prove that a front of temperature $\theta = \theta_0$ may arise. This type of phenomenon is known as *slow diffusion*; heat spends a positive time to spread over the neighborhood.

2. If $p > q$, the opposite effect occurs: the conductivity dominates over the specific heat. In this case the phenomenon is called *fast diffusion*. We shall prove that in some cases, θ stabilizes toward a constant value in a finite time.

Definition 9.1. The pair $(\mathbf{u}, \theta)$ is said to be a *weak solution* of (9.2) if

(i) We have

$$\mathbf{u} \in L^2(0,T; W^{1,2}_\sigma(\Omega)) \cap L^\infty(0,T; L^2_\sigma(\Omega)),$$
$$\varphi(\theta) \in \varphi_D + L^2(0,T; H^1_0(\Omega)), \quad \theta \in L^\infty(Q_T), \quad \text{and} \quad \theta \geq 0.$$

(ii) $\mathbf{u}(\cdot,0) = \mathbf{u}_0$ and for any test function $\mathbf{w} \in W^{1,2}_\sigma(\Omega) \cap L^N_\sigma(\Omega)$, we have

$$\int_\Omega (\mathbf{u}_t \cdot \mathbf{w} + (\mathbf{u}\cdot\nabla)\mathbf{u}\cdot\mathbf{w} + \mu(\theta)\mathbf{D}(\mathbf{u}) : \nabla\mathbf{w}) = \int_\Omega \mathbf{F}(\theta)\cdot\mathbf{w} \tag{9.3}$$

for a.e. $t \in (0,T)$.

(iii) $\theta_t \in L^2(0,T; H^{-1}(\Omega))$ and for any test functions $\zeta \in L^2(0,T; H^1_0(\Omega))$ and $\psi \in L^2(0,T; H^1_0(\Omega)) \cap W^{1,1}(0,T; L^2(\Omega))$ with $\psi(T) = 0$, we have

$$\int_0^T \langle\theta_t, \zeta\rangle + \int_0^T\int_\Omega (\nabla\varphi(\theta) - \theta\mathbf{u})\cdot\nabla\zeta = 0 \tag{9.4}$$

and

$$\int_0^T \langle\theta_t, \psi\rangle + \int_0^T\int_\Omega (\theta - \theta_0)\,\psi_t = 0, \tag{9.5}$$

with $\langle\cdot,\cdot\rangle$ denoting the duality product between $H^{-1}(\Omega)$ and $H^1_0(\Omega)$.

It is proved in [132] that under the assumptions

$$\begin{cases} \varphi \in \mathcal{C}([0,\infty)) \cap \mathcal{C}^1((0,\infty)), \quad \varphi(0) = 0, \quad \varphi \text{ nondecreasing}, \\ \mathbf{F} \in \mathcal{C}^{0,1}_{\mathrm{loc}}([0,\infty); \mathbb{R}^N), \\ \mu \in \mathcal{C}^{0,1}_{\mathrm{loc}}([0,\infty)) \quad \text{satisfying } 0 < m_0 \leq \mu(s) \leq m_1 \quad \forall s \in [0,\infty), \\ \quad \text{and if } \mu' \neq 0 \text{ or } \mathbf{F}' \neq 0, \text{ then } \varphi^{-1} \text{ is Hölder continuous of exponent } \alpha, \end{cases}$$

there exists a weak solution of (9.2).

9.2 Spatial localization. Let us assume $\varphi(s) := s^m$, with $m > 1$. Given a weak solution $(\mathbf{u}, \theta)$ of (9.2), we perform the change of unknown $\hat\theta := \theta^m$. Writing $q := 1/m$, introducing the expression of $\hat\theta$ in the second equation of (9.2) and omitting the hats, we are led to the form of the heat equation

$$\left(\theta^q\right)_t + \mathbf{u}\cdot\nabla\theta^q - \Delta\theta = 0.$$

In order to define the characteristics corresponding to $\mathbf{u}$ we need to assume certain regularity on the velocity field, namely $\mathbf{u} \in \mathcal{C}([0,T]; \mathcal{C}^1_\sigma(\bar\Omega))$. Since $\mathbf{u} = \mathbf{0}$ in $\partial\Omega$, we can extend $\mathbf{u}$ by zero to all $\mathbb{R}^N$ to get $\mathbf{u} \in \mathcal{C}([0,T]; \mathcal{C}^1_\sigma(\mathbb{R}^N))$. It is well known that with this regularity, the characteristics χ given as a solution of the problem

$$\begin{cases} \dfrac{\partial\chi}{\partial t}(\mathbf{x},t) = \mathbf{u}(\chi(\mathbf{x},t),t) & \text{in } \mathbb{R}^N \times (0,T), \\ \chi(\mathbf{x},0) = \mathbf{x} & \text{in } \mathbb{R}^N, \end{cases} \tag{9.6}$$

are uniquely defined with the regularity $\chi \in C^1(\mathbb{R}^N \times [0, T])$.

In the following, we shall suppose that the initial data θ_0 vanishes in some ball $\mathbf{B}_{\rho_0}$ of radius ρ_0 centered in $\mathbf{x}_0$ and compactly imbedded in Ω. Then the following property is a consequence of the continuity of χ:

$$\begin{cases} \text{there exist } \hat{t} > 0 \text{ and } \rho_1 > \rho_0 \text{ such that if } t < \hat{t} \text{ and } \rho < \rho_1, \\ \text{then } \chi(\mathbf{B}_\rho, t) \subset \Omega. \end{cases} \tag{9.7}$$

We first state the result on the finite speed of propagation and waiting time property *along the characteristics*.

Theorem 9.1. *Suppose that $q < 1$ and $\theta_0 \equiv 0$ in $\mathbf{B}_{\rho_0}$. Then there exists $t^* \in (0, \hat{t})$ and a continuous function $r(\tau)$ defined in $(0, t^*)$, with $r(0) = \rho_0$, such that any second component of a weak solution of (9.2) satisfies*

$$\theta(\mathbf{x}, t) \equiv 0 \quad a.e.\ in\ \left\{(\mathbf{x}, t) : \mathbf{x} \in \chi(\mathbf{B}_{r(t)}, t),\ t \in (0, t^*)\right\}.$$

Besides, if the initial data satisfies the following flatness condition

$$\int_{\mathbf{B}_\rho} |\theta_0|^{q+1} \leq \delta_0(\rho - \rho_0)_+^{1/(1-\kappa)} \quad for\ a.e.\ \rho < \rho_1, \tag{9.8}$$

for some $\delta_0 > 0$ and with $\kappa < 1$ given by (9.24) then there exists $t_* \in (0, \hat{t})$ such that

$$\theta(\mathbf{x}, t) \equiv 0 \quad a.e.\ in\ \left\{(\mathbf{x}, t) : \mathbf{x} \in \chi(\mathbf{B}_{\rho_0}, t),\ t \in (0, t_*)\right\}.$$

Finally, we state a result on finite speed of propagation for which weaker requirements on the regularity of the velocity field are needed. We shall only assume $\mathbf{u} \in L^\infty_\sigma(Q_T)$. Although in this case we cannot define the characteristics corresponding to $\mathbf{u}$ in a classical sense, the boundedness of the velocity field still allows us to use ideas similar to those of Theorem 9.1.

Theorem 9.2. *Suppose that $q < 1$ and $\theta_0 \equiv 0$ in $\mathbf{B}_{\rho_0}$. Then there exist $t^* > 0$ and a nonnegative function $r(\tau)$ defined in $(0, t^*)$, with $r(0) = \rho_0$, such that any second component of a weak solution of (9.2) satisfies*

$$\theta(\mathbf{x}, t) \equiv 0 \quad a.e.\ in\ \left\{(\mathbf{x}, t) : \mathbf{x} \in \mathbf{B}_{r(t)},\ t \in (0, t^*)\right\}.$$

9.3 Proofs of Theorems 9.1 and 9.2. In the proof of both theorems, we shall use special test functions to localize the natural energies corresponding to the problem in suitable sets of the form

$$\mathcal{P} \equiv \mathcal{P}(\rho, t) := \left\{(\mathbf{x}, \tau) \in \mathbb{R}^N \times \mathbb{R}_+ : \mathbf{x} \in \chi(\mathbf{B}_R, \tau),\ \tau \in (0, t)\right\}, \tag{9.9}$$

with $\mathbf{B}_R$ the ball of $\mathbb{R}^N$ centered at the origin and of radius $R \equiv R(\rho, \tau)$, for certain nonnegative arguments ρ, τ, and with χ given by (9.6). We introduce the time sections of $\mathcal{P}$,

$$P(s) := \mathcal{P} \cap \{\tau = s\} \subset \mathbb{R}^N, \quad s \in (0, t), \tag{9.10}$$

so $\mathcal{P} = \bigcup_{s\in(0,t)} P(s)$, and the lateral boundary of $\mathcal{P}$,

$$\partial_l \mathcal{P} := \left\{ (\mathbf{x}, \tau) \in \mathbb{R}^N \times \mathbb{R}_+ : \mathbf{x} \in \partial\chi(\mathbf{B}_R, \tau),\ \ \tau \in (0, t) \right\},$$

so the parabolic boundary of $\mathcal{P}$ is given by $\partial\mathcal{P} = P(0) \cup P(t) \cup \partial_l\mathcal{P}$. Let us formulate some basic properties related to this collection of sets. If $(n_t, \mathbf{n}_x)$ is the unitary outward normal vector to $\mathcal{P}$, then we have

$$n_t + \mathbf{u} \cdot \mathbf{n}_x = -\frac{\partial R}{\partial \tau} \det(\frac{D\chi}{D\mathbf{x}}), \tag{9.11}$$

and for any integrable function g, it holds that

$$\frac{d}{d\rho} \int_{\mathcal{P}(\rho,t)} g(\mathbf{x}, \tau) d\mathbf{x} d\tau = \int_{\partial_l \mathcal{P}(\rho,t)} \frac{dR}{d\rho} g(\mathbf{x}, \tau) d\mathbf{x} d\tau, \tag{9.12}$$

where we used the fact that, $\chi(\cdot, \tau)$ being a diffeomorphism, we have $\partial\chi(\mathbf{B}_R, \tau) = \chi(\partial\mathbf{B}_R, \tau)$. In this domain, we consider the *local energy functions* defined by

$$E(\rho, t) := \int_{\mathcal{P}} |\nabla\theta(\mathbf{x}, \tau)|^2 \quad \text{and} \quad b(\rho, t) := \operatorname*{ess\,sup}_{\tau\in(0,t)} \int_{P(\tau)} |\theta(\mathbf{x}, \tau)|^{q+1}. \tag{9.13}$$

Next, we shall show how the local energies come into our problem. If θ is the second component of any weak solution of (9.2), we choose the following test function in (9.4) and (9.5):

$$\xi_{\mathcal{P}}(\theta(\mathbf{x}, t)) := \begin{cases} \theta(\mathbf{x}, t) & \text{if } (\mathbf{x}, t) \in \mathcal{P}, \\ 0 & \text{otherwise.} \end{cases}$$

In [149], it is proven that $\xi_{\mathcal{P}}$ is an admissible test function. After applying the divergence theorem, we get

$$\frac{q}{q+1} \int_{\partial\mathcal{P}} (\theta^{q+1} n_t + \theta^{q+1} \mathbf{u} \cdot \mathbf{n}_x)\, d\mathbf{x}\, d\tau + \int_{\mathcal{P}} |\nabla\theta|^2\, d\mathbf{x}\, d\tau = \int_{\partial\mathcal{P}} \theta\nabla\theta \cdot \mathbf{n}_x\, d\mathbf{x}\, d\tau.$$

Using the decomposition $\partial\mathcal{P} = P(0) \bigcup P(t) \bigcup \partial_l\mathcal{P}$ and taking into account that $n_t = 1$ and $\mathbf{n}_x = 0$ in $P(t)$ and $n_t = -1$ and $\mathbf{n}_x = 0$ in $P(0)$, we obtain

$$\begin{aligned} &\frac{q}{q+1} \int_{P(t)} |\theta(t)|^{q+1}\, d\mathbf{x} + \int_{\mathcal{P}} |\nabla\theta|^2\, d\mathbf{x}\, d\tau \\ &\quad \le \int_{\partial_l\mathcal{P}} |\nabla\theta|\, |\theta|\, |\mathbf{n}_x|\, d\mathbf{x}\, d\tau - \frac{q}{q+1} \int_{\partial_l\mathcal{P}} |\theta|^{q+1} (n_t + \mathbf{u} \cdot \mathbf{n}_x)\, d\mathbf{x}\, d\tau \\ &\quad + \frac{q}{q+1} \int_{P(0)} |\theta_0|^{q+1}\, d\mathbf{x} \\ &\quad := j_1 + j_2 + j_3. \end{aligned} \tag{9.14}$$

The proofs of the theorems share a common scheme: First, we perform estimates of the terms j_i of (9.14) by using well-known inequalities, such as those of Hölder, Young, and Poincaré, together with an interpolation-trace inequality (see [149]), which allows us to deduce a differential inequality for the local energy E. Second, a direct integration of such inequality leads to the desired result.

We start with the proof of Theorem 9.2. Notice that for this result we do not need to consider the characteristics corresponding to $\mathbf{u}$, so we simply fix $\chi = Identity$ in the definition of $\mathcal{P}$.

In the proof of Theorem 9.2, we proceed in two steps.

Step 1. We consider the domain of integration in (9.14) as the collection of truncated cones given by

$$\mathcal{P}(\rho, t) := \{(\mathbf{x}, \tau) : \mathbf{x} \in \mathbf{B}_R, \ \tau \in (0, t)\},$$

with $R \equiv R(\rho, \tau) := \rho - \tau u$, $u := \|\mathbf{u}\|_{L^\infty(\mathbf{B}_{\rho_0} \times (0,T))}$, $\rho \in I := (tu + \varepsilon, \rho_0]$, and $t < t_1 := \min\{\hat{t}, \frac{\rho_0 - \varepsilon}{u}\}$, with $\hat{t}$ given by (9.7). Notice that I is nonempty since by construction $tu + \varepsilon < \rho_0$. We begin by estimating the term j_1. On the one hand, the function $\rho \to \int_{\partial_l \mathcal{P}} |\nabla\theta|^2 \, d\Gamma$ is well defined for a.e. $\rho \in I$ due to the regularity $\theta \in L^2(0, T; H^1(\Omega))$. We can use Hölder's inequality to get

$$j_1 \le \left(\int_{\partial_l \mathcal{P}} |\theta|^2 \, d\Gamma \right)^{1/2} \left(\int_{\partial_l \mathcal{P}} |\nabla\theta|^2 \, d\Gamma \right)^{1/2}. \tag{9.15}$$

On the other hand, by (9.12) we have $\frac{\partial E}{\partial \rho} = \int_{\partial_l \mathcal{P}} |\nabla\theta|^2 \, d\Gamma$ a.e. $\rho \in I$, and from (9.15) we get

$$j_1 \le \|\theta\|_{L^2(\partial_l \mathcal{P}(\rho))} \left(\frac{\partial E}{\partial \rho}(\rho, t) \right)^{1/2}. \tag{9.16}$$

Taking into account $\partial_l \mathcal{P} = \bigcup_{\tau \in (0,t)} \partial P(\tau)$ with $\partial P(\tau) \equiv \partial \mathbf{B}_R$ and applying the interpolation-trace inequality (see [149]), we obtain

$$\|\theta\|_{L^2(\partial P(\tau))} \le C_1 \left(\|\nabla\theta\|_{L^2(P(\tau))} + R^{-\delta} \|\theta\|_{L^{q+1}(P(\tau))} \right)^\gamma \|\theta\|^{1-\gamma}_{L^{q+1}(P(\tau))}, \tag{9.17}$$

with $C_1 > 0$ a universal constant and with

$$\gamma := \frac{2N - (q+1)(N-1)}{2(N+q+1) - N(q+1)} \in (0, 1) \quad \text{and} \quad \delta := 1 + \frac{N(1-q)}{2(q+1)}.$$

Notice that since $\rho \in I$,

$$R(\tau)^{-\delta} \le \left(\min_{\tau \in (0,t)} (\rho - u\tau) \right)^{-\delta} \le \varepsilon^{-\delta}.$$

Defining $L := C_1 \max\{1, \varepsilon^{-\gamma\delta}\}$ and integrating the expression (9.17) in $(0, t)$, we obtain

$$\int_0^t \|\theta\|^2_{L^2(\partial P(\tau))} \, d\tau \le L \int_0^t \left(\|\nabla\theta\|_{L^2(P(\tau))} + \|\theta\|_{L^{q+1}(P(\tau))} \right)^{2\gamma} \|\theta\|^{2(1-\gamma)}_{L^{q+1}(P(\tau))} \, d\tau,$$

and thanks to the inequality $(a+b)^2 \le 2(a^2+b^2)$ and to Hölder's inequality with exponent $1/\gamma$, we deduce

$$\int_0^t \|\theta\|^2_{L^2(\partial P(\tau))}\,d\tau \le 4L^2 \int_0^t \left(\|\nabla\theta\|^2_{L^2(P(\tau))} + \|\theta\|^2_{L^{q+1}(P(\tau))}\right)^{\gamma} d\tau \\ \times \left(\int_0^t \|\theta\|^2_{L^{q+1}(P(\tau))}\,d\tau\right)^{1-\gamma}. \tag{9.18}$$

Using $q < 1$ and b nondecreasing with respect to ρ leads to

$$\left(\int_0^t \|\theta\|^2_{L^2(\partial P(\tau))}\,d\tau\right)^{1/2} \le 2Lt^{(1-\gamma)/2}\left(E(\rho,t) + tb(\rho_0,t)^{\frac{2}{q+1}-1} b(\rho,t)\right)^{\gamma/2} \\ \times b(\rho,t)^{(1-\gamma)/(q+1)},$$

with E and b given in (9.13). Defining

$$K_0(t) := 2Lt^{(1-\gamma)/2}\max\left\{1, t_1 b(\rho_0,t_1)^{\frac{2}{q+1}-1}\right\}, \tag{9.19}$$

we get

$$\left(\int_0^t \|\theta\|^2_{L^2(\partial P(\tau))}\,d\tau\right)^{1/2} \le K_0(t)\,(b(\rho,t)+E(\rho,t))^{\nu}$$

with

$$\nu := \frac{\gamma}{2} + \frac{1-\gamma}{q+1} < 1. \tag{9.20}$$

We deduce from (9.16)

$$j_1 \le K_0(t)\left(\frac{\partial E(\rho,t)}{\partial\rho}\right)^{1/2} (b(\rho,t)+E(\rho,t))^{\nu}. \tag{9.21}$$

Finally, by assumption, $j_3 = 0$ and the estimate of j_2 is simple because of the choice of the domain of integration. We have

$$\mathbf{n}_t + \mathbf{u}\cdot\mathbf{n}_x = \|\mathbf{u}\|_{L^\infty} + \mathbf{u}\cdot\mathbf{n}_x \ge 0,$$

and therefore $j_2 \le 0$.

Step 2. From (9.14) and the estimates obtained in Step 1 for j_1, j_2, and j_3, we have

$$2M\left(\int_{P(t)} |\theta|^{q+1}\,dt + E(\rho,t)\right) \le K_0(t)\left(\frac{\partial E(\rho,t)}{\partial\rho}\right)^{1/2}(b(\rho,t)+E(\rho,t))^{\nu}, \tag{9.22}$$

with $M = q/2(q+1)$. Since functions E, b and $\frac{\partial E}{\partial \rho}$ are nondecreasing in time, (9.22) remains valid if we change $\int_{P(t)} |\theta|^{q+1} dt$ by $b(\rho, t)$ and $2M$ by M. We get

$$M\left(b(\rho,t) + E(\rho,t)\right) \leq K_0(t) \left(\frac{\partial E(\rho,t)}{\partial \rho}\right)^{1/2} \left(b(\rho,t) + E(\rho,t)\right)^{\nu},$$

and therefore

$$M^2 \left(b(\rho,t) + E(\rho,t)\right)^{\kappa} \leq K^2(t) \frac{\partial E(\rho,t)}{\partial \rho} \tag{9.23}$$

for a.e. $\rho \in I$ and $t < t_1$ and with

$$\kappa := 2(1-\nu), \tag{9.24}$$

ν given by (9.20) and

$$K(t) := \frac{K_0(t)}{M} = \frac{C_1}{M} \max\left\{1, \varepsilon^{\gamma\delta}\right\}^{1/2} \max\left\{1, t_1 M^{\frac{2}{q+1}-1}\right\} t^{\frac{1-\gamma}{2}}. \tag{9.25}$$

Due to the crucial assumption $q < 1$, we have $\kappa < 1$, and a direct integration of (9.23) in (ρ, ρ_0) with $\rho \in I$ leads to

$$E^{1-\kappa}(\rho,t) \leq E^{1-\kappa}(\rho_0,t) - K^{-2}(t)(\rho_0 - \rho). \tag{9.26}$$

We define

$$z(t) := \rho_0 - K^2(t) E^{1-\kappa}(\rho_0, t), \quad \text{for } t > 0.$$

Using the definitions of E and K, we easily deduce that z is continuous and decreasing with $z(0) = \rho_0$ and $\lim_{t\to\infty} z(t) = -\infty$. Therefore, (9.26) implies that

$$E(\rho,t) = 0 \quad \text{a.e. in } P_0 := \left\{(\rho,t) : t \in (0,t^*),\ \rho \in (ut+\varepsilon, z(t))\right\}$$

with $t^* \in (0, \min\{t_0, t_1\})$. Notice that P_0 is nonempty if

$$\rho_0 > K^2(t) E^{1-\kappa}(\rho_0,t) + ut + \varepsilon \quad \text{for } t \in (0, T^*),$$

which we may ensure by taking t^*, ε small enough. Finally, from the definition of $E(\rho,t)$, we deduce $\theta(\mathbf{x},t) \equiv 0$ if $|\mathbf{x}| \leq \rho - ut \leq z(t) - ut := r(t)$ with $r(0) = \rho_0$ and $r(t) > 0$ in $(0, t^*)$. □

In the proof of Theorem 9.1, we closely follow the proof of Theorem 9.2 but take advantage of the possibility of defining the characteristics corresponding to $\mathbf{u}$. We shall only prove the waiting time property, the proof of the finite speed of propagation being a straightforward modification of the former.

Step 1. We consider the domain of integration given by

$$\mathcal{P} \equiv \mathcal{P}(\rho,t) := \{(\mathbf{x},\tau) : \mathbf{x} \in \chi(B_R,\tau),\ \tau \in (0,t)\}, \tag{9.27}$$

with χ defined as the unique solution of (9.6), $R(\tau) \equiv \rho$ with $\rho \in (\rho_0, \rho_1)$, and $t \in (0, \hat{t})$ with $\hat{t}$ given by (9.7). Using (9.12) and Hölder's inequality, we estimate j_1 as in (9.15):

$$\begin{aligned} j_1 &\leq \int_{\partial_l \mathcal{P}} |u|\,|\nabla u|\,|\mathbf{n}_x| \leq \left(\int_{\partial_l \mathcal{P}} |\nabla u|^2\, d\Gamma\right)^{1/2} \left(\int_{\partial_l \mathcal{P}} |u|^2\,|\mathbf{n}_x|^2\, d\Gamma\right)^{1/2} \\ &\leq \left(\frac{\partial E}{\partial \rho}\right)^{1/2} \left(\int_{\partial_l \mathcal{P}} |u|^2\, d\Gamma\right)^{1/2}, \end{aligned} \tag{9.28}$$

where we used $|\mathbf{n}_x| \leq 1$. Since χ is a diffeomorphism, the interpolation-trace inequality remains valid for the same exponents γ and δ although with, in general, a different universal constant C_1 depending on the Jacobian of χ,

$$\|\theta\|_{L^2(\partial P(\tau))} \leq C_1 \left(\|\nabla\theta\|_{L^2(P(\tau))} + \rho^{-\delta}\, \|\theta\|_{L^{q+1}(P(\tau))}\right)^{\gamma} \|\theta\|^{1-\gamma}_{L^{q+1}(P(\tau))}.$$

Now in the same way as in the previous proof, we arrive at the estimate

$$j_1 \leq K_0(\rho, t) \left(\frac{\partial E}{\partial \rho}(\rho, t)\right)^{1/2} (b(\rho, t) + E(\rho, t))^{\nu}, \tag{9.29}$$

with ν given by (9.20) and K_0 given by (9.19) but with L replaced by $L(\rho) := C_1 \max\{1, \rho^{-\gamma\delta}\}$. Applying Young's inequality with exponent $1/\nu$ to (9.29), we have for all $\varepsilon > 0$

$$j_1 \leq C_\varepsilon K_0^{\frac{1}{1-\nu}} \left(\frac{\partial E(\rho, t)}{\partial \rho}\right)^{1/\kappa} + \varepsilon\, (b(\rho, t) + E(\rho, t)), \tag{9.30}$$

with $C_\varepsilon = \varepsilon^{\nu/(\nu-1)}$ and $\kappa := 2(1-\nu)$. The estimate of j_2 follows directly from (9.11) and the choice of the domain of integration. We have

$$\mathbf{n}_t + \mathbf{u} \cdot \mathbf{n}_x = -\frac{dR}{d\tau} \det\left(\frac{D\chi}{D\mathbf{x}}\right) \equiv 0,$$

because R does not depend on τ. Therefore, $j_2 \equiv 0$. Finally, by assumption, $j_3 \leq \delta_0 (\rho - \rho_0)_+^{\frac{1}{1-\kappa}}$.

Step 2. Taking $M = q/2(q+1)$, we get from (9.14)

$$\begin{aligned} \frac{M}{2} (b(\rho, t) + E(\rho, t)) &\leq C_\varepsilon K_0^{1/1-\nu}(\rho, t) \left(\frac{\partial E(\rho, t)}{\partial \rho}\right)^{1/\kappa} \\ &\quad + \varepsilon\, (b(\rho, t) + E(\rho, t)) + \delta_0 (\rho - \rho_0)_+^{1/(1-\kappa)}, \end{aligned} \tag{9.31}$$

with $\rho \in (\rho_0, \rho_1)$ and $t \in (0, \hat{t})$. Choosing $\varepsilon < M/4$ and raising both sides of (9.31) to the power κ leads to

$$E(\rho, t)^{\kappa} \leq \hat{C}_1(t) \frac{\partial E(\rho, t)}{\partial \rho} + \hat{C}_2 (\rho - \rho_0)_+^{\frac{\kappa}{1-\kappa}} \tag{9.32}$$

for a.e. $\rho \in (\rho_0, \rho_1)$ and $t \in (0, \hat{t})$ with

$$\hat{C}_1(t) := c\left(\frac{4C_\varepsilon}{M} \max_{\rho\in(\rho_0,\rho_1)} K_0^{1/1-\nu}(\rho, t)\right)^\kappa =: \tilde{c}t^{1-\gamma}, \quad \hat{C}_2 := c\left(\frac{4C_\varepsilon\delta_0}{M}\right)^\kappa$$

and c the constant of the inequality $(a+b)^\kappa \leq c(a^\kappa + b^\kappa)$. Notice that we have again the equivalency $q < 1 \iff \kappa < 1$. To finish, we construct an upper bound for E, which will imply the result. We consider the problem

$$\begin{cases} z^\kappa(\rho) = \tilde{c}t_*^{1-\gamma}\dfrac{dz}{d\rho}(\rho) + \hat{C}_2(\rho-\rho_0)_+^{\frac{\kappa}{1-\kappa}} & \text{in } \rho \in (\rho_0, \rho_1), \\ z(\rho_1) \geq E(\rho_1, t_*), \end{cases}$$

where t_* is still to be chosen. The function $z(\rho) := A(\rho - \rho_0)_+^{\frac{1}{1-\kappa}}$ solves the problem if we choose the positive constant A such that

$$A^\kappa = \frac{\tilde{c}t_*^{1-\gamma}}{1-\kappa}A + \hat{C}_2 \quad \text{and} \quad A \geq E(\rho_1, t_*)(\rho_1 - \rho_0)^{\frac{-1}{1-\kappa}}. \tag{9.33}$$

It is not difficult to check the existence of such a constant A fulfilling (9.33) for t_* small enough when we impose the additional restriction $A^\kappa > \hat{C}_2$. By the monotonicity of the problem, we deduce $E(\rho, t) \leq z(\rho)$ in $\rho \in (\rho_0, \rho_1)$, $t \in (0, t_*)$ and therefore $E(\rho_0, t) = z(\rho_0) = 0$ in $t \in (0, t_*)$, from which the assertion follows. □

9.4 Time localization. The property of time localization or extinction in finite time in evolution problems is associated to parabolic singularity and it may appear both in linear and nonlinear problems; see, e.g., [126] and the references therein. The techniques to study this property are similar to those used to prove the existence of free boundaries in parabolic degenerated equations, and they involve either a comparison principle for the problem or the use of energy methods. We introduce more precisely the notion of extinction in finite time for problem (9.2).

Definition 9.2. Given a weak solution $(\mathbf{u}, \theta)$ of (9.2), we say that θ has the *property of extinction in finite time* if there exists $t_f > 0$ such that $\theta(\cdot, t) = 0$ a.e. in Ω for all $t \geq t_f$.

In problem (9.2), extinction in finite time occurs when the heat conduction is fast, which is related to the class of functions φ possesing a Lipschitz continuous inverse (and other additional properties), and it is a *global* property of solutions, i.e., a property involving the structure of the differential equation as well as the behavior of the auxiliary conditions that solutions satisfy.

In the previous section we showed that the properties of localization in space of the support of solutions of (9.2) are local and therefore independent of the boundedness of the domain. On the contrary, since the property of localization in time is global, it could be expected that the behavior of solutions is different when the domain under consideration is bounded or unbounded. And this is the case. For instance, for equations of the type

$$\theta_t - \Delta\varphi(\theta) = 0 \quad \text{in } \mathbb{R}^N \times (0, \infty) \tag{9.34}$$

with $\varphi(s) := |s|^m \operatorname{sign}(s)$, in [67] it is proven that the property holds if $0 < m < (N-2)/N$ and $N \geq 3$ with some other additional conditions. However, this same equation in a bounded domain has the extinction property if $0 < m < 1$. More precisely, a sufficient condition for the occurrence of this property for solutions of (9.2), which includes (9.34) in the particular case $\mathbf{u} = \mathbf{0}$, is given in the following theorem.

Theorem 9.3. *Let* $(\mathbf{u}, \theta)$ *be a weak solution of problem* (9.2) *with boundary data satisfying* $\varphi_D \equiv 0$ *a.e. on* Σ_T. *Assume that*

$$\varphi'(s) \geq cs^{m-1} \quad \textit{with } m \in (0, 1) \tag{9.35}$$

and $c > 0$. *Then* θ *has the extinction in finite time property.*

The proof is inspired by that given in [67]. The idea is to use θ^p, for a suitable p, as a test function for the second equation of problem (9.2). Doing this formally, we obtain

$$\frac{d}{dt}\int_\Omega \theta^{p+1}dx + \int_\Omega \mathbf{u}\cdot\nabla\theta^{p+1}dx + \frac{p+1}{p}\int_\Omega \nabla\varphi(\theta)\cdot\nabla\theta^p dx = 0. \tag{9.36}$$

Using (9.35) and $\operatorname{div}\mathbf{u} = 0$, we obtain

$$\frac{d}{dt}\int_\Omega \theta^{p+1}dx + \frac{4cp}{p+1}\int_\Omega \left|\nabla\theta^{\frac{p+m}{2}}\right|^2 dx \leq 0. \tag{9.37}$$

Sobolev's theorem implies that

$$\int_\Omega \left|\nabla\theta^{\frac{p+m}{2}}\right|^2 dx = \left\|\theta^{\frac{p+m}{2}}\right\|^2_{H^1_0(\Omega)} \geq C(q,\Omega)\left\|\theta^{\frac{p+m}{2}}\right\|^2_{L^q(\Omega)}$$

with $q \in [1, 2^*]$, and 2^* is the critical Sobolev exponent of the imbedding $H^1(\Omega) \subset L^{2^*}(\Omega)$. When $q = 2\frac{p+1}{p+m}$, we obtain

$$\int_\Omega \left|\nabla\theta^{\frac{p+m}{2}}\right|^2 dx \geq C(q,\Omega)\,\|\theta(t)\|^{p+m}_{L^{p+1}(\Omega)}. \tag{9.38}$$

We ensure that $q \in [1, 2^*]$ by taking $p \geq (2 - 2^*m)/(2^* - 2)$. Notice that in the case of $\Omega := \mathbb{R}^N$, Sobolev's theorem is only valid for $q = 2^*$, forcing us to take $p = \frac{2-2^*m}{2^*-2}$ and therefore limiting the range of m to $(0, (N-2)/N)$; see [67]. Gathering (9.37) and (9.38), we deduce

$$\frac{d}{dt}\|\theta(t)\|^{p+1}_{L^{p+1}(\Omega)} + C\,\|\theta(t)\|^{p+m}_{L^{p+1}(\Omega)} \leq 0,$$

with $C := \frac{4cpC(q,\Omega)}{p+1}$. Defining $E(t) := \|\theta(t)\|^{p+1}_{L^{p+1}(\Omega)}$ and $\alpha = \frac{p+m}{p+1}$, we formulate the above inequality as

$$\begin{cases} \dfrac{dE}{dt}(t) + CE(t)^{\alpha} \leq 0, \\ E(0) = \|\theta_0\|^{p+1}_{L^{p+1}(\Omega)} := E_0 \geq 0. \end{cases} \tag{9.39}$$

Due to the assumption that $m < 1$, we have $\alpha < 1$. The solution of the differential equation associated to (9.39) is an upper bound for any solution of the differential inequality. A direct integration shows that any solution of (9.39) must therefore satisfy

$$E(t) \leq \max\left\{\left(E_0^{1-\alpha} - C(1-\alpha)t\right)^{\frac{1}{1-\alpha}}, 0\right\},$$

and then $E(t) = 0$ for $t \geq t_f := E_0^{1-\alpha}/C(1-\alpha)$, from which the assertion follows.

10 Simultaneous motion in the surface channel and the underground water

10.1 System of equations: Statement of the problem. The mathematical models of simultaneous flows of surface and underground water were proposed in [47, 37] (see also [40, 44, 48, 43]). These models were based on the equations of hydraulics for open channels (the diffusion analogues of the Saint-Venant equations or the equations of diffusion waves) and the planar filtration equation (Boussinesq's equation).

In the simplest case, where the channel has a rectangular cross-section and is of constant width and the confining layer and the bottom of the channel are horizontal, the corresponding system of equations and the internal compatibility conditions are

$$\frac{\partial H}{\partial t} = \nabla(H\nabla H) + f_{\Omega}, \quad x \in \Omega/\Pi, \quad \Omega \subset \mathbb{R}^2, \quad t \in (0,T), \tag{10.1}$$

$$\begin{aligned} \frac{\partial u}{\partial t} &= \frac{\partial}{\partial s}\left(\psi(s,u)\left|\frac{\partial u}{\partial s}\right|^{-1/2}\frac{\partial u}{\partial s}\right) + \left[H\frac{\partial H}{\partial \mathbf{n}}\right]_{\Pi} + f_{\Pi}, \\ & x \in \Pi, \quad t \in (0,T), \end{aligned} \tag{10.2}$$

$$H\frac{\partial H}{\partial \mathbf{n}}\bigg|_{\pm} = \sigma(u - H_{\pm}), \quad x \in \Pi, \quad t \in (0,T). \tag{10.3}$$

Here $H(x,t)$ is the level of the underground water in the region $\Omega \subset \mathbb{R}^2$, $u(s,t)$ is the water level in the channel which corresponds to the plane curve Π, s is the arc length along Π, $\mathbf{n}$ is the normal to Π, $H_{\pm}$ are the values of H on the opposite sides of H, $[H]_{\Pi} = H_+ - H_-$, and $f_{\Omega}(x,t)$, $f_{\Pi}(s,t)$ represent the prescribed external flux of water—the "sources." The function ψ is defined by the formula

$$\psi(s,u) = R^{2/3}\omega/k \tag{10.4}$$

where $R(s, u)$ is the hydraulic radius (the ratio of the real section area and the wetted channel parameter), $\omega(s, u)$ is the cross-section area of the real channel (i.e., the cross-section of the area occupied by the water), $k(s)$ is the coefficient of roughness. Under the assumptions imposed, the function ψ satisfies the relations

$$\psi(s, u) = \psi_0(s, u)|u|^{5/3}, \quad |\ln \psi_0(s, u)| \leq C < \infty. \tag{10.5}$$

Let us point out that the equations of system (10.1), (10.2) are posed in different domains. Boussinesq's equation (10.1) is defined in the two-dimensional domain $\Omega \setminus \Pi$, which, generally speaking, is multiply connected. The diffusion analogue of the Saint-Venant equation (10.2) is posed on the curve Π. These equations are coupled by the jump across the curve of the term $[H\frac{\partial H}{\partial \mathbf{n}}]_\Pi$ on the right-hand side of (10.2) and through equation (10.3). Moreover, both equations degenerate with respect to H, u, and u_s. System (10.1)–(10.3) endowed with the boundary and initial conditions

$$\left(\sigma_1 \frac{\partial H}{\partial n} + \sigma_2 H\right) = g, \quad x \in \Gamma = \partial\Omega, \quad t \in (0, T), \tag{10.6}$$

$$\left(\kappa_1 \psi(s, u) \left|\frac{\partial u}{\partial s}\right|^{-1/2} \frac{\partial u}{\partial s} + \kappa_2 u\right) = g, \quad x \in \Pi \cap \Gamma, \quad t \in (0, T), \tag{10.7}$$

$$H(x, 0) = H_0(x), \quad u(x, 0) = u_0(x), \quad x \in \Omega, \tag{10.8}$$

for the functions $H(x, t)$, $u(s, t)$ is complete. The existence of a weak solution of system (10.1)–(10.3) $\mathbf{V} = (H, u) \in W$,

$$W = \Big\{(H, u) : 0 \leq (H, u) \leq M, \; H^{1/2}\nabla H \in L^2\left(0, T; L^2(B_\rho)\right),$$
$$\psi^{2/3}\left|\frac{\partial u}{\partial s}\right| \in L^{3/2}\left(0, T; L^{3/2}(\Pi_\rho)\right)\Big\}, \quad 0 < \rho < \infty,$$

was proved in [47, 37, 48, 39] for the basic initial-boundary value problems (including (10.6)–(10.8)).

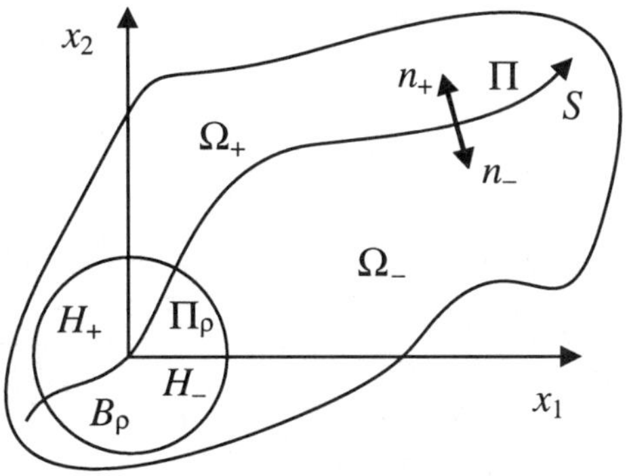

Figure 10.1: The horizontal cross-section of the flow domain.

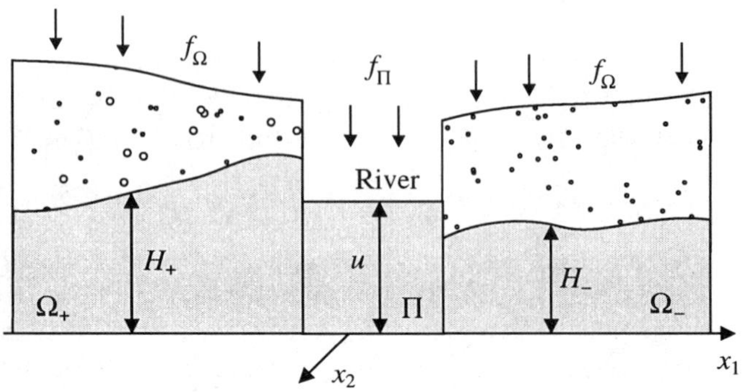

Figure 10.2: The vertical cross-section of the flow domain.

We study the local properties of the solutions $\mathbf{v}(x, t)$ of system (10.1)–(10.3) in a ball $B_\rho(x_0) = \{x : |x - x_0| < \rho\}, x_0 \in \Pi$ regardless of the boundary conditions on $\partial\Omega$. Without losing generality, we assume that $x_0 = 0, s = x_1$,

$$\Pi_\rho = \{x \in \Omega : x_2 = 0,\ x_1 < \rho\}, \qquad B_\rho^{\pm} = \{x \in B_\rho : 0 < \pm x_2\}.$$

In our study, we follow [42, 41, 38, 37].

10.2 Localization effects in the equations of diffusion waves. The solutions of the equation of diffusion waves (equation (10.2) without the second term on the right-hand side) can describe both the processes of fast and slow diffusion. Let us consider the following initial-boundary value problem:

$$\frac{\partial u}{\partial t} = \frac{\partial}{\partial s}\left(|u|^\alpha \left|\frac{\partial u}{\partial s}\right|^{-1/2} \frac{\partial u}{\partial s}\right) + f_\Pi \quad \text{for } s \in (0,1), t \in (0,T), \tag{10.9}$$

$$\begin{aligned} &u(i,t) = u^i(t) \quad (i = 1,2) \quad \text{or } \frac{\partial u}{\partial s} = 0 \text{ at the level } s = 0,\\ &u(s,0) = u_0(s) \quad \text{for } s \in (0,1), \end{aligned} \tag{10.10}$$

$$0 \le (u^i(t), u_0(s)) \le C_0. \tag{10.11}$$

Let us assume that $f_\Pi(s,t) \ge 0$ and

$$u^0(t) = u^1(t) = u^0 = \text{const} > 0, \quad \delta \le u_0(s) < \infty. \tag{10.12}$$

Then according to the maximum principle,

$$\min(\delta, u^0) \le u(x,t)$$

so that the motion is caused by the degeneracy of equation (10.9) with respect to the derivative $\frac{\partial u}{\partial s}$. Introducing the new function $v = u(s,t) - u^0$, we rewrite problem (10.9), (10.10) in the form

$$\frac{\partial v}{\partial t} = \frac{\partial}{\partial s}\left(a_0 \left|\frac{\partial v}{\partial s}\right|^{-1/2} \frac{\partial v}{\partial s}\right) + f_\Pi, \quad s \in (0,1), \quad t \in (0,T)$$

$$v(i,t) = 0 \quad (i = 1,2) \quad \left(\text{or } \left.\frac{\partial v}{\partial s}\right|_{s=0} = 0\right),$$

$$v(s,0) = u_0(s) - u^0 \geq 0, \quad s \in (0,1),$$

where

$$C_2 \leq a_0 \leq C_1, \quad p = \frac{3}{2} < 1 + \gamma = 2.$$

Results of Chapter 2 (Theorem 2.1) allow us to formulate the following assertion.

Theorem 10.1 (Localization in finite time). *Let conditions* (10.12) *be fulfilled with* $f_\Pi(s,t) \equiv 0$. *Then the solution of problem* (10.9)–(10.11) *is constant beginning with a finite time* t^*:

$$u(s,t) \equiv u^0 \quad \textit{for } s \in (0,1),\ t \geq t^*.$$

If $f_\Pi \geq 0$, *and for some* $t_f < t^*$

$$\|f_\Pi\|_{L^2(\Omega)}^{p/(p-1)} \leq \varepsilon \left(1 - \frac{t}{t_f}\right)_+^{1/(1-\nu)}, \quad \nu = p/2, \tag{10.13}$$

with a constant ε, *then the following estimate holds:*

$$\left\|u(\cdot,t) - u^0\right\|_{L^2(\Omega)}^2 \leq C\left(1 - \frac{t}{t_f}\right)_+^{1/(1-\nu)}.$$

Specifically,

$$u(s,t) \equiv u^0 \quad \textit{in } \Omega \times \{t \geq t_f\}.$$

In physical terms, this assertion means that the water level in the channel becomes constant in a finite time provided the external source f_Π is absent.

If $f \not\equiv 0$ and condition (10.13) is fulfilled, one can point out the source intensity $\varepsilon > 0$ such that the water level in the channel stabilizes at the same instant t_f when the source disappears.

As opposed to this situation, the character of motion in problem (10.9)–(10.11) with zero initial data is also defined by the term containing $|u|^\alpha$. Introducing the new function

$$v = u^{1/\gamma}, \quad \gamma = \frac{1}{1+2\alpha},$$

we come to the equation

$$\frac{\partial v^\gamma}{\partial t} = \frac{\partial}{\partial s}\left(|\gamma|^{1/2} \left|\frac{\partial v}{\partial s}\right|^{p-2} \frac{\partial v}{\partial s}\right) + f_\Pi(s,t), \quad (s,t) \in (0,1) \times (0,T), \quad p = 3/2.$$

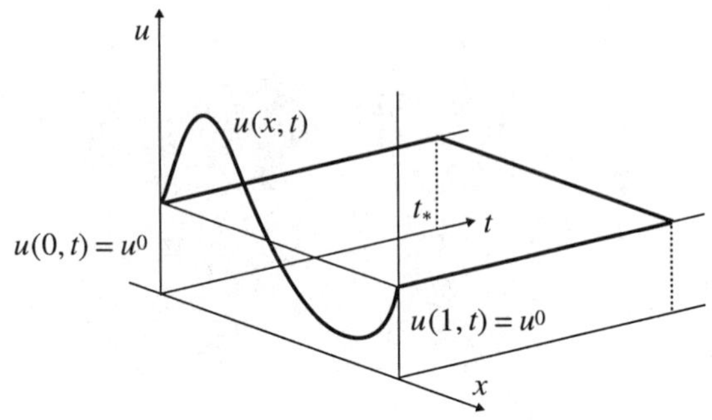

Figure 10.3: Stabilization to a constant level at a finite time.

If

$$p = \frac{3}{2} > 1 + \gamma = \frac{2(1+\alpha)}{1+2\alpha} \Longleftrightarrow \alpha > 1/2,$$

results of Chapter 3 (slow diffusion) are once again applicable.

Theorem 10.2 (Finite speed of wetting of a dry bottom). *Let $u(s,t) \geq 0$ be a weak solution of equation* (10.9) *with $\alpha > 1/2$ and let*

$$f_\Pi = 0, \quad u_0(s) = u(s,0) = 0 \quad \textit{for } |s| \leq \rho_0,\, t \in (0,T). \tag{10.14}$$

Then there exists

$$u(s,t) = 0 \quad \textit{for } |s| \leq \rho(t), \quad \theta = \theta(\alpha) > 0, \tag{10.15}$$

where $\rho(t)$ is defined by the formula

$$\rho^{1+\sigma}(t) = \rho_0^{1+\sigma} - Ct^\theta$$

with constants $C = C(C_0, \alpha)$, $\theta = \theta(\alpha)$, $\sigma = \sigma(\alpha)$. If, additionally to (10.14),

$$\int_{\Pi_\rho} |u_0(s)|^2 ds + \int_0^T \int_{\Pi_\rho} |f|^2 ds dt \leq \varepsilon(\rho - \rho_0)_+^{1/(1-\nu)}, \quad \rho_0 \leq \rho,$$

then there exists $t^* \in (0,T)$ such that

$$u(s,t) = 0 \quad \textit{in } (-\rho_0, \rho_0) \times (0, t^*).$$

10.3 Finite speed of propagation for simultaneous flows. Let us return to the model (10.1)–(10.3) describing simultaneous flows. We will consider weak solutions $\mathbf{V} \in W$ of this system in the domain $B_\rho \times (0,T)$. Let us point out that all nonnegative bounded solutions of equation (10.1) can only describe processes of slow flow with finite speed of propagation.

Following the general scheme presented in Section 3, we introduce the energy functions

$$
\begin{aligned}
b(t,\rho) &= \| H(\cdot,t) \|^2_{L^2(B_\rho)} + \| u(\cdot,t) \|^2_{L^2(\Pi_\rho)},\\
b(\rho,t) &= b_H(\rho,t) + b_u(\rho,t),\\
E(\rho,t) &= E_H(\rho,t) + E_u(\rho,t),\\
\overline{b}(\rho,t) &= \operatorname*{ess\,sup}_{0\le\tau\le t} b(\rho,\tau),\\
E(\rho,t) &= \int_0^t \left((H\nabla H, \nabla H)_{B_\rho} + \left(\psi \left|\frac{\partial u}{\partial s}\right|^{3/2}, 1\right)_{\Pi_\rho} \right) d\tau
\end{aligned}
$$

under the usual notation

$$
(u,v)_{B_\rho} = \int_{B_\rho} uv dx, \quad (u,v)_\Pi = \int_{\Pi_\rho} uv ds.
$$

It is easy to check that the energy functions possess the properties

$$
C_2\tilde{E} \le E(\rho,t) \le C_1\tilde{E}, \quad C_2\tilde{E}_1 \le \frac{\partial E}{\partial \rho} \le C_1\tilde{E}_1,
$$

where

$$
\begin{aligned}
\tilde{E} &= \int_0^t \left(\|\sqrt{|H|}\nabla H\|^2_{L^2(B_\rho)} + \left\| |u|^{\alpha/2} \left|\frac{\partial u}{\partial s}\right|^{3/4} \right\|^2_{L^2(\Pi_\rho)} \right) d\tau,\\
\tilde{E}_1 &= \int_0^t \left(\|\sqrt{|H|}\nabla H\|^2_{L^2(S_\rho)} + \left[|u|^{\alpha} \left|\frac{\partial u}{\partial s}\right|^{3/2} \right]\Bigg|_{-\rho}^{\rho} \right) d\tau,\\
[a]|_{s=-\rho}^{s=\rho} &= a(-\rho) + a(\rho).
\end{aligned}
$$

Our aim is to prove the following result.

Theorem 10.3. *Let* $\mathbf{v} = (H,u) \in W$ *be a weak solution of problem* (10.1)–(10.3) *under the assumptions*

$$
\begin{aligned}
H(x,0) &= 0, \quad f_\Omega = 0 \quad in\ B_{\rho_0} \times (0,T),\\
u(s,0) &= 0, \quad f_\Pi = 0 \quad in\ \Pi_{\rho_0} \times (0,T).
\end{aligned}
$$

Then there exist $t^* \in (0,T)$ *and* $\rho(t)$ *such that*

$$
H(x,t) = 0 \quad x \in B_{\rho(t)}, \quad u(s,t) = 0 \quad s \in \Pi_{\rho(t)}, \quad t \in (0,t^*)
$$

with $\rho(t)$ *defined by the formula*

$$
\rho^{1+\sigma}(t) = \rho_0^{1+\sigma} - Ct^\theta,
$$

$C = C(C_0, b(\rho_0, t), E(\rho_0, t))$. *If, moreover,*

$$\|H(\cdot,0)\|^2_{L^2(B_\rho)} + \|u(\cdot,0)\|^2_{L^2(\Pi_\rho)} + \int_0^T \left(\|f_\Omega\|^2_{L^2(B_\rho)} + \|f_\Pi\|^2_{L^2(\Pi_\rho)}\right) d\tau$$
$$\leq \varepsilon\,(\rho-\rho_0)_+^{\vartheta}, \quad \rho > \rho_0, \quad \vartheta(\alpha) > 0,$$

then there exists $t^* > 0$ *such that* $W \equiv (H, u) = 0$ *in* $B_{\rho_0} \times (0, t^*)$.

Proof. It is convenient to deal with equations written in the form (1.8) of Chapter 2, i.e., to preserve the nonlinear terms (with respect to H and u) on the right-hand sides.

Let us multiply equation (10.1) by $H(x,t)$, equation (10.2) by $u(s,t)$ and then formally integrate the results over the domains $B_\rho \times (0,t)$ and $\Pi_\rho \times (0,t)$. Let us add the resulting relations, taking into account the internal conjugate conditions (10.3). This leads to the energy relation (the formula of integration by parts)

$$b(\rho,\tau)|_{\tau=0}^{\tau=t} + E(\rho,t) = \sum_{i=1}^{4} I_i, \tag{10.16}$$

where

$$I_1 = \int_0^t \int_{S_\rho} H^2 \nabla H \mathbf{n}\, dS\, d\tau, \qquad I_2 = \int_0^t \psi |u_s|^{-1/2} u_s u \Big|_{s=-\rho}^{s=\rho} d\tau,$$
$$I_3 = \int_0^t \int_{B_\rho} H f_\Omega\, dx\, d\tau, \qquad I_4 = \int_0^t \int_{\Pi_\rho} u f_\Pi\, ds\, d\tau.$$

Applying the interpolation-trace inequality, we obtain the estimates

$$\int_{S_\rho} |H|^3 dS \leq C\left[\left(\int_{B_\rho} |H||\nabla H|^2 dx\right)^{1/2} + \rho^{\delta_1}\left(\int_{B_\rho} |H|^2 dx\right)^{3/4}\right]^{4/3}$$
$$\times \left[\int_{B_\rho} |H|^2 dx\right]^{1/2} \leq \tilde{C}\left(\frac{\partial E_H}{\partial t} + b_H\right)^{4/3} b_H^{1/2} \rho^{4\delta_1/3},$$

$$|u|^{(\alpha+3/2)} \leq C\left[\left(\int_{\Pi_\rho} |u|^\alpha \left|\frac{\partial u}{\partial s}\right|^{3/2} ds\right)^{2/3} + \rho^{\delta_2}\left(\int_{\Pi_\rho} |u|^2 dx\right)^{(3+2\alpha)/6}\right]^{3\theta/2}$$
$$\times \left[\int_{\Pi_\rho} |u|^2 dx\right]^{(1-\theta)(3+2\alpha)/4}$$
$$\leq \tilde{C}\left(\frac{\partial E_H}{\partial t} + b_h\right)^{\theta} b_H^{(1-\theta)(3+2\alpha)/4} \rho^{3\theta\delta_2/2}$$

with the parameters $C = C(\alpha)$, $\tilde{C} = \tilde{C}(C, \overline{b}(\rho_0, T))$, $\delta_1 = -3/2$, $\delta_2 = -(3 + 2\alpha)/6$, and

$$\frac{3+2\alpha}{6} \geq \frac{2}{3} \Longleftrightarrow \alpha \geq \frac{1}{2}.$$

The terms I_1, I_2 can be estimated in the following way:

$$
\begin{aligned}
I_1 &\le \left(\frac{\partial E_H}{\partial \rho}\right)^{1/2}\left(\int_0^t\int_{S_\rho}|H|^3 dS d\tau\right)^{1/2}\\
&\le \tilde{C}t^{1/3}\rho^{\delta_1/3}\left(\frac{\partial E}{\partial \rho}\right)^{1/2}(E+\overline{b})^{7/12} \qquad (10.17)\\
&\le \frac{1}{8}(E+\overline{b})+\tilde{C}t^{4/5}\rho^{4\delta_1/5}\left(\frac{\partial E}{\partial \rho}\right)^{6/5},\\
I_2 &\le \left(\frac{\partial E_u}{\partial \rho}\right)^{1/3}\left(\int_0^t [|u|^{(\alpha+3/2)}]d\tau\right)^{2/3}\\
&\le \tilde{C}t^{2(1-\theta)/3}\rho^{\theta\delta_2/3}\left(\frac{\partial E}{\partial \rho}\right)^{1/3}(E+\overline{b})^{\kappa} \qquad (10.18)\\
&\le \frac{1}{8}(E+\overline{b})+\tilde{C}t^{36(1-\theta)/15}\rho^{18\theta\delta_2/5}\left(\frac{\partial E}{\partial \rho}\right)^{6/5}.
\end{aligned}
$$

In (10.18) we used the inequality $18\kappa/13 \ge 1$. The terms I_3 and I_4 are estimated in the standard way. Gathering (10.16), (10.17), (10.18) (with $t \le t^*$ and t^* chosen sufficiently small), we come to the ordinary differential inequality which is either of the form (1.16) of Chapter 3 in the case of the first assertion of Theorem 10.3) or of the form (1.20) of Chapter 3 in the case of the second assertion. The proof is completed by application of Lemmas 2.2 and 2.3. □

Reverting to the original physical problem, we interpret the results as follows. If the domain B_{ρ_0} was dry at the initial moment, i.e., the levels of the surface and ground water were zero therein, then the first assertion of the theorem gives estimates on the location of the free boundaries $H(x,t)$ and $u(s,t)$. The second assertion states that whatever the flux outside B_{ρ_0}, this domain can be swamped only in a finite nonzero time.

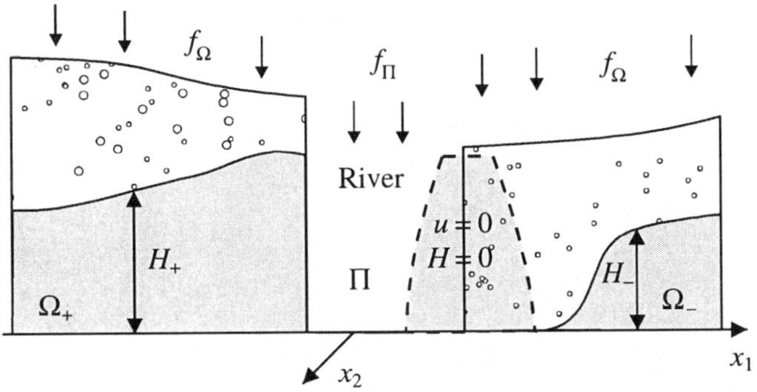

Figure 10.4: Finite speed of wetting.

11 Solute transport through a porous medium with micro and macro structures

Following [20], we study the properties of weak solutions to the initial-boundary value problem

$$\frac{\partial}{\partial t}\left(\vartheta u+\rho f_1(u)\right)-\nabla\cdot\left(D\nabla u-qu\right)$$
$$=\int_{\partial\Omega'}\gamma h\left(x,x'\right)\left(u(x,t)-u'\left(x,x',t\right)\right)ds'+f(x,t), \qquad (11.1)$$
$$u=u(x,t), \quad (x,t)\in Q_T=\Omega\times(0,T), \quad \Omega\subset R^N, \quad N\geq 1;$$
$$\frac{\partial}{\partial t}\left(\theta' u'+\rho' f_2\left(u'\right)\right)-\nabla'\left(D'\nabla' u'\right)=f'\left(x,x',t\right), \qquad (11.2)$$
$$u'=u'\left(x,x',t\right), \quad (x,t)\in Q_T,$$
$$(x',t)\in Q'_T=\Omega'\times(0,T), \quad \Omega'\in R^M, \quad M\geq 1;$$
$$\left(D'\nabla' u'\cdot\mathbf{n}'+h\left(u'-u\right)\right)\big|_{S'_T}=0, \quad S'_T=\partial\Omega'\times(0,T); \qquad (11.3)$$
$$u(x,x',0)=u'_0\left(x,x'\right), \quad x\in\Omega, \quad x'\in\Omega'; \qquad (11.4)$$
$$u(x,0)=u_0(x), \quad x\in\Omega; \qquad (11.5)$$
$$(D\nabla u\cdot\mathbf{n}+\mu u)|_{S_T}=0, \quad S_T=\partial\Omega\times(0,T). \qquad (11.6)$$

The system under consideration consists of two degenerate parabolic equations, one of them with a functional nonlocal term. The differential operators ∇, ∇' act in the variables x and x', respectively. $\mathbf{n}$ and $\mathbf{n}'$ are the unit normal vectors to Γ and Γ'; ρ, ρ', γ, and μ are positive constants. System (11.1), (11.2) describes the solute transport through a porous medium, where the chemical species undergoes a chemical reaction on the surface of the porous skeleton [166, 280]. The temperature in the fluid and solid is constant. $u(x,t)$ stands for the concentration of the reactant in the fluid, and $u'(x,x',t)$ is the reactant concentration in the solid catalyst pellet.

A detailed description of such models is given in [280]. System (11.1), (11.2) can be obtained as a by-product of the main balance laws (2.1), (2.2) and the empirical laws. Let Ω denote the domain occupied by a porous medium. Let θ and θ' be the porousness of the medium, let ρ and ρ' be the densities of the reactants, and let $f_1(u)$ and $f_2(u')$ be the reactant masses forming in the result of the chemical reactions in the domains Ω and Ω'. Then the total reactant densities in Ω and Ω' are defined by the formulas

$$\overline{\rho}=\theta u+\rho f_1(u), \quad \overline{\rho}'=\theta' u'+\rho' f_2(u'). \qquad (11.7)$$

Furthermore, the distribution of the density of the mass sources brought into Ω from the surface Ω' is given by the formula

$$h(x,t)=\int_{\partial\Omega'}\gamma h(x,x')\left(u(x,t)-u'(x,x',t)\right)ds'+f(x,t).$$

In each domain Ω and Ω', the momentum balance law (2.2) is replaced by the Fick law

$$\rho \mathbf{v} = -\left(D\nabla u - \mathbf{q}u\right), \qquad \rho'\mathbf{v}' = -D'\nabla' u', \tag{11.8}$$

where $\mathbf{q}$ is the velocity of the fluid. Substituting (11.7)–(11.8) into the mass balance laws (2.1) corresponding to Ω and Ω', we arrive at system (11.1), (11.2).

The existence and uniqueness of a solution to problem (11.1)–(11.6) and some properties of the moving boundary (the boundary of the set where the concentrations are positive) were established in [166]. Besides, for the problem in the one-dimensional formulation ($N = 1$) certain properties of the moving boundary were studied by means of comparison methods.

The local energy method allows us to extend results of [166] to the multidimensional case $N > 1$ and under less restrictive assumptions on the problem data.

We are interested in the local properties of the solution $W(x, x', t) = (u(x, t), u'(x, x', t)$ so that no information about the boundary conditions for $u(x, t)$ on Γ is used. However, these arguments do not apply to study $u'(x, x', t)$ because equation (11.1) involves the integral of $u'(x, x', t)$ over Γ.

Let us assume the following:

$$|\theta(x, t), \theta'(x, x'), \mathbf{q}(x, t), \theta_t| \leq \frac{1}{\nu} < \infty, \qquad \nu = \text{const}, \tag{11.9}$$

$$\forall\,(x, x', t) \in \overline{\Omega} \times \overline{\Omega}' \times [0, T], \quad \left(\rho, \rho', \theta, \theta'\right) \geq 0,$$

$$\forall\,(x, t) \in \overline{Q}_T, \quad \xi \in \mathbb{R}^N, \quad \nu|\xi|^2 \leq (\xi^*, D\xi) \leq \frac{1}{\nu}, \tag{11.10}$$

$$\forall \xi \in \mathbb{R}^M, \quad (x, x', t) \in \overline{Q} \times \overline{Q}', \quad 0 \leq (\xi^*, D'\xi), \tag{11.11}$$

For all $(x, t) \in \overline{Q}_T$, $u \in \mathbb{R}$

$$G_0|u|^{p+1} \leq G(x, t, u) \equiv \theta\frac{u^2}{2} + \rho_1\left(uf_1(u) - \int_0^u f_1(\tau)d\tau\right) \tag{11.12}$$

with $0 < p < 1$, $G_0 = \text{const}$. For all $(x, x', t,) \in \overline{Q}'_T \cup \overline{Q}_T$, $u' \in \mathbb{R}$,

$$0 \leq G'(x, x', t, u') \equiv \theta'\frac{u'^2}{2} + \rho_1'\left(u'f_2(u') - \int_0^{u'} f_2(\tau)d\tau\right) \tag{11.13}$$

with

$$\gamma = \text{const} > 0, \qquad h(x, x') \geq 0. \tag{11.14}$$

Conditions (11.12) and (11.13) are fulfilled if, say,

$$0 \leq (\theta, \theta', \rho'), \quad 0 < \nu \leq \rho, \quad f_1(u) = |u|^{p-1}u, \quad f_2(u') = |u'|^{q-1}u'.$$

We shall consider the weak solutions of system (11.1), (11.2) satisfying conditions (11.3)–(11.5) and such that

$$E(t,\rho) \equiv \int_0^t \int_{B_\rho} D\nabla u \cdot \nabla u dx dt < \infty,$$

$$E'(t,\rho) \equiv \int_0^t \int_{B_\rho} \int_{\Omega'} D'\nabla' u' \cdot \nabla' u' dx dx' dt < \infty,$$

$$b(t,\rho) = \sup_{0\le\tau\le t} \text{ess} \int_{B_\rho} G(x,\tau,u(x,\tau)dx < \infty, \tag{11.15}$$

$$b'(t,\rho) = \sup_{0\le\tau\le t} \text{ess} \int_{B_\rho} \int_{\Omega'} G'(x,x',u'(x,x',\tau))dx'dx, \tag{11.16}$$

where $\overline{B}_\rho \subset \Omega$. We refer to [166] for the theorems of existence and uniqueness of such solutions.

Let us consider first the case where $0 \le h(x,x')$, $0 \le D'$, and $0 \le G'$. The aim is to prove the following assertions.

Theorem 11.1. *Let* $W(x,x',t) = (u(x,t),\ u'(x,x',t))$ *be a weak solution of problem* (11.1)–(11.5) *with* $f \equiv f' \equiv 0$ *and such that*

$$W(x,x',0) = 0 \quad \text{for } x \in B_{\rho_0},\ \overline{B}_{\rho_0} \subset \Omega,\ x' \in \Omega'.$$

Let conditions (11.9)–(11.14) *be fulfilled. Then there exists* t^* *depending on* T, $E(T,\rho_0)$, $b(T,\rho_0)$ *such that*

$$\forall x' \in \Omega', \quad x \in B_\rho, \quad u(x,t) = u'(x,x',t) = 0,$$

where $0 \le \rho \le \rho(t) \le \rho_0$, $t \in (0,t^*)$, *and the function* $\rho(t)$ *is defined by the formula*

$$\rho(t) = \rho_0 - Ct^{1-\sigma} E^\beta(t,\rho_0)$$

with

$$\sigma = \frac{N(1-p)+1+p}{N(1-p)+2(p+1)}, \qquad \beta = \frac{(1-\sigma)}{1+p};$$

the constant C *is defined by* T, $E(T,\rho_0)$, $b(T,\rho_0)$, β, *and* $\rho(t^*) = 0$.

Theorem 11.2. *Let* $W(x,x',t)$ *be a weak solution of problem* (11.1)–(11.5) *with* $f \not\equiv 0$, $f' \not\equiv 0$. *Let us assume that conditions* (11.9)–(11.14) *hold and, moreover,*

$$Q(\rho,0) + \int_0^T \|f\|^{(p+1)/p}_{(p+1)/p,B_\rho} dt \le \varepsilon(\rho-\rho_0)_+^{1/\beta}, \tag{11.17}$$

where $v_+ = max(0,v)$, $\rho \in (\rho_0,\rho_1)$,

$$Q(\rho,t) \equiv \int_{B_\rho} \left[G(x,t,u(x,t)) + \gamma \int_{\Omega'} G'(x,x',t,u'(x,x',t))dx' \right] dx$$

for some ε *(small in comparison with* $E(T,\rho_1)$*). Then there exists* t^* *depending on* $E(T,\rho_1)$, $b(T,\rho_1)$ *such that*

$$W(x,x',t)=0 \quad \forall x\in B_{\rho_0}, \quad x'\in\Omega', \quad t\le t^*.$$

If condition (11.14) fails and the function $h(x,x')$ is allowed to change sign, conditions (11.11), (11.12), (11.14) should be replaced by

$$\nu|\xi|^2\le(\xi^*,D'\xi)\le\frac{1}{\nu} \tag{11.18}$$

for all $\xi\in\mathbb{R}^M$, $(x,x',t)\in\overline{Q}\times\overline{Q}'$,

$$G_0'|u|^{q+1}\le G'(x,x',t,u), \quad 0<q<1, \quad |h|\le\frac{1}{\nu}. \tag{11.19}$$

Theorem 11.3. *Let conditions* (11.9)–(11.12), (11.17) *be fulfilled and* $W=(u,u')$ *be a weak solution of problem* (11.1)–(11.5) *with* $f\equiv f'\equiv 0$ *such that*

$$\forall x\in B_{\rho_0}, \quad \overline{B}_{\rho_0}\subset\Omega, \quad x'\in\Omega', \quad W(x,x',0)=0.$$

Then the assertion of Theorem 11.1 *holds.*

Let $f\not\equiv 0$, $f'\not\equiv 0$, *and, additionally,*

$$Q(\rho,0)+\int_0^T\left(\|f\|^{(p+1)/p}_{(p+1)/p,B_\rho}+\|f'\|^{(q+1)/q}_{(q+1)/q,\Omega'\cup B_\rho}\right)dt\le\varepsilon(\rho-\rho_0)_+^{1/\beta},$$

for some $\varepsilon>0$. *Then there exists* t^* *depending on* $E(T,\rho_1)$, $b(T,\rho_1)$ *such that*

$$W(x,x',t)=0 \quad \forall x\in B_{\rho_0}, \quad x'\in\Omega', \quad t\le t^* W(x,x',t)=0.$$

The proof of the above theorems follows the routine scheme of analysis of the nonlinear ordinary differential inequalities for suitably chosen energy functions. The principal ingredient is the derivation of the energy equality

$$\begin{aligned} I_0 &= E(\rho,t)+\gamma E'(\rho,t)+\int_{B_\rho}\left(G+\gamma\int_{\Omega'}G'dx'\right)dx\Bigg|_{\tau=0}^{\tau=t} \\ &= I_1+I_2+I_3+I_4. \end{aligned} \tag{11.20}$$

In this relation,

$$\begin{aligned} I_1 &= -\gamma\int_0^t\int_{B_\rho}\left(\int_{\partial\Omega'}G(x,x',\tau)dx'\right)dxd\tau, \\ G(x,x',\tau) &= h\big(u(x,\tau)-u'(x,x',\tau)\big)^2, \\ I_2 &= \int_0^t\int_{B_\rho}u\mathbf{q}\cdot\nabla u dxd\tau, \end{aligned}$$

$$I_3 = \int_0^t \int_{B_\rho} \left(\theta_t \frac{u^2}{2}\right) dx d\tau,$$

$$I_4 = \int_0^t \int_{S_\rho} (D\nabla u - qu)\, \nabla u\, \mathbf{n}\, dS d\tau,$$

$$I_5 = \int_0^t \int_{B_\rho} fudxd\tau,$$

$$I_6 = \gamma \int_0^t \int_{B_\rho} \int_{\Omega'} f'u'dx'dxd\tau, \ S_\rho = \partial B_\rho.$$

Equality (11.20) is obtained by multiplying equation (11.1) by $u(x,t)$, multiplying equation (11.2) by $\gamma u'(x,x',t)$, integrating by parts, and summing the results. It can be derived directly from the integral identities in the definition of the weak solution introduced in [166].

Under the conditions of Theorem 11.1, the relations

$$I_1 \le 0, \quad I_5 = I_6 = 0, \quad G_0 \|U\|^{p+1}_{p+1,B_\rho} \le b(\rho,t), \tag{11.21}$$

$$Q(\rho,0) = \int_{B_\rho} \left[G + \gamma \int_{\Omega'} G'dx'\right]\bigg|_{t=0} dx = 0, \quad \rho \le \rho_0. \tag{11.22}$$

hold.

The terms I_2, I_3, I_4 obey the estimates

$$\begin{aligned}
|I_2| &\le CE^{1/2} b^{\frac{1-\sigma}{p+1}} t^{\frac{1-\sigma}{2}} \left(E + \rho^{2\sigma} b^{2/(p+1)}\right)^{\sigma/2} \le Ct^{(1-\sigma)/2}(E+b),\\
|I_3| &\le Cb^{\frac{2(1-\sigma)}{p+1}} t^{1-\sigma} \left(E + \rho^{2\sigma}\, b^{2/(p+1)}\right)^{\sigma} \le Ct^{(1-\sigma)}(E+b),\\
|I_4| &\le Ct^{(1-\sigma)/2}(E+b)^{\sigma/2+(1-\sigma)/p+1}\left({E_\rho}^{1/2} + (E+b)^{\sigma/2+(1-\sigma)/p+1}\right) b^{\frac{2(1-\sigma)}{p+1}}\\
&\le Ct^{\kappa(1-\sigma)} {E_\rho}^{\kappa} + (1/4 + Ct^{1-\sigma})(E+b),
\end{aligned} \tag{11.23}$$

where

$$\frac{\kappa - 1}{\kappa} = \beta = \frac{(1-\sigma)(1-p)}{1+p} > 0.$$

Here and throughout the section, C denotes the various constants that depend only on T, $E(T,\rho_0)$, $b(T,\rho_0)$, ν, and N. Choosing t small enough in (11.23) and using (11.15), (11.16), (11.20)–(11.22), we come to the ordinary differential inequality for the energy function E,

$$E \le E + \gamma E' \le I_0 \le Ct^{\kappa(1-\sigma)} {E_\rho}^{\kappa}, \quad \rho \le \rho_0.$$

Integrating it with respect to ρ, we obtain the inequality

$$E^\beta(\rho) \le E^\beta(\rho_0) - \frac{1}{C\beta} t^{\sigma-1}(\rho_0 - \rho), \tag{11.24}$$

and the assertion of Theorem 11.1 follows. Under the conditions of Theorem 11.2,

$$0 \le Q(\rho, 0), \quad \rho_0 \le \rho, \quad I_5 \ne 0,$$

but $Q(\rho, 0)$ satisfies (11.17) so that for I_5, we obtain the estimate

$$\begin{aligned} |I_5| &\le \int_0^t \|f\|_{(p+1)/p, B_\rho} \|U\|_{p+1, B_\rho} d\tau \le t^{p+1} b(\rho, t) + C \int_0^t \|f\|_{p+1/p, B_\rho}^{(p+1)/p} d\tau \\ &\le t^{p+1} b(\rho, t) + C\varepsilon(\rho - \rho_0)_+^{1/\beta}. \end{aligned} \tag{11.25}$$

Gathering (11.17), (11.20), (11.23), (11.25), we have the inhomogeneous ordinary differential inequality for the energy function E,

$$E \le C\left(t^{\kappa(1-\sigma)} (E_\rho)^\kappa + \varepsilon(\rho - \rho_0)_+^{\frac{\kappa}{\kappa - 1}} \right), \quad \rho_0 \le \rho \le \rho_1, \tag{11.26}$$

whence the assertion of Theorem 11.2 by virtue of Lemma 2.4 of Section 1.

Under the conditions of Theorem 11.3, relations (11.21) and (11.22) are still valid, but the term I_1 in (11.20) does not have a definite sign. It now admits the estimate

$$|I_1| \le C\left(\int_0^t \left(\|U\|_{2, B_\rho}^2 + \|U'\|_{2, B_\rho \cup \Omega'}^2 \right) d\tau \right) = I'_1 + I''_1.$$

The term I_1'' can be evaluated in the same way as I_3 in (11.23). Under the conditions (11.18), (11.19), for I''_1 we have the estimate

$$|I''_1| \le Ct^{1-\sigma}(E' + b').$$

Thus for small t, we once again obtain inequality (11.24), and the first assertion of Theorem 11.3 follows.

To establish the second assertion of Theorem 11.3, we estimate the term I_5 as we did I_6 in (11.26):

$$I_5 \le t^{q+1} b' + C\varepsilon(\rho - \rho_0)_+^{1/\beta},$$

which gives the differential inequality (11.26) for the energy function E.

12 A quasilinear degenerate system arising in semiconductor theory

In solid-state physics, the drift-diffusion equations are today the most widely used model to describe semiconductor devices. The drift-diffusion models describe the flow of electrons in the conduction band of the semiconductor material and of holes (or defect electrons) in the valence band of the crystal, influenced by the electric field. Mathematically, they form a system of parabolic equations for the electron density n and the hole density p and the Poisson equation for the electric potential V:

$$\frac{\partial n}{\partial t} - \nabla \cdot (\nabla r(n) - n\nabla V) = -R(n, p), \tag{12.1}$$

$$\frac{\partial p}{\partial t} - \nabla \cdot (\nabla r(p) + p\nabla V) = -R(n, p), \tag{12.2}$$

$$\Delta V = n - p - C(x) \quad \text{in } Q_T = \Omega \times (0, T), \tag{12.3}$$

where $\Omega \subset \mathbb{R}^d$ $(1 \leq d \leq 3)$ is the (bounded) domain occupied by the semiconductor crystal. Here, $C = C(x)$ denotes the doping profile (fixed charged background ions) characterizing the semiconductor under consideration, $r(s)$ is the pressure function, and $R(n, p)$ the recombination-generation rate. The process of transfering an electron of the conduction band of the semiconductor into the lower energetic valence band is called recombination of electron–hole pairs. The inverse process, i.e., the transfer of a valence electron to the conduction band, is termed generation of electrons and holes. If recombination of electron–hole pairs exceeds generation, then $R(n, p) > 0$; in the opposite case, we have $R(n, p) < 0$.

In the standard drift-diffusion model,

$$r(s) = s \quad \text{and} \quad R(n, p) = q(n, p)(np - n_i^2),$$

where $q(n, p)$ is a positive function and $n_i = n_i(x) > 0$ is the so-called *intrinsic density*. The standard model can be derived from Boltzmann's equation under the assumption that the semiconductor device is in the low injection regime (i.e., for small absolute values of the applied voltage). It is shown in [200] that in the high injection regime, the diffusion terms $\nabla r(n)$, $\nabla r(p)$ are no longer linear, and the function $r(s)$ has to be taken as

$$r(s) = s^\alpha, \quad \alpha = \frac{5}{3}.$$

With this pressure function, the equations (12.1), (12.2) become of degenerate type, and solutions may exist for which $n = 0$ or $p = 0$ holds locally (so-called *vacuum solutions*).

The function r can be interpreted in the language of gas dynamics. We assume that the particles behave—thermodynamically speaking—as an ideal gas such that the gas law $r = nT$ holds (T denotes the particle temperature). In the isothermal case $T =$ const, the pressure turns out to be linear: $r(n) = n$. In the isentropic case, however, the temperature (only) depends on the concentrations. Then $T(n) = n^{2/3}$ holds for particles without spin in adiabatic and hence for isentropic states [116], which implies that $r(n) = n^{5/3}$ (and similarly for the holes).

The equations are supplemented with physically motivated boundary conditions. The boundary $\partial\Omega$ consists of two disjoint subsets Γ_D and Γ_N. The carrier densities and the potential are fixed at Γ_D (Ohmic contacts), whereas Γ_N models the union of insulating boundary segments:

$$n = n_D, \quad p = p_D, \quad V = V_D \quad \text{on } \Gamma_D \times (0, T), \tag{12.4}$$

$$\nabla r(n) \cdot \mathbf{n} = \nabla r(p) \cdot \mathbf{n} = \nabla V \cdot \mathbf{n} = 0 \quad \text{on } \Gamma_N \times (0, T), \tag{12.5}$$

where $\mathbf{n}$ denotes the exterior normal vector of $\partial\Omega$, which is assumed to exist a.e. We assume that the densities at time $t = 0$ are known:

$$n(0) = n_I, \quad p(0) = p_I \quad \text{in } \Omega. \tag{12.6}$$

The standard (low injection) model has been mathematically and numerically investigated in many papers (see [249, 254] and references therein). The existence and uniqueness of weak solutions have been shown. The isentropic (high injection) model is analyzed in [133, 198, 199, 200, 201]. The existence of weak solutions (satisfying $\nabla r(n), \nabla r(p) \in L^2$ and $n, p \in L^\infty$) has been proved. The uniqueness of solutions is shown in some special situations [133, 199, 201]. For the derivation of the model, we refer to [200, 202, 260].

Denote by $W^{s,p}(X)$ the space $W^{s,p}(0, T; X)$ if X is a Banach space. Furthermore, introduce $\mathcal{V} = \{u \in H^1(\Omega) : u = 0 \text{ on } \Gamma_D\}$. In the following, we assume that

$$r(s) = s^\alpha, \quad \alpha > 1,$$

and that there exists a solution (n, p, V) to (12.1)–(12.6) satisfying

$$n, p \in L^\infty(Q_T) \cap H^1(\mathcal{V}^*), \quad r(n), r(p) \in L^2(H^1), \quad V \in L^\infty(H^1).$$

The existence of a solution with these regularity properties is shown in [133, 201]. We have the following theorems.

Theorem 12.1 (Finite speed of propagation). *Let $x_0 \in \Omega$, $0 < \rho_0 < \operatorname{dist}(x_0, \partial\Omega)$ and $T > 0$. Assume that*

$$n_I = 0, \quad p_I = 0 \quad \text{in } B_{\rho_0}(x_0)$$

and

$$R(u, v)(u^\alpha + v^\alpha) \geq -\kappa_R(u^{\alpha+1} + v^{\alpha+1}) \quad \text{for all } u, v \geq 0 \tag{12.7}$$

with $\kappa_R \geq 0$ holds. Then there exist $T_1 > 0$ and a nonincreasing function ρ satisfying $\rho(\tau) > 0$, $0 \leq \tau < T_1$, and $\rho(0) = \rho_0$ such that

$$n(x, t) = 0, \quad p(x, t) = 0 \quad \text{for a.e. } x \in B_{\rho(t)}(x_0), \; t \in (0, T_1).$$

Proof. Using local elliptic regularity theory (cf., e.g., [183]) and noting that $n, p \in L^\infty(B_{\rho_0}(x_0))$, we see that $\nabla V \in L^\infty(B_{\rho_0}(x_0) \times (0, T))$. Let

$$M = \|\nabla V\|_{0,\infty,B_{\rho_0}(x_0)\times(0,T)}, \quad \varepsilon \in (0, \rho_0), \quad t_1 = \varepsilon/2M,$$

and consider the cone

$$P = P(\rho, t) = \{(x, \tau) : x \in B_r(x_0),\ \ \tau \in (0, t)\},$$

where $\rho \in (\varepsilon, \rho_0]$, $t \in (0, t_1)$, and $r = r(\rho, \tau) = \rho - M\tau$. For almost all ρ and τ,

$$\begin{aligned}\int_P n\nabla V \cdot \nabla n^\alpha dx d\tau &= \frac{\alpha}{\alpha+1}\int_P \nabla V \cdot \nabla n^{\alpha+1} dx d\tau \\ &= -\frac{\alpha}{\alpha+1}\int_P \Delta V n^{\alpha+1} dx d\tau \\ &\quad + \frac{\alpha}{\alpha+1}\int_{\partial_l P} (\nabla V \cdot \mathbf{n}_x) n^{\alpha+1} d\sigma d\tau,\end{aligned}$$

and therefore, using the local integration-by-parts formula,

$$\begin{aligned}&\frac{1}{\alpha+1}\int_{P\cap\{\tau=t\}} n(t)^{\alpha+1} dx + \int_P |\nabla n^\alpha|^2 dx d\tau \\ &\quad \le \frac{1}{\alpha+1}\int_{P\cap\{\tau=0\}} n(0)^{\alpha+1} dx + \int_{\partial_l P} (\nabla n^\alpha \cdot \mathbf{n}_x) n^\alpha d\sigma d\tau \\ &\quad - \frac{1}{\alpha+1}\int_{\partial_l P} (\mathbf{n}_\tau + \nabla V \cdot \mathbf{n}_x) n^{\alpha+1} d\sigma d\tau - \frac{\alpha}{\alpha+1}\int_P \Delta V n^{\alpha+1} dx d\tau \\ &\quad - \int_P R(n, p) n^\alpha dx d\tau \\ &\quad = I_1 + \cdots + I_5.\end{aligned} \tag{12.8}$$

Since $n(0)$ vanishes in $B_{\rho_0}(x_0)$, we have $I_1 = 0$. For the estimate of I_2, observe that in spherical coordinates with center x_0 (cf. [33, 149])

$$\begin{aligned}\frac{\partial E_n}{\partial \rho}(\rho, t) &= \frac{\partial}{\partial \rho}\int_P |\nabla n^\alpha|^2 dx d\tau = \frac{\partial}{\partial \rho}\int_0^t \int_0^{r(\rho,t)} \int_{S^{d-1}} |\nabla n^\alpha|^2 \tilde{r}^{d-1} d\omega d\tilde{r} d\tau \\ &= \int_{\partial_l P} |\nabla n^\alpha|^2 d\sigma d\tau.\end{aligned}$$

Hence

$$I_2 \le \left(\int_{\partial_l P} |\nabla n^\alpha|^2\right)^{1/2} \left(\int_{\partial_l P} n^{2\alpha}\right)^{1/2} = \left(\frac{\partial E_n}{\partial \rho}\right)^{1/2} \|n^\alpha\|_{0,2,\partial_l P}.$$

We use the interpolation-trace lemma

$$\|n^\alpha\|_{0,2,\partial B_r} \le c_0 \left(\|\nabla n^\alpha\|_{0,2,B_r} + r^{-\delta}\|n^\alpha\|_{0,1+1/\alpha,B_r}\right)^\theta \|n^\alpha\|_{0,1+1/\alpha,B_r}^{1-\theta}, \tag{12.9}$$

where

$$\theta = \frac{d(\alpha-1) + (\alpha+1)}{d(\alpha-1) + 2(\alpha+1)} \in (0, 1), \qquad \delta = \frac{2(\alpha+1) + d(\alpha-1)}{2(\alpha+1)} > 1. \tag{12.10}$$

By the definition of r, we have

$$r^{-\delta} = (\rho - M\tau)^{-\delta} \leq (\varepsilon - Mt_1)^{-\delta} = (2/\varepsilon)^{\delta}.$$

Thus applying Hölder's inequality with exponent $1/\theta$ and setting $K_1 = c_0^2 \max(1, (2/\varepsilon)^{\delta})$, we obtain

$$\begin{aligned}
&\int_0^t \|n^\alpha\|_{0,2,\partial B_r} d\tau \\
&\quad \leq 2K_1 \int_0^t \left(\|\nabla n^\alpha\|_{0,2,B_r}^2 + \|n^\alpha\|_{0,1+1/\alpha,B_r}^2 \right)^\theta \|n^\alpha\|_{0,1+1/\alpha,B_r}^{2(1-\theta)} d\tau \\
&\quad \leq 2K_1 \left(\int_0^t \|\nabla n^\alpha\|_{0,2,B_r}^2 d\tau + \int_0^t \|n^\alpha\|_{0,1+1/\alpha,B_r}^2 d\tau \right)^\theta \\
&\quad \quad \times \left(\int_0^t \|n^\alpha\|_{0,1+1/\alpha,B_r}^2 d\tau \right)^{1-\theta} \\
&\quad \leq 2K_1 t^{1-\theta} \left(E_n(\rho, t) + t_1 b_n(\rho_0, t_1)^{(\alpha-1)/(\alpha+1)} b_n(\rho, t) \right)^\theta \\
&\quad \quad \times b_n(\rho, t)^{2\alpha(1-\theta)/(\alpha+1)},
\end{aligned}$$

where

$$b_n(\rho, t) = \sup_{\tau \in (0,t)} \int_{B_{r(\rho,\tau)}(x_0)} n(x, \tau)^{\alpha+1} dx.$$

This yields

$$\begin{aligned}
\|n^\alpha\|_{0,2,\partial_l P} &\leq K_2 t^{(1-\theta)/2} (E_n(\rho, t) + b_n(\rho, t))^{\theta/2} b_n(\rho, t)^{\alpha(1-\theta)/(\alpha+1)} \\
&\leq K_2 t^{(1-\theta)/2} (E_n(\rho, t) + b_n(\rho, t))^\mu,
\end{aligned}$$

where $K_2^2 = 2K_1 \max(1, t_1 b_n(\rho_0, t_1)^{(\alpha-1)/(\alpha+1)})$ and

$$\mu = \frac{\theta}{2} + \frac{\alpha}{\alpha+1}(1-\theta) \in (1/2, 1).$$

We conclude that

$$I_2 \leq K_2 t^{(1-\theta)/2} \left(\frac{\partial E_n}{\partial \rho} \right)^{1/2} (E_n(\rho, t) + b_n(\rho, t))^\mu.$$

Thanks to the special structure of $r = r(\rho, \tau)$ and the definition of M, we have

$$\mathbf{n}_\tau + \nabla V \cdot \mathbf{n}_x = \frac{M + \nabla V \cdot \mathbf{e}_x}{\sqrt{1+M^2}} \geq 0$$

so that $I_3 \leq 0$. Furthermore,

$$I_4 \leq K_3 \int_P n^{\alpha+1} dx d\tau,$$

where $K_3 = \frac{\alpha}{\alpha+1}\|\Delta V\|_{0,\infty,Q_T}$.

For p we get an analogous inequality to (12.8) and similar estimates involving the local energies E_p and b_p defined by

$$E_p(\rho,t) = \int_{P(\rho,t)} |\nabla p^\alpha|^2 dx d\tau, \quad b_p(\rho,t) = \sup_{\tau\in(0,t)} \int_{B_{r(\rho,t)}(x_0)} p(x,\tau)^{\alpha+1} dx.$$

Therefore, we have the estimate

$$\begin{aligned}
&\frac{1}{\alpha+1}\int_{B_r(x_0)} (n(t)^{\alpha+1} + p(t)^{\alpha+1}) dx + \int_P (|\nabla n^\alpha|^2 + |\nabla p^\alpha|^2) dx d\tau \\
&\quad \le K_4 t^{(1-\theta)/2}\left(\left(\frac{\partial E_n}{\partial\rho}\right)^{1/2}(E_n+b_n)^\mu + \left(\frac{\partial E_p}{\partial\rho}\right)^{1/2}(E_p+b_p)^\mu\right) \\
&\quad\quad + K_3\int_P (n^{\alpha+1}+p^{\alpha+1}) dx d\tau - \int_P R(n,p)(n^{\alpha+1}+p^{\alpha+1}) dx d\tau,
\end{aligned}$$

where

$$K_4^2 = 2K_1 \max\left(1, t_1 b_n(\rho_0,t_1)^{(\alpha-1)/(\alpha+1)}, t_1 b_p(\rho_0,t_1)^{(\alpha-1)/(\alpha+1)}\right).$$

Introduce

$$E = E_n + E_p, \qquad b = b_n + b_p.$$

Employing the assumption on $R(n,p)$ gives

$$\begin{aligned}
&\frac{1}{\alpha+1}\int_{B_r(x_0)} (n(t)^{\alpha+1} + p(t)^{\alpha+1}) dx + E(\rho,t) \\
&\quad \le 2K_4 t^{(1-\theta)/2}\left(\frac{\partial E}{\partial\rho}\right)^{1/2}(E+b)^\mu + (K_3+\kappa_R)\int_P (n^{\alpha+1}+p^{\alpha+1}) dx d\tau \\
&\quad \le 2K_4 t^{(1-\theta)/2}\left(\frac{\partial E}{\partial\rho}\right)^{1/2}(E+b)(\rho,t)^\mu + tK_5 b(\rho,t),
\end{aligned}$$

with $K_5 = K_3 + \kappa_R$. Since the right-hand side of the above inequality is nondecreasing in t, we can write

$$(E+b)(\rho,t) \le 2(\alpha+1)K_4 t^{(1-\theta)/2}\left(\frac{\partial E}{\partial\rho}\right)^{1/2}(b+E)^\mu + (\alpha+1)K_5 t b.$$

Choosing $t < t_2 = \min(t_1, (2(\alpha+1)K_5)^{-1})$, we get

$$b + E \le 4(\alpha+1)K_4 t^{(1-\theta)/2}\left(\frac{\partial E}{\partial\rho}\right)^{1/2}(b+E)^\mu \tag{12.11}$$

and

$$E(\rho,t)^{2(1-\mu)} \le (b+E)(\rho,t)^{2(1-\mu)} \le K_6 t^{1-\theta}\frac{\partial}{\partial\rho}E(\rho,t),$$

where $K_6 = 16(\alpha+1)^2 K_4^2$. Integrating this differential inequality for E in (ρ, ρ_0) gives (note that $\mu > 1/2$)

$$E(\rho, t)^{2\mu-1} \le E(\rho_0, t)^{2\mu-1} - K_6^{-1} t^{\theta-1}(\rho_0 - \rho).$$

Let

$$\tilde{\rho}(t) = \rho_0 - K_6 t^{1-\theta} E(\rho_0, t)^{2\mu-1}.$$

Then $\tilde{\rho}(0) = \rho_0$ and $\tilde{\rho}$ is nonincreasing. Choose $T_1 \in (0, t_2)$ such that $\tilde{\rho}(T_1) > \varepsilon$. Then for $t \in (0, T_1)$ and $\rho \in (\varepsilon, \tilde{\rho}(t)]$,

$$\begin{aligned} E(\rho, t)^{2\mu-1} &\le E(\tilde{\rho}(t), t)^{2\mu-1} \\ &\le E(\rho_0, t)^{2\mu-1} - K_6^{-1} t^{\theta-1}(\rho_0 - \tilde{\rho}(t)) = 0. \end{aligned}$$

Thus (see (12.11)) for $\rho = \tilde{\rho}(t)$,

$$n(x, t) = p(x, t) = 0 \quad \text{for a.e. } t \in (0, T_1), \ x \in B_{\tilde{\rho}(t)}(x_0). \qquad \square$$

13 Blowup in solutions of the thermistor problem

In this section, the energy method is applied to study the blowup regimes in solutions of the mathematical model of the heat conduction in a conductor in the presence of Joule heating. The process is described by the system (see [22, 23, 24, 111, 114, 115, 112, 195])

$$\frac{\partial \theta}{\partial t} = \operatorname{div}(k(\theta)\nabla\theta) + \sigma(\theta)|\nabla\varphi|^2, \tag{13.1}$$

$$0 = \operatorname{div}(\sigma(\theta)\nabla\varphi), \quad (x, t) \in Q = \Omega \times (0, T). \tag{13.2}$$

Here $\theta(x, t)$ is the temperature inside the conductor (a fluid), $\varphi(x, t)$ is the electric potential, $k(\theta) > 0$ is the thermal conductivity, and $\sigma(\theta) > 0$ is the electric conductivity.

System (13.1), (13.2) can be inferred from the energy balance law (2.3) completed by assumptions on the character of motion of the fluid medium and the laws of electrodynamics. Indeed, let us accept the following hypotheses:

(A1) The fluid (the conductor) is motionless,

$$\mathbf{v}(x, t) \equiv 0, \qquad \mathbf{D}(x, t) \equiv 0, \qquad \mathbf{S} : \mathbf{D} \equiv 0. \tag{13.3}$$

(A2) Its specific inner energy U depends only on the temperature: $U = U(\theta)$.

(A3) The heat flux obeys the Fourier law

$$\mathbf{q} = -k(\theta)\nabla\theta. \tag{13.4}$$

(A4) The electric current density $\mathbf{I}$ follows the Ohm law

$$\mathbf{I} = -\sigma(\theta)\,\nabla\varphi \tag{13.5}$$

and the conservation law

$$\operatorname{div}\mathbf{I} = 0. \tag{13.6}$$

(A5) The distribution of the energy sources in equation (2.3) is defined by the formula

$$Q = -\mathbf{E}\mathbf{I} = \sigma(\theta)|\nabla\varphi|^2, \tag{13.7}$$

where $\mathbf{E} = -\nabla\varphi$ is the electric field.

Under these assumptions, (2.3), (13.3)–(13.7) yield the system of equations

$$\begin{aligned} \rho\frac{\partial U(\theta)}{\partial\theta}\frac{\partial\theta}{\partial t} &= \operatorname{div}(k(\theta)\,\nabla\theta) + \sigma(\theta)\,|\nabla\varphi|^2, \\ 0 &= \operatorname{div}(\sigma(\theta)\,\nabla\varphi). \end{aligned}$$

Under the additional condition

$$\rho\, c \equiv \rho\,\frac{dU}{d\theta} = 1,$$

the last system coincides with system (13.1), (13.2).

System (13.1), (13.2) consists of a parabolic equation for the temperature $\theta(x,t)$ and an elliptic equation for the electric potential $\varphi(x,t)$. Having been written in the equivalent form

$$\frac{\partial\theta}{\partial t} = \operatorname{div}(k(\theta)\nabla\theta + \sigma(\theta)\varphi\,\nabla\varphi), \quad \operatorname{div}(\sigma(\theta)\,\nabla\varphi) = 0,$$

it becomes analogous to the system of equations of the two-phase filtration for the concentration s and the reduced pressure p considered in Section 4.

We will assume that the data of problem (13.1), (13.2) satisfy the conditions

$$\begin{gathered} 0 \le \theta_0(x) \le M < \infty \quad \text{in } \Omega, \\ \forall\, s \ge 0, \quad k(s) > 0, \quad \sigma(s) < \infty, \quad k(s),\ \sigma(s) \in C(\mathbb{R}^+). \end{gathered}$$

The following boundary conditions can be imposed:

1. the Neumann condition

$$\frac{\partial u}{\partial\mathbf{n}} = 0 \quad \text{on } \Gamma \times (0,T),$$

where $\mathbf{n}$ is the outward unit normal to Γ;

2. the mixed boundary conditions

$$u = 0 \quad \text{on } \Gamma_0 \times (0, T), \qquad \frac{\partial u}{\partial n} = 0 \quad \text{on } \Gamma_1 \times (0, T), \tag{13.8}$$

where Γ_0, Γ_1 are two relatively open subsets of Γ such that

$$\Gamma_0 \cap \Gamma_1 = \emptyset, \quad \overline{\Gamma_0} \cup \overline{\Gamma_1} = \Gamma;$$

3. the Dirichlet boundary condition

$$u = 0 \quad \text{on } \Gamma \times (0, T).$$

Remark 13.1. We might consider as well the nonhomogeneous boundary conditions, but for the sake of presentation we do not address this question here.

In these three situations, we can divide the boundary conditions for φ. However, to produce an electric current in Ω, the boundary data should be nonhomogeneous. That is why we assume that

$$\varphi = \varphi_0 \quad \text{on } \Gamma_0', \qquad \sigma(u)\frac{\partial \varphi}{\partial \mathbf{n}} = \varphi_1 \quad \text{on } \Gamma_1',$$

where Γ_0', Γ_1' are two relatively open subsets of Γ such that

$$\Gamma_0' \cap \Gamma_1' = \emptyset, \quad \overline{\Gamma_0'} \cup \overline{\Gamma_1'} = \Gamma.$$

The functions φ_0, φ_1 depend on x and t. Note that we are not excluding the cases $\Gamma_0' = \emptyset$ (the Neumann conditions) or $\Gamma_1' = \emptyset$ (the Dirichlet conditions). We will always assume that our problem has a sufficiently smooth solution. The existence results and various complements on the problem can be found in [22, 23, 24, 111, 114, 115, 112, 195, 113].

In all of the above-listed cases, we aim to show how the blowup of u could be induced by φ_0, φ_1, provided these two functions are chosen large enough.

13.1 The Neumann boundary condition for $u(x,t)$.

13.1.1 *Blowup driven by φ_0.* Let (u, φ) be a solution of the problem

$$u_t = \operatorname{div}(\kappa(u)\nabla u) + \sigma(u)|\nabla\varphi|^2 \quad \text{in } \Omega \times (0, T), \tag{13.9a}$$

$$u(x, 0) = u_0, \tag{13.9b}$$

$$\frac{\partial u}{\partial \mathbf{n}} = 0 \quad \text{on } \Gamma \times (0, T), \tag{13.9c}$$

$$\operatorname{div}(\sigma(u)\nabla\varphi) = 0 \quad \text{in } \Omega \times (0, T), \tag{13.9d}$$

$$\varphi = \varphi_0 \quad \text{on } \Gamma_0' \times (0, T), \qquad \sigma(u)\frac{\partial \varphi}{\partial \mathbf{n}} = \varphi_1 \quad \text{on } \Gamma_1' \times (0, T), \tag{13.9e}$$

where $\Gamma_0' \neq \emptyset$ and Γ_1' are admitted to be empty. We assume that the solution (u, φ) exists and is sufficiently smooth.

Theorem 13.1. *Let f be a smooth, positive function of two variables such that*

$$u \to f(\sigma(u), \kappa(u)) \text{ is nonincreasing,} \tag{13.10}$$

$$\int^{+\infty} f(\sigma(s), \kappa(s))\, ds < +\infty,$$

$$\exists C > 0 \quad \text{such that} \quad f(\sigma(u), \kappa(u))\sigma(u) \geq C \quad \forall u \in \mathbb{R}. \tag{13.11}$$

Denote by C' the best of such constants for which

$$\int_{\Gamma_0'} |\varphi - \overline{\varphi}|^2\, d\gamma \leq C' \int_\Omega |\nabla\varphi|^2\, dx \quad \forall \varphi \in H^1(\Omega),$$

where $\overline{\varphi}$ is the average of φ on Γ_0'; that is, if $d\gamma$ denotes the area of the surface element on Γ, then

$$\overline{\varphi} = \frac{1}{|\Gamma_0'|} \int_{\Gamma_0'} \varphi\, d\gamma, \quad |\Gamma_0'| \text{ is the surface measure of } \Gamma_0'.$$

Then if

$$\int_\Omega \int_{u_0}^{+\infty} f(\sigma(s), \kappa(s))\, ds dx < \frac{C}{C'} \int_0^{+\infty} \int_{\Gamma_0'} |\varphi_0 - \overline{\varphi}_0|^2\, d\gamma dt, \tag{13.12}$$

problem (13.9a)–(13.9e) *cannot have a global-in-time "smooth" solution.*

Proof. Set

$$Y(t) = \int_\Omega \int_u^{+\infty} f(\sigma(s), \kappa(s))\, ds dx. \tag{13.13}$$

Differentiating Y with respect to t leads to

$$\frac{dY}{dt} = -\int_\Omega f(\sigma(u), \kappa(u)) u_t\, dx. \tag{13.14}$$

Using equation (13.9a), we obtain

$$\begin{aligned}
\frac{dY}{dt} &= -\int_\Omega f(\sigma(u), \kappa(u)) \nabla \cdot \kappa(u) \nabla u\, dx - \int_\Omega f(\sigma(u), \kappa(u))\sigma(u)|\nabla\varphi|^2\, dx \\
&= \int_\Omega f(\sigma(u), \kappa(u))' \kappa(u) |\nabla u|^2\, dx - \int_\Omega f(\sigma(u), \kappa(u))\sigma(u)|\nabla\varphi|^2\, dx \\
&\leq -C \int_\Omega |\nabla\varphi|^2\, dx.
\end{aligned} \tag{13.15}$$

Here we used (13.10) and (13.11). It is now easy to see that the value $\overline{\varphi}_0$ delivers the minimum to the function

$$\lambda \to \int_{\Gamma_0'} |\varphi_0 - \lambda|^2 \, d\gamma, \tag{13.16}$$

whence

$$\int_{\Gamma_0'} |\varphi_0 - \overline{\varphi}_0|^2 \, d\gamma \le \int_{\Gamma_0'} |\varphi_0 - \varphi_\Omega|^2 \, d\gamma \le \int_{\Gamma} |\varphi - \varphi_\Omega|^2 \, d\gamma \le C' \int_{\Omega} |\nabla \varphi|^2 \, dx \tag{13.17}$$

for all $\varphi \in H^1(\Omega)$; φ_Ω denotes the average of φ on Ω and C' is the best of the constants such that

$$\int_{\Gamma_0'} |\varphi - \overline{\varphi}|^2 \, d\gamma \le C' \int_{\Omega} |\nabla \varphi|^2 \, dx \quad \forall \varphi \in H^1(\Omega). \tag{13.18}$$

Going back to (13.15), we obtain

$$\frac{dY}{dt} \le -\frac{C}{C'} \int_{\Gamma_0'} |\varphi_0 - \overline{\varphi}_0|^2 \, d\gamma,$$

and integrating in t

$$Y(t) \le Y(0) - \frac{C}{C'} \int_0^t \int_{\Gamma_0'} |\varphi_0 - \overline{\varphi}_0|^2 \, d\gamma dt.$$

It now follows that if (13.12) holds, then $Y(t)$ becomes negative for all t sufficiently large, which is impossible. □

Remark 13.2. In the case where φ_0 is independent of t, and under the conditions of the theorem on σ and κ, the solution $u(x,t)$ blows up provided that φ_0 is not constant on Γ_0'. The sharpness of this condition was studied in [23].

Remark 13.3. The function f can be taken as follows:

1.
$$f(\sigma,\kappa) = \frac{1}{\sigma}, \quad \sigma' \ge 0, \quad \int^{+\infty} \frac{ds}{\sigma(s)} < +\infty.$$

2.
$$\begin{cases} f(\sigma,\kappa) = \dfrac{1}{\sigma\kappa}, & (\sigma\kappa)' \ge 0, \\ \displaystyle\int^{+\infty} \frac{ds}{\sigma(s)\kappa(s)} < +\infty, & \kappa \text{ is bounded.} \end{cases}$$

3.
$$\begin{cases} f(\sigma,\kappa) = \dfrac{\kappa}{\sigma}, & \left(\dfrac{\kappa}{\sigma}\right)' \le 0, \\ \displaystyle\int^{+\infty} \frac{\kappa(s)}{\sigma(s)} ds < +\infty, & \kappa \text{ is bounded away from zero.} \end{cases}$$

13.1.2 *Blowup driven by* φ_1. Let (u, φ) be the solution to problem (13.9a)–(13.9e). We assume that $\Gamma_1' \neq \emptyset$ while Γ_0' is admitted to be empty. If this is the case, then by virtue of equation (13.9d), one has to impose the compatibility condition

$$\int_\Gamma \varphi_1(x,t)\,d\gamma = 0 \quad \forall t \geq 0, \tag{13.19}$$

whence φ is determined up to a constant. Notice that our choice to impose the Neumann boundary condition is guided by the fact that in this case the existence of a solution is easy to establish. In the other case, some additional work would be required.

Theorem 13.2. *Let f be a smooth, positive function of two variables such that*

$$u \to f(\sigma(u), \kappa(u)) \quad \text{is nonincreasing}, \tag{13.20}$$

$$\int^{+\infty} f(\sigma(s), \kappa(s))\,ds < +\infty, \tag{13.21}$$

$$\exists C > 0 \quad \text{such that } \sigma(u) \leq C f(\sigma(u), \kappa(u)) \quad \forall u \geq 0. \tag{13.22}$$

Set

$$V_0 = \{\xi \in H^1(\Omega) |\ \xi = 0 \text{ on } \Gamma_0'\}, \tag{13.23}$$

$$\Phi_1(t) = \sup\left\{ \left| \int_{\Gamma_1} \varphi_1(x,t)\xi(x)\,d\gamma(x) \right| \hat{E};\ \xi \in V_0,\ \int_\Omega |\nabla \xi|^2\,dx = 1 \right\}. \tag{13.24}$$

Then if

$$\int_\Omega \int_{u_0}^{+\infty} f(\sigma(s), \kappa(s))\,ds dx < \frac{1}{C} \int_0^{+\infty} \Phi_1(t)^2 dt, \tag{13.25}$$

problem (13.9a)–(13.9e) *cannot have a global-in-time "smooth" solution.*

Remark 13.4. In the case where $\Gamma_0' = \emptyset$, V_0 coincides with the whole of $H^1(\Omega)$. Note that $\Phi_1(t)$ is nothing but the norm of the continuous linear form

$$\xi \to \int_{\Gamma_1} \varphi_1(x,t)\xi(x)\,d\gamma(x).$$

Proof of Theorem 13.2. Set

$$Y(t) = \int_\Omega \int_u^{+\infty} f(\sigma(s), \kappa(s)) ds dx.$$

Arguing as in (13.15) and using (13.20) and (13.22), we have

$$\begin{aligned}\frac{dY}{dt} &= -\int_\Omega f(\sigma(u),\kappa(u))\,\mathrm{div}(\kappa(u)\nabla u)\,dx\\ &\quad -\int_\Omega f(\sigma(u),\kappa(u))\sigma(u)|\nabla\varphi|^2\,dx\\ &= \int_\Omega f(\sigma(u),\kappa(u))'\kappa(u)|\nabla u|^2\,dx - \int_\Omega f(\sigma(u),\kappa(u))\sigma(u)|\nabla\varphi|^2\,dx\\ &\le -\frac{1}{C}\int_\Omega \sigma(u)^2|\nabla\varphi|^2\,dx.\end{aligned} \tag{13.26}$$

For every function $\xi \in V_0$,

$$0 = \int_\Omega \mathrm{div}(\sigma(u)\nabla\varphi)\xi\,dx = \int_\Omega \mathrm{div}(\sigma(u)\nabla\varphi\xi)\,dx - \int_\Omega \sigma(u)\nabla\varphi\cdot\nabla\xi\,dx.$$

Integrating by parts, we have that

$$\int_\Omega \sigma(u)\nabla\varphi\cdot\nabla\xi\,dx = \int_{\Gamma_1'}\sigma(u)\frac{\partial\varphi}{\partial n}\xi\,d\gamma = \int_{\Gamma_1'}\varphi_1(x,t)\xi\,d\gamma,$$

and applying the Cauchy–Schwarz inequality, we obtain

$$\left|\int_{\Gamma_1'}\varphi_1\xi\,d\gamma\right|^2 \le \int_\Omega \sigma(u)^2|\nabla\varphi|^2\,dx\int_\Omega|\nabla\xi|^2\,dx.$$

Taking the supremum in $\xi \in V_0$, $\int_\Omega |\nabla\xi|^2\,dx = 1$, we get the inequality

$$\Phi_1(t)^2 \le \int_\Omega \sigma(u)^2|\nabla\varphi|^2\,dx.$$

Reverting to (13.26), we obtain the inequality

$$\frac{dY}{dt} \le -\frac{1}{C}\Phi_1(t)^2,$$

and integrating in t then yields

$$Y(t) \le Y(0) - \frac{1}{C}\int_0^t \Phi_1(s)^2\,ds.$$

It follows from (13.25) that $Y(t)$ is negative for all t large enough, a contradiction. □

Remark 13.5. Changing φ_1 into $\lambda\varphi_1$, if necessary, we see that the blowup occurs whenever φ_1 is "large" enough. This means that so is the electric current through Γ.

Remark 13.6. The following are examples of the admissible functions f:

1. $$f(\sigma,\kappa)=\sigma, \qquad \sigma'\le 0, \qquad \int^{+\infty}\sigma(s)\,ds<+\infty.$$

2. $$f(\sigma,\kappa)=\frac{\sigma}{\kappa}, \quad \left(\frac{\sigma}{\kappa}\right)'\le 0, \quad \int^{+\infty}\frac{\sigma(s)}{\kappa(s)}ds<+\infty, \quad \kappa \text{ is bounded.}$$

3. $$\begin{cases} f(\sigma,\kappa)=\sigma\kappa, \quad (\sigma\kappa)'\le 0, \\ \displaystyle\int^{+\infty}\sigma(s)\kappa(s)\,ds<+\infty, \quad \kappa \text{ is bounded away from } 0. \end{cases}$$

Remark 13.7. The first examples in Remarks 13.3 and 13.6 are, in a certain sense, counterparts. This could be explained in space dimension 2. Let (u,φ) be a solution of problem (13.9a)–(13.9e). By virtue of (13.9d), one can find a function ψ such that

$$(\psi_y,-\psi_x)=\sigma(u)(\varphi_x,\varphi_y).$$

Then we have

$$\frac{1}{\sigma}|\nabla\psi|^2=\sigma|\nabla\varphi|^2, \qquad \operatorname{div}\left(\frac{1}{\sigma}\nabla\psi\right)=\operatorname{div}(-\varphi_y,\varphi_x)=0.$$

Moreover, the boundary condition for φ on Γ_1' reads

$$\sigma\varphi_x n_x+\sigma\varphi_y n_y=\varphi_1$$

($\mathbf{n}=(n_x,n_y)$ stands for the outward unit normal to Γ) or

$$\psi_y n_x-\psi_x n_y=\frac{\partial\psi}{\partial\tau}=\varphi_1,$$

where τ is the unit vector orthogonal to $\mathbf{n}$. Integrating the last condition, we see that (u,ψ) satisfies (13.9a)–(13.9e) with $1/\sigma$ instead of σ and with the Neumann boundary condition replaced by the Dirichlet condition.

13.2 Mixed boundary condition for $u(x,t)$. We will assume throughout this section that (13.8) holds. Under this assumption our technique fails when integrating by parts in (13.15) or (13.26) because now $\frac{\partial u}{\partial n}$ does not vanish on the whole of the boundary Γ. In order to circumvent this difficulty we introduce an auxiliary function into the definition of Y. We introduce the first eigenfunction of the Laplace operator. Let us denote by λ the first eigenvalue of the problem

$$\begin{cases} -\Delta\Psi=\lambda\Psi & \text{in } \Omega, \\ \Psi=0 & \text{on } \Gamma_0, \qquad \dfrac{\partial\Psi}{\partial n}=0 \quad \text{on } \Gamma_1. \end{cases} \tag{13.27}$$

Lemma 13.1. *Let λ be the first eigenvalue of problem* (13.27). *Then*

1. $\lambda > 0$;
2. *the eigenspace associated with λ is one dimensional;*
3. *the solution of problem* (13.27) *does not change sign in Ω, so there is a unique function Ψ that solves* (13.27) *and is such that*

$$\int_\Omega \Psi(x)\,dx = 1; \tag{13.28}$$

4. *Ψ is positive in* $\Omega \cup \Gamma_1$.

Proof. Let V_0 be the space

$$V_0 = \{v \in H^1(\Omega) \mid v = 0 \text{ on } \Gamma_0\}.$$

It is easy to show that λ is obtained as

$$\lambda = \inf\left\{\int_\Omega |\nabla u(x)|^2\,dx > 0 : v \in V_0, \int_\Omega v^2 = 1\right\}. \tag{13.29}$$

This infimum is achieved and generates the eigenfunction $u(x) \in V_0$ such that

$$\int_\Omega \nabla u \cdot \nabla \xi\,dx = \lambda \int_\Omega u\xi\,dx \quad \forall \xi \in V_0. \tag{13.30}$$

It is asserted that u does not change sign in Ω. Indeed, assuming the contrary, we have that $u^+ = \max(u, 0)$ vanishes in Ω. Moreover, using $\xi = u^+$ in (13.30), we get

$$\int_\Omega |\nabla u^+|^2\,dx = \lambda \int_\Omega (u^+)^2\,dx,$$

and $u^+/|u^+|_2$ ($|\cdot|_2$ is the usual norm on $L^2(\Omega)$) is a point where the infimum (13.29) is achieved, i.e., u^+ is a solution of (13.27) and satisfies

$$-\Delta u^+ = \lambda u^+, \quad u^+ \geq 0 \quad \text{in } \Omega.$$

By virtue of the strong maximum principle (see [272, 183]), this yields $u^+ = 0$, which is impossible if u changes sign. By rescaling, we can always assume that

$$\int_\Omega u\,dx = 1,$$

and due to the maximum principle, $u > 0$ in $\Omega \cup \Gamma_1$. Let us check that λ is a simple eigenvalue. Let v be an element of the eigenspace $\{v > 0,\ v \perp u\}$. Then

$$0 = \int_\Omega \nabla u \nabla v\,dx = \lambda \int_\Omega uv\,dx,$$

whence $v = 0$. □

13.2.1 *Blowup driven by* φ_0. Let (u, φ) be a solution of the problem

$$u_t = \operatorname{div}(\kappa(u)\nabla u) + \sigma(u)|\nabla\varphi|^2 \quad \text{in } \Omega \times (0, T), \tag{13.31a}$$

$$u(x, 0) = u_0, \tag{13.31b}$$

$$\begin{aligned} u &= 0 && \text{on } \Gamma_0 \times (0, T),\\ \frac{\partial u}{\partial \mathbf{n}} &= 0 && \text{on } \Gamma_1 \times (0, T), \end{aligned} \tag{13.31c}$$

$$\operatorname{div}(\sigma(u)\nabla\varphi) = 0 \quad \text{in } \Omega \times (0, T), \tag{13.31d}$$

$$\begin{aligned} \varphi &= \varphi_0 && \text{on } \Gamma_0' \times (0, T),\\ \sigma(u)\frac{\partial \varphi}{\partial \mathbf{n}} &= \varphi_1 && \text{on } \Gamma_1' \times (0, T) \end{aligned} \tag{13.31e}$$

where $\Gamma_0' \neq \emptyset$ and Γ_1' can be empty. We assume that (u, φ) exists and is sufficiently smooth.

Lemma 13.2. *Let* Ψ *be the unique solution to* (13.31a)–(13.31e) *such that*

$$\int_\Omega \Psi(x)\, dx = 1.$$

Then for any subdomain of positive superficial measure $\widetilde{\Gamma}_1 \Subset \Gamma_1$, *there exists a constant* $C = C(\Psi, \widetilde{\Gamma}_1)$ *such that*

$$\int_\Omega \Psi(x)|\nabla\varphi(x)|^2\, dx \geq C \int_{\widetilde{\Gamma}_1} |\varphi - \overline{\varphi}|^2\, d\gamma \quad \forall \varphi \in H^1(\Omega), \tag{13.32}$$

where $\overline{\varphi} = \frac{1}{|\widetilde{\Gamma}_1|}\int_{\widetilde{\Gamma}_1} \varphi\, d\gamma$.

Proof. We know that $\Psi > 0$ in $\Omega \cup \Gamma_1$. Let Ω' be an open set such that its closure $\overline{\Omega'}$ satisfies $\widetilde{\Gamma}_1 \subset \overline{\Omega'} \subset \Omega \cup \Gamma_1$. Then there is a positive constant α such that $\Psi \geq \alpha$ on $\overline{\Omega'}$. Thus

$$\int_\Omega \Psi|\nabla\varphi|^2\, dx \geq \int_{\Omega'} \Psi|\nabla\varphi|^2\, dx \geq \alpha \int_{\Omega'} |\nabla\varphi|^2\, dx \geq C \int_{\widetilde{\Gamma}_1} |\varphi - \overline{\varphi}|^2\, d\gamma$$

for every $\varphi \in H^1(\Omega)$ (see (13.17)), which completes the proof. □

Remark 13.8. In the one-dimensional case, we can assume without loss of generality that $\widetilde{\Gamma}_1 = \Gamma_1$.

Theorem 13.3. *Let* f *be a smooth, positive function of two variables such that*

$$\begin{gathered} u \to f(\sigma(u), \kappa(u)) \quad \textit{is nonincreasing},\\ \int^{+\infty} f(\sigma(s), \kappa(s))\, ds < +\infty,\\ \exists C > 0 \quad \textit{such that} \quad f(\sigma(u), \kappa(u))\sigma(u) \geq C \quad \forall u \in \mathbb{R}^+,\\ \int_{\mathbf{R}} f(\sigma(s), \kappa(s))\kappa(s)\, ds = M < +\infty. \end{gathered} \tag{13.33}$$

Moreover, assume that there exists a subset $\widetilde{\Gamma}_1$ *of positive superficial measure such that*

$$\widetilde{\Gamma}_1 \Subset \Gamma_0' \cap \Gamma_1.$$

Denote by C' *the best constant such that*

$$\int_{\widetilde{\Gamma}_1} |\varphi - \overline{\varphi}|^2 \, d\gamma \leq C' \int_\Omega \Psi |\nabla \varphi|^2 \, dx \quad \forall \varphi \in H^1(\Omega),$$

where $\overline{\varphi}$ *is the average of* φ *on* $\widetilde{\Gamma}_1$ *and* Ψ *the solution to* (13.27) *satisfying* (13.28). *Then if for* t *large enough*

$$\int_\Omega \int_{u_0}^{+\infty} f(\sigma(s), \kappa(s)) \, ds dx < \int_0^t \left\{ \frac{C}{C'} \int_{\widetilde{\Gamma}_1} |\varphi_0 - \overline{\varphi}_0|^2 \, d\gamma - \lambda M \right\} dt, \tag{13.34}$$

problem (13.31) *cannot have a global-in-time "smooth" solution.*

Proof. Set

$$Y(t) = \int_\Omega \Psi(x) \int_u^{+\infty} f(\sigma(s), \kappa(s)) \, ds dx. \tag{13.35}$$

Differentiating Y with respect to t leads to

$$\frac{dY}{dt} = -\int_\Omega \Psi(x) f(\sigma(u), \kappa(u)) u_t \, dx. \tag{13.36}$$

Using equation (13.9a), we obtain

$$\begin{aligned}
\frac{dY}{dt} &= -\int_\Omega \Psi f(\sigma(u), \kappa(u)) \nabla \cdot \kappa(u) \nabla u \, dx - \int_\Omega \Psi f(\sigma(u), \kappa(u)) \sigma(u) |\nabla \varphi|^2 \, dx \\
&= \int_\Omega \Psi f(\sigma(u), \kappa(u))' \kappa(u) |\nabla u|^2 \, dx + \int_\Omega f(\sigma(u), \kappa(u)) \kappa(u) \nabla u \nabla \Psi \, dx \\
&\quad - \int_\Omega \Psi f(\sigma(u), \kappa(u)) \sigma(u) |\nabla \varphi|^2 \, dx \\
&\leq \int_\Omega f(\sigma(u), \kappa(u)) \kappa(u) \nabla u \cdot \nabla \Psi \, dx - C \int_\Omega \Psi |\nabla \varphi|^2 \, dx \\
&= \int_\Omega \nabla \Big(\int_0^u f(\sigma(s), \kappa(s)) \kappa(s) \, ds \Big) \cdot \nabla \Psi \, dx - C \int_\Omega \Psi |\nabla \varphi|^2 \, dx \\
&= \int_\Omega \int_0^u f(\sigma(s), \kappa(s)) \kappa(s) \, ds (-\Delta \Psi) \, dx - C \int_\Omega \Psi |\nabla \varphi|^2 \, dx \\
&\leq M\lambda \int_\Omega \Psi \, dx - C \int_\Omega \Psi |\nabla \varphi|^2 \, dx.
\end{aligned}$$

With the use of (13.28), (13.32), we get

$$\frac{dY}{dt} \leq M\lambda - \frac{C}{C'} \int_{\widetilde{\Gamma}_1} |\varphi_0 - \overline{\varphi}_0|^2 \, d\gamma. \tag{13.37}$$

Integrating in t, we arrive at the inequality

$$Y(t) \leq \int_{\Omega}\int_{u_0}^{+\infty} f(\sigma(s),\kappa(s))\,ds dx - \int_0^t \left\{\frac{C}{C'}\int_{\widetilde{\Gamma}_1} |\varphi_0-\overline{\varphi}_0|^2\,d\gamma - \lambda M\right\} dt. \tag{13.38}$$

which contradicts the positivity of Y for large t if (13.34) holds. □

Remark 13.9. In the one-dimensional case, one can assume that $\widetilde{\Gamma}_1 = \Gamma_1 \subset \Gamma_0'$ (see Remark 13.8). In the case where φ_0 is independent of t and under the conditions of the theorem for σ and κ, u blows up provided that

$$\int_{\widetilde{\Gamma}_1} |\varphi_0-\overline{\varphi}_0|^2\,d\gamma > \lambda M.$$

For f, we may take the function

$$f(\sigma,\kappa) = \frac{1}{\sigma} \quad \sigma' \geq 0, \quad \int^{+\infty}\frac{ds}{\sigma(s)} < +\infty$$

together with (13.33).

13.2.2 *Blowup driven by* φ_1. Let (u,φ) be the solution to problem (13.31a)–(13.31e). We assume that $\Gamma_1' \neq \emptyset$, while Γ_0' may be empty. Due to equation (13.9d), we have to impose the compatibility condition (13.19).

Theorem 13.4. *Let f be a smooth, positive function of two variables such that*

$$u \to f(\sigma(u),\kappa(u)) \quad \textit{is nonincreasing,} \tag{13.39}$$

$$\int^{+\infty} f(\sigma(s),\kappa(s))\,ds < +\infty,$$

$$\exists C>0 \quad \textit{such that} \quad \sigma(u) \leq Cf(\sigma(u),\kappa(u)) \quad \forall u \in \mathbb{R}^+, \tag{13.40}$$

$$\int_{\mathbf{R}} f(\sigma(s),\kappa(s))\kappa(s)\,ds = M < +\infty. \tag{13.41}$$

Let $\widetilde{\Gamma}_1 \Subset \Gamma_1$ be a subset of positive superficial measure and Ω' be an open subset of Ω such that

$$\widetilde{\Gamma}_1 \subset \overline{\Omega'} \subset \Omega \cup \Gamma_1, \quad \widetilde{\Gamma}_1 \subset \Gamma_1'. \tag{13.42}$$

Then for some positive constant α,

$$\Psi \geq \alpha \quad \textit{on } \overline{\Omega'}. \tag{13.43}$$

Set

$$V_0 = \{\xi \in H^1(\Omega');\ \xi = 0 \textit{ on } \partial\Omega'\backslash\widetilde{\Gamma}_1\}, \tag{13.44}$$

$$\Phi_1(t) = \sup\left\{\left|\int_{\widetilde{\Gamma}_1}\varphi_1(x,t)\xi(x)\,d\gamma(x)\right| \hat{E} \mid \xi \in V_0, \int_{\Omega'}|\nabla\xi|^2\,dx = 1\right\}.$$

Then if for t large enough one has

$$\int_\Omega \Psi(x)\int_{u_0}^{+\infty} f(\sigma(s),\kappa(s))\,ds dx < \int_0^t \left\{\frac{\alpha}{C}\Phi_1(s)^2 - \lambda M\right\} ds, \tag{13.45}$$

problem (13.31a)–(13.31e) *cannot have a global-in-time "smooth" solution.*

Proof. We argue in the standard way and therefore omit some details. Choosing Y as in (13.35) and using (13.39), (13.40), (13.41), we have that

$$\begin{aligned}\frac{dY}{dt} &= -\int_\Omega \Psi f(\sigma(u),\kappa(u))\nabla\cdot\kappa(u)\nabla u\,dx \\ &\quad -\int_\Omega \Psi f(\sigma(u),\kappa(u))\sigma(u)|\nabla\varphi|^2\,dx \\ &= \int_\Omega \Psi f(\sigma(u),\kappa(u))'\kappa(u)|\nabla u|^2\,dx + \int_\Omega f(\sigma(u),\kappa(u))\kappa(u)\nabla u\nabla\Psi\,dx \\ &\quad -\int_\Omega \Psi f(\sigma(u),\kappa(u))\sigma(u)|\nabla\varphi|^2\,dx \\ &\le \int_\Omega f(\sigma(u),\kappa(u))\kappa(u)\nabla u\cdot\nabla\Psi\,dx - \frac{1}{C}\int_\Omega \Psi\sigma(u)^2|\nabla\varphi|^2\,dx \\ &= \int_\Omega \nabla\left(\int_0^u f(\sigma(s),\kappa(s))\kappa(s)\,ds\right)\cdot\nabla\Psi\,dx - \frac{1}{C}\int_\Omega \Psi\sigma(u)^2|\nabla\varphi|^2\,dx \\ &= \int_\Omega\int_0^u f(\sigma(s),\kappa(s))\kappa(s)\,ds(-\Delta\Psi)\,dx - \frac{1}{C}\int_\Omega \Psi\sigma(u)^2|\nabla\varphi|^2\,dx \\ &\le M\lambda - \frac{\alpha}{C}\int_{\Omega'} \sigma(u)^2|\nabla\varphi|^2\,dx.\end{aligned} \tag{13.46}$$

Next, for any function $\xi \in V_0$ extended by zero to the whole of Ω,

$$0 = \int_\Omega \nabla\cdot(\sigma(u)\nabla\varphi)\xi\,dx = \int_{\Omega'} \nabla\cdot(\sigma(u)\nabla\varphi\xi)\,dx - \int_{\Omega'} \sigma(u)\nabla\varphi\cdot\nabla\xi\,dx.$$

Integrating by parts, we get

$$\int_{\Omega'} \sigma(u)\nabla\varphi\cdot\nabla\xi\,dx = \int_{\widetilde{\Gamma}_1} \sigma(u)\frac{\partial\varphi}{\partial\mathbf{n}}\xi\,d\gamma = \int_{\widetilde{\Gamma}_1} \varphi_1\xi\,d\gamma.$$

By the Cauchy–Schwarz inequality,

$$\left|\int_{\widetilde{\Gamma}_1} \varphi_1(x,t)\xi\,d\gamma\right|^2 \le \int_{\Omega'} \sigma(u)^2|\nabla\varphi|^2\,dx \int_{\Omega'} |\nabla\xi|^2\,dx.$$

Taking the supremum in $\xi \in V_0$, $\int_{\Omega'} |\nabla\xi|^2\,dx = 1$ leads to

$$\Phi_1(t)^2 \le \int_{\Omega'} \sigma(u)^2|\nabla\varphi|^2\,dx.$$

Returning to (13.46) and integrating in t, we have

$$Y(t) \leq Y(0) - \int_0^t \left\{ \frac{\alpha}{C} \Phi_1(s)^2 - \lambda M \right\} ds,$$

which is a contradiction to (13.45). □

Remark 13.10. Substituting φ_1 by $\lambda\varphi_1$, if necessary, we see that the blowup occurs whenever φ_1 is "large" enough on $\widetilde{\Gamma}_1$. An admissible function f is, for instance,

$$f(\sigma, \kappa) = \sigma, \quad \sigma' \leq 0, \quad \int^{+\infty} \sigma(s)\, ds < +\infty$$

together with (13.41).

13.3 The Dirichlet boundary condition for *u*. In the one-dimensional problem, the case of the Dirichlet boundary conditions for u and φ was studied in $[L_1]$ and [L2]. We consider here a simple one-dimensional model, but with a blowup that could be driven by the Neumann boundary condition. We denote $\Omega = (0, 1)$ and consider the problem

$$u_t = (\kappa(u)u_x)_x + \sigma(u)\varphi_x^2 \quad \text{in } \Omega \times (0, T), \tag{13.47a}$$

$$u(0, t) = u(1, t) = 0 \quad \text{on } (0, T), \tag{13.47b}$$

$$u(., 0) = u_0, \tag{13.47c}$$

$$(\sigma(u)\varphi_x)_x = 0 \quad \text{in } \Omega \times (0, T), \tag{13.47d}$$

$$\varphi(0, t) = \varphi_0 \quad \text{or} \quad \sigma(0)\varphi_x(0, t) = \varphi_1, \quad \sigma(0)\varphi_x(1, t) = \varphi_1 \quad \text{on } (0, T). \tag{13.47e}$$

Remark 13.11. If the Neumann boundary conditions for φ are adopted, the compatibility condition (13.19) holds.

Let λ be the first eigenvalue of the problem

$$\begin{cases} -\Psi_{xx} = \lambda\Psi & \text{in } \Omega, \\ \Psi(0) = \Psi(1) = 0. \end{cases} \tag{13.48}$$

The function Ψ is assumed to be normalized as

$$\int_\Omega \Psi(x)\, dx = 1.$$

It is well known that, in fact, $\lambda = \pi^2$ and $\Psi(x) = \frac{\pi}{2}\sin(\pi x)$.

Theorem 13.5. *Let σ be a nonincreasing function so that*

$$\int^{+\infty} \sigma(s)\, ds < +\infty, \quad M = \int_{\mathbf{R}} \sigma(s)\kappa(s)\, ds < +\infty. \tag{13.49}$$

Then if for all t large enough

$$\int_{\Omega} \Psi(x) \int_{u_0}^{+\infty} \sigma(s) ds dx < \int_0^t \{\varphi_1(s)^2 - \lambda M\} ds, \tag{13.50}$$

problem (13.47a)–(13.47e) *cannot have a global-in-time "smooth" solution.*

Proof. It follows from (13.47d) that

$$\sigma(u)\varphi_x = Cst = C(t) \quad \text{in } \Omega.$$

Hence by (13.47e),

$$\sigma(u)\varphi_x = \sigma(0)\varphi_x(1,t) = \varphi_1(t),$$

whence

$$\varphi_x = \frac{\varphi_1(t)}{\sigma(u)}.$$

Using this equality in (13.47a), we see that u satisfies

$$u_t = (\kappa(u)u_x)_x + \frac{\varphi_1(t)^2}{\sigma(u)}.$$

Multiplying this equation by $\Psi\sigma(u)$ and integrating over Ω, we get

$$\begin{aligned}
\int_{\Omega} \Psi\sigma(u)u_t dx &= \int_{\Omega} \Psi\sigma(u)(k(u)u_x)_x dx + \varphi_1^2(t) \\
&= \int_{\Omega} (\Psi\sigma(u)k(u)u_x)_x dx - \int_{\Omega} (\Psi\sigma(u))_x k(u)u_x dx + \varphi_1^2(t) \\
&= -\int_{\Omega} \Psi\sigma'(u)k(u)u_x^2 dx - \int_{\Omega} \Psi_x\sigma(u)k(u)u_x dx + \varphi_1^2(t).
\end{aligned}$$

Using the fact that σ is nonincreasing, we obtain

$$\begin{aligned}
\int_{\Omega} \Psi\sigma(u)u_t dx &\geq -\int_{\Omega} \Psi_x \left(\int_0^u \sigma(s)k(s)\, ds\right)_x dx + \varphi_1^2(t) \\
&= -\int_{\Omega} \Psi_{xx} \left(\int_0^u \sigma(s)k(s)\, ds\right) dx + \varphi_1^2(t) \qquad (13.51) \\
&= -\int_{\Omega} \lambda\Psi_{xx} \left(\int_0^u \sigma(s)k(s)\, ds\right) dx + \varphi_1^2(t).
\end{aligned}$$

Let us set (see (13.49))

$$Y(t) = \int_{\Omega} \Psi \int_u^{+\infty} \sigma(s)\, ds dx. \tag{13.52}$$

It is clear that (13.51) reads

$$-\frac{dY}{dt} \geq -\int_\Omega \lambda\Psi\left(\int_0^u \sigma(s)\kappa(s)\,ds\right)dx + \varphi_1(t)^2$$

or (see (13.49))

$$\frac{dY}{dt} \leq \lambda M - \varphi_1(t)^2.$$

Integrating between 0 and t, we obtain

$$Y(t) \leq Y(0) - \int_0^t \{-\lambda M + \varphi_1(s)^2\}\,ds.$$

It follows from (13.50), (13.52) that Y is negative for t large enough, which is impossible. □

13.3.1 *Generalization to the case of a moving nonhomogeneous incompressible medium.* The technique used here can be applied to the study of a more complicated mathematical model describing the heat conduction in a moving nonhomogeneous incompressible medium. In this case, the main equations have the form

$$\begin{aligned} \rho\frac{dU(u)}{dt} &\equiv \rho C(u)\left(\frac{\partial u}{\partial t} + (\mathbf{v}\cdot\nabla)u\right) \\ &= \operatorname{div}(\kappa(u)\nabla u) + \sigma(u)|\nabla\varphi|^2 + \mathbf{S}:\mathbf{D}, \end{aligned} \tag{13.53}$$

$$\operatorname{div}(\sigma(u)\nabla\varphi) = 0, \tag{13.54}$$

$$\frac{d\rho}{dt} \equiv \frac{\partial\rho}{\partial t} + \mathbf{v}\cdot\nabla\rho = 0, \quad \operatorname{div}\mathbf{v} = 0, \tag{13.55}$$

where $U(u)$ is the specific inner energy, $\mathbf{v}$ is the velocity,

$$\mathbf{D}(x,t) = \mathbf{D}^{ij} = \frac{1}{2}\left(\frac{\partial v_i}{\partial x_j} + \frac{\partial v_j}{\partial x_i}\right)$$

is the rate of strain tensor, ρ is the density of the medium, $\mathbf{S} = \mathbf{S}^{ij} = \mathbf{S}^{ji}$ is the stress tensor, and $\mathbf{S}:\mathbf{D} = \sum_{ij}\mathbf{S}^{ij}\mathbf{D}^{ij}$. We assume that $U(u)$, $\rho(x,t)$, $\mathbf{v}(x,t)$, $\mathbf{S}(x,t)$, and $\mathbf{D}(x,t)$ are prescribed functions. System (13.1), (13.2) represents a special case of (13.53)–(13.55) under the assumptions

$$\mathbf{v} \equiv 0, \qquad \mathbf{S}:\mathbf{D} = 0,$$

and

$$\rho = \text{const}, \quad \rho C(u) \equiv \rho\frac{dU(u)}{du} = 1.$$

Let us assume that

$$0 < \rho < \infty, \qquad 0 < U'(u), \qquad \mathbf{P}:\mathbf{D} \geq 0,$$

and

$$\mathbf{v} \cdot \mathbf{n}|_\Gamma = 0,$$

where $\mathbf{n}$ is the outward unit normal to Γ. For the sake of simplicity, we also assume that $(\mathbf{v}, \rho) \in C^1$. Let us restrict our considerations to the case of the boundary conditions of Theorem 13.1 and derive the main differential inequality, an analogue of (13.15).

Assuming that all requested operations make sense, let us set

$$\theta(x,t) = U(u(x,t)), \qquad Y(t) = \int_\Omega \rho \int_{\theta(x,t)}^{\infty} f(\underline{\sigma}(s), \underline{\kappa}(s)) dx ds,$$

where $\underline{\sigma}(\theta) = \sigma(u(\theta))$, $\underline{\kappa}(\theta) = \kappa(u(\theta))$. Then it follows from (13.53) that

$$\rho \frac{d\theta}{dt} = \operatorname{div}(\overline{\kappa}(\theta)\nabla\theta) + \underline{\sigma}(\theta)|\nabla\varphi|^2 + \mathbf{S} : \mathbf{D},$$

where $\overline{\kappa} = \underline{\kappa}/U'(u)$. Differentiating $Y(t)$ leads to the equality

$$\frac{dY}{dt} = -\int_\Omega \rho f \frac{d\theta}{dt} dx = -\int_\Omega f\{\operatorname{div}(\overline{\kappa}(\theta)\nabla\theta) + \underline{\sigma}(\theta)|\nabla\varphi|^2 + \mathbf{P} : \mathbf{D}\} dx,$$

whence

$$\frac{dY}{dt} \leq -\int_\Omega f\{\operatorname{div}(\overline{\kappa}(\theta)\nabla\theta) + \underline{\sigma}(\theta)|\nabla\varphi|^2\} dx.$$

Thus under the assumptions of Theorem 13.1, the problem does not admit any global solution.

Appendix

1 The function spaces

We provide here some reference information concerning the function spaces, the embedding theorems and frequently used inequalities. We retain the definitions and notations introduced in the monographs [235, 233].

Let $\Omega \subset \mathbb{R}^N$ be a bounded domain. It is always assumed that the boundary $\Gamma = \partial\Omega$ is smooth. The cases where Ω is unbounded are always specially indicated.

The space $L^p(\Omega)$, $1 < p < \infty$, is constituted by the measurable functions integrable on Ω with the power p (in the sense of Lebesgue). The norm in $L^p(\Omega)$ is defined by the formula

$$\|u\|_{L^p(\Omega)} \equiv \|u\|_{p,\Omega} = \left(\int_\Omega |u(x)|^p dx \right)^{1/p}. \tag{1.1}$$

If $p = \infty$, the space $L^\infty(\Omega)$ is constituted by the functions essentially bounded with the finite norm

$$\|u\|_{L^\infty(\Omega)} \equiv \|u\|_{\infty,\Omega} = \overline{\lim}_{p\to\infty} \|u\|_{p,\Omega}. \tag{1.2}$$

The space $L^p_{\text{loc}}(\Omega)$ is the set of measurable functions locally integrable in Ω with the power p.

By $W^{l,p}(\Omega)$ with $l \in \mathbb{N}$, $1 < p < \infty$, we understand the so-called Sobolev space of functions that possesses weak derivatives up to the order l that are integrable in

Ω with the power p. The norm in $W^{l,p}(\Omega)$ is defined as

$$\|u\|_{W^{l,p}(\Omega)} \equiv \|u\|_{p,\Omega}^{(l)} = \left(\sum_{k=0}^{l} \|D^k u\|_{p,\Omega}^p\right)^{1/p} \tag{1.3}$$

under the notation

$$D^\alpha = \frac{\partial^{|\alpha|}}{\partial x_1^{\alpha_1}, \dots, \partial x_N^{\alpha_N}};$$

$\alpha = (\alpha_1, \dots, \alpha_N)$ is the multiindex, $|\alpha| = \alpha_1 + \dots + \alpha_N$, and

$$D^1 u = Du = \left(\frac{\partial u}{\partial x_1}, \dots, \frac{\partial u}{\partial x_N}\right), \qquad D^0 u = u.$$

For the vector-valued functions $\mathbf{u}(x) = (u_1(x), \dots, u_m(x))$ with the components $u_i \in W^{l,p}(\Omega)$, we use the notation $\mathbf{u} \in W^{l,p}(\Omega)$ and set

$$\| \mathbf{u} \|_{W^{l,p}(\Omega)} = \left(\sum_{i=1}^{m} \sum_{k=0}^{l} \| D^k u_i \|_{W^{l,p}(\Omega)}^p\right)^{1/p} \tag{1.4}$$

The "anisotropic" spaces $W^{l,\mathbf{p}}(\Omega)$, $\vec{p} = (p_1, \dots, p_N)$ are constituted by the vector-valued functions whose components are integrable with different powers. All these spaces are Banach spaces, while for $p = 2$ they are Hilbert spaces. In the latter case, we use the notation $W^{l,2} = H^l$.

The spaces $\mathring{W}^{l,p}(\Omega)$ are defined as the completion of $C_0^\infty(\Omega)$ (the set of infinite differentiable functions with compact supports in Ω) with respect to the norm of $W^{1,p}(\Omega)$.

We recall the formula of integration by parts

$$\int_\Omega \frac{\partial u}{\partial x_i} v \, dx = -\int_\Omega u \frac{\partial v}{\partial x_i} \, dx + \int_\Gamma uv \cos(\mathbf{n}, x_i) \, d\Gamma, \tag{1.5}$$

where $\mathbf{n}$ is the unit outer normal vector to Γ. Formula (1.5) is safely correct for the functions $u, v \in C^1(\Omega)$ on any bounded domain Ω with $\partial\Omega \in C^1$.

These requirements can be relaxed: once the embedding theorems are formulated, it becomes clear that (1.5) holds for $(u(x), v(x)) \in W^{1,p}(\Omega)$. Formula (1.5) becomes especially simple if $u \in \mathring{W}^{1,p}(\Omega)$, $v \in W^{1,p}(\Omega)$:

$$\int_\Omega \frac{\partial u}{\partial x_i} v \, dx = -\int_\Omega u \frac{\partial v}{\partial x_i} u \, dx. \tag{1.6}$$

The space $C^\alpha(\Omega)$, $\alpha \in (0, 1)$, of Hölder-continuous functions with the exponent α and the norm

$$\|u\|_{C^\alpha(\overline{\Omega})} = \|u\|_{C(\overline{\Omega})} + H^\alpha(u), \tag{1.7}$$

$$H^\alpha(u) = \sup_{(x_1,x_2)\in\overline{\Omega}} \frac{|u(x_1) - u(x_2)|}{|x_1 - x_2|^\alpha}.$$

Correspondingly, the space $H^{l+\alpha}(\Omega)$ ($l \in \mathbb{N}$, $0 \le \alpha \le 1$) is constituted by the functions possessing all ordinary derivatives up to the order l, being Hölder-continuous with the exponent α, and with the norm

$$\|u\|_{C^{l+\alpha}(\Omega)} = \sum_{k=1}^{l} \left\| D^k u \right\|_{C(\Omega)} + \sum_{k=0}^{l} H^{\alpha}(D^k u). \tag{1.8}$$

We denote by $L^q(0, T; L^p(\Omega))$ the function space obtained as the completion of $C^{\infty}(\Omega \times (0, T))$ in the norm

$$\|u\|_{L^q(0,T;L^p(\Omega))} = \left(\int_0^T \|u\|^q_{L^p(\Omega)}\, dt \right)^{1/q}.$$

Correspondingly,

$$\|u\|_{L^{\infty}(0,T;L^p(\Omega))} = \overline{\lim}_{q\to\infty} \|u\|_{L^q(0,T;L^p(\Omega))}.$$

2 Elementary inequalities

2.1 Algebraic inequalities. For every $\mathbf{a} = (a_1, \ldots, a_k)$, $|\mathbf{a}| \neq 0$, and $s > 0$, the inequality

$$C_1(a_1^s + \cdots + a_k^s) \le (a_1 + \cdots + a_k)^s \le C_2(a_1^s + \cdots + a_k^s)$$

holds with finite constants C_1, C_2 depending only on k and s.

The *Cauchy inequality*

$$|a_{ij}| \le \sqrt{a_{ij}\xi_i\xi_j}\sqrt{a_{ij}\eta_i\eta_j} \tag{2.1}$$

holds for every positive defined symmetric form $\{a_{ij}\}$ and the arbitrary vectors $\vec{\xi}, \vec{\eta} \in \mathbb{R}^N$, $|\vec{\xi}| \neq 0$, $|\vec{\eta}| \neq 0$.

The *Young inequality*

$$ab \le \frac{1}{p}(\varepsilon a)^p + \frac{1}{p'}\left(\frac{b}{\varepsilon}\right)^{p'}, \quad p' = \frac{p}{p-1}, \tag{2.2}$$

holds for every $a, b \ge 0$, $\varepsilon > 0$, $1 < p < \infty$. For $p = 2$, (2.2) is also termed the Cauchy inequality.

2.2 Integral inequalities. The direct and inverse *Hölder inequalities*,

$$\left| \int_{\Omega} uv\, dx \right| \le \| u \|_{p,\Omega} \| v \|_{p',\Omega}, \quad p \ge 1, \quad p' = \frac{p}{p-1}, \tag{2.3}$$

$$\int_{\Omega} |u||v|\, dx \ge \left(\int_{\Omega} |u|^p\, dx \right)^{\frac{1}{p}} \left(\int_{\Omega} |v|^{p/(p-1)}\, dx \right)^{\frac{p-1}{p}}, \quad p \in (0, 1), \tag{2.4}$$

hold, provided the corresponding norms and integrals are bounded.

The *Jensen inequality* [261]: Let g be a nonnegative function measurable and integrable on $\Omega \subset \mathbb{R}^N$ and Φ be a continuous concave function defined on the whole of $\mathbb{R}$. Then the inequality

$$\Phi\left(\int_\Omega g\,dx\right) \le \frac{1}{M}\int_\Omega \Phi(Mg)\,dx, \qquad M = \operatorname{meas}\Omega$$

holds.

3 Embedding theorems

Here we present several frequently used assertions from the theory of embedding of the function spaces [233, 235]. These facts can also be found in [2, 251].

3.1 Interpolation inequalities.

Lemma 3.1. *Let $\Omega \subset \mathbb{R}^N$ be a bounded domain with the piecewise smooth boundary Γ, and let Γ_r be the intersection of Ω with some r-dimensional smooth hypersurface, $r \le N$. In particular, if $r = N$, then $\Gamma = \Omega$; for $r = N-1$ one may take $\Gamma = \partial\Omega$.*

For every function $u(x) \in W^{l,p}(\Omega)$ ($l \ge 1$ is natural, $1 < p < \infty$), $N \le pl$, $r > N - pl$, there exists the trace $u(x)|_{\Gamma_r}$ of $u(x)$ on Γ_r; besides

$$u|_{\Gamma_r} \in L^q(\Gamma_r), \quad \|u\|_{q,\Gamma} \le C\|u\|_{W^{l,p}(\Omega)}. \tag{3.1}$$

(If $q \le pr/(N - pl)$, for $N = pl$ the number $q < \infty$ is arbitrary). If $pl > N$, then $u(x)$ is a Hölder-continuous function of the class $C^{k+\alpha}(\overline{\Omega})$, where $k = l - 1 - [N/p]$, $\alpha = 1 + [N/p] - N/p$ if N/p is not integer, and $0 < \alpha < 1$ is arbitrary if N/p is an integer. Besides,

$$\|u\|_{C^{k+\alpha}(\Omega)} \le C\|u\|_{W^{l,p}(\Omega)}. \tag{3.2}$$

The symbol $[N/p]$ stands for the integer part of the number N/p. The constants in (3.1), (3.2) are independent of $u(x)$.

Observe that the embedding $W^{l,p}(\Omega) \hookrightarrow L^q(\Gamma_r)$ is compact for $N \ge pl$ if $q < pr/(N - lp)$. If $pl > N$, it is compactly embedded into $C^k(\Omega)$. The more detailed dependence between the spaces $W^{l,p}(\Omega)$ and $L^q(\Gamma_r)$ is reflected by the so-called *interpolation* (or *multiplicative*) inequalities.

To begin with, let us consider the spaces $W^{l,p}(\Omega)$ and $\mathring{W}^{l,p}(\Omega)$.

Lemma 3.2. *Let $u(x) \in \mathring{W}^{1,p}(\Omega)$, $1 < p < \infty$. The interpolation inequalities*

$$\|u\|_{q,\Omega} \le C\big(\|Du\|_{p,\Omega}\big)^{\theta}\big(\|u\|_{r,\Omega}\big)^{1-\theta} \tag{3.3}$$

hold, where

$$\theta = (1/r - 1/q)/(1/r - 1/p^*), \quad p^* = Np/(N - p).$$

Here

$$C = \left(\frac{N-1}{N}p^*\right)^\theta, \quad \theta \in [0,1], \quad q \in [r, p^*] (or\, q \in [p^*, r]),$$

if $p < N$;

$$C = \max\left(\frac{N-1}{N}q,\ 1 + \frac{p-1}{p}r\right)^\theta, \quad \theta \in [0,1], \quad q \in [r, \infty),$$

if $N \le p$. *If* $p > N$, *then* (3.3) *is true for* $q = \infty$ *with* $\theta = Np/(Np + r(p-N))$ *and some constant* $C < \infty$ *not depending on* Ω.

Remark 3.1. Inequality (3.3) holds as well for all $u \in W^{1,p}(\Omega)$ such that

$$u(x) = 0 \quad \text{in } \Omega_1 \subset \Omega \quad \text{with} \quad \operatorname{meas} \Omega_1 > 0 \qquad \left(\text{or} \int_{\Omega_1} u\,dx = 0\right);$$
$$\text{or if} \quad u(x) = 0 \quad \text{on } \Gamma_1 \subset \Gamma \quad \text{with} \quad \operatorname{meas} \Gamma_1 > 0.$$

The constant C from (3.3) may depend on Ω.

Remark 3.2. Let $u \in W^{1,p}(\Omega)$. Introduce the function

$$v(x) = u(x) - M, \qquad M \operatorname{meas} \Omega = \int_\Omega u(x)\,dx.$$

Obviously, $\int_\Omega v(x)dx = 0$. Hence, according to Remark 3.1, (3.3) holds. Making use of Young's inequality and observing that $|M| \operatorname{meas} \Omega \le \|u\|_{1,\Omega}$, we conclude that

$$\|u\|_{q,\Gamma} \le C\big(\|Du\|_{p,\Omega} + \|u\|_{1,\Omega}\big)^\theta \big(\|u\|_{r,\Gamma}\big)^{1-\theta}. \tag{3.4}$$

Remark 3.3. It is known [233] that the standard norm $\|u\|^{(1)}_{p,\Omega}$ (cf. with (1.4)) is equivalent to the norms

$$\|Du\|_{p,\Omega} + s(\operatorname{meas} \Omega_1)^{1/p-1/N-1/\gamma}\|u\|_{\gamma,\Omega}$$

for all $\gamma \in [1, p]$, $s = \text{const} > 0$, and $\Omega_1 \subset \Omega$ such that $\operatorname{meas} \Omega_1 > 0$.

Lemma 3.3. *Let* $u(x) \in W^{1,p}(\Omega)$, $1 < p < \infty$. *The interpolation inequalities*

$$\|u\|_{q,\Omega} \le C\left(\|u\|^{(1)}_{p,\Omega}\right)^\theta \|u\|^{1-\theta}_{r,\Omega}, \tag{3.5}$$

$$\|u\|_{q,\Gamma} \le C\left(\|u\|^{(1)}_{p,\Omega}\right)^\theta \|u\|^{1-\theta}_{r,\Omega}. \tag{3.6}$$

hold.

The parameters q, p, r, *and* θ *in* (3.5) *satisfy the conditions of Lemma* 3.2, *and those of* (3.6) *satisfy the conditions*

$$\theta = \frac{qN - r(N-1)}{p(N+r) - Nr}\frac{p}{q} \in (0,1),$$

$$1 \le r < \frac{Np}{N-p}, \quad 1 \le q < \frac{p(N-1)}{N-p} \quad \text{if } N > p,$$

$$1 \le r < \infty, \quad 1 \le q < \infty \quad \text{if } p = N,$$

$$1 \le r \le \infty, \quad 1 \le q \le \infty \quad \text{if } p > N.$$

The constant C *in* (3.5), (3.6) *may depend on* Ω.

Remark 3.4. By Remark 3.3, the norm $\|u\|^{(1)}_{p,\Omega}$ in (3.5), (3.6) can be replaced by any equivalent one. For instance, one may use (3.5), (3.6) in the form

$$\|u\|_{q,\Omega} \le C\big(\|Du\|_{p,\Omega} + \|u\|_{\gamma,\Omega}\big)^{\theta}\|u\|^{1-\theta}_{r,\Omega}, \tag{3.7}$$

$$\|u\|_{q,\Gamma} \le C\big(\|Du\|_{p,\Omega} + \|u\|_{\gamma,\Omega}\big)^{\theta}\|u\|^{1-\theta}_{r,\Omega} \tag{3.8}$$

with $1 \le \gamma < \infty$.

Accept the notation

$$B_\rho(x_0) = \{x : \quad |x - x_0| < \rho\}, \quad S_\rho(x_0) = \partial B_\rho(x_0).$$

Lemma 3.4 ([149]). *Let* $\Omega = B_\rho(x_0)$, $\rho > 0$, $x_0 \in \Omega$. *For all* $u \in W^{1,p}(B_\rho(x_0))$, *the inequalities*

$$\|u\|_{q,B_\rho(x_0)} \le C\big(\|Du\|_{p,B_\rho(x_0)} + \rho^{\delta}\|u\|_{\gamma,B_\rho(x_0)}\big)^{\theta}\|u\|^{1-\theta}_{r,B_\rho(x_0)}, \tag{3.9}$$

$$\|u\|_{q,S_\rho} \le C\big(\|Du\|_{p,B_\rho(x_0)} + \rho^{\delta}\|u\|_{\gamma,B_\rho(x_0)}\big)^{\theta}\|u\|^{1-\theta}_{r,B_\rho(x_0)} \tag{3.10}$$

hold, where $\gamma \in [1,p]$ *and the constants* q, p, r, *and* θ (*which are different in* (3.9) *and* (3.10)) *are defined in Lemmas* 3.2 *and* 3.3, *and*

$$\delta = -\left(1 + \frac{p-\gamma}{\gamma p}N\right). \tag{3.11}$$

Remark 3.5. All the inequalities above hold for the vector-valued functions whose components belong to the corresponding function space. The proof amounts to checking the validity of the above inequalities for every component of the considered vector-function and applying (2.3).

For instance, it is easy to get from (3.3) that

$$\begin{aligned}
\|\mathbf{u}\|^q_{q,\Omega} &= \sum_{i=1}^m \|u_i\|^q_{q,\Omega} \le \sum_{i=1}^m C^q \|Du_i\|^{q\theta}_{q,\Omega}\|u_i\|^{q(1-\theta)}_{r,\Omega} \\
&\le C\left(\sum_{i=1}^m \|Du_i\|^p_{q,\Omega}\right)^{q\theta/p}\left(\sum_{i=1}^m \|u_i\|^r_{r,\Omega}\right)^{q(1-\theta)/r} \\
&\le C\big(\|D\vec{u}\|_{q,\Omega}\big)^{q\theta}\big(\|\mathbf{u}\|_{r,\Omega}\big)^{q(1-\theta)}.
\end{aligned}$$

Lemma 3.5. *Let $\Omega \in \mathbb{R}^N$ and $u \in W^{l,p}(\Omega) \cap L^r(\Omega)$, $1 \le l$, $1 \le r < \infty$, $1 < p < \infty$. Then*

$$\|u\|_{W^{k,q}(\Omega)} \le C\|u\|^{\theta}_{W^{l,q}(\Omega)}\|u\|^{1-\theta}_{L^r(\Omega)}, \tag{3.12}$$

where

$$\theta = \frac{q(kr+N)Nr}{p(N+rl)-Nr}\frac{p}{q}, \quad \frac{k}{l} \le \theta \le 1;$$

the constant C depends only on l, p, r, k, q, θ, and Ω and may be unbounded, in particular, if $\Omega = \mathbb{R}^N$.

3.2 Anisotropic function spaces.

Lemma 3.6 ([229]). *Let $\Omega \subset \mathbb{R}^N$ be a bounded domain with the smooth boundary Γ,*

$$u \in W^{1,\infty}(\Omega), \qquad \left|\frac{\partial u}{\partial x_i}\right| \le M, \quad i = 1, \dots, N.$$

Then

$$\|u\|_{q,\Gamma} \le C(M,\Omega)\left(\sum_{k=1}^{N}\left\|\frac{\partial u}{\partial x_i}\right\|_{p_i,\Omega} + \|u\|_{r,\Omega}\right)^{\theta}\|u\|^{1-\theta}_{r,\Omega}, \tag{3.13}$$

where

$$\frac{N}{p^*} \le q, \quad 1 < r \le q, \quad 1 < p_i < \infty, \quad p^* = \sum_{i=1}^{N}\frac{1}{p_i}, \tag{3.14}$$

$$\frac{1}{N(1-p^*)} + \frac{1}{r} > 0, \quad \theta = \left(\frac{1}{r} - \frac{N-1}{Nq}\right) / \left(\frac{1}{N} - \frac{1}{Np^*} + \frac{1}{r}\right). \tag{3.15}$$

Remark 3.6. Inequality (3.13) is invariant with respect to the scaling transformation $u(x) \to u(x)/M$. Hence for every $u \in L^r(\Omega)$ such that $D_{x_i}u \in L^{p_i}(\Omega)$, it holds with the constant $C(1,\Omega)$.

Lemma 3.7. *Let $\Omega = B_\rho(x_0)$ $(\rho > 0)$, $u(x) \in L^r(B_\rho(x_0))$, and $D_{x_i}u \in L^{p_i}(B_\rho(x_0))$. Then*

$$\|u\|_{q,S_\rho} \le C\left(\sum_{i=1}^{N}\rho^{\delta_i}\left\|\frac{\partial u}{\partial x_i}\right\|_{p_i,B_\rho(x_0)} + \rho^{\delta}\|u\|_{r,B_\rho(x_0)}\right)^{\theta}\|u\|^{1-\theta}_{r,B_\rho(x_0)}, \tag{3.16}$$

$$1 \le \gamma < \infty,$$

where q, p, r, and θ satisfy the conditions of Lemma 3.6,

$$\delta = ((N-1)/q - N/q)/\theta, \qquad \delta_i = \delta + 1 + N/r - N/p_i. \tag{3.17}$$

References

[1] U. G. ABDULLAEV, *Instantaneous shrinking and exact local estimations of solutions in nonlinear diffusion absorption*, Adv. Math. Sci. Appl., 8 (1998), pp. 483–503.

[2] R. ADAMS, *Sobolev Spaces*, Academic Press, New York, 1975.

[3] H. W. ALT AND S. LUCKHAUS, *Quasilinear elliptic-parabolic differential equations*, Math. Z., 183 (1983), pp. 311–341.

[4] L. ALVAREZ, *On the behaviour of the free boundary of some nonhomogeneous elliptic problems*, Appl. Anal., 36 (1990), pp. 131–144.

[5] L. ALVAREZ AND J. I. DÍAZ, *On the behaviour near the free boundary of solutions of some nonhomogeneous elliptic problems*, in Actas del IX CEDYA, J. M. Sanz-Serna, ed., Valladolid, Spain, 1987, Univ. de Valladolid, pp. 55–59.

[6] F. ANDREU, V. CASELLES, J. I. DÍAZ, AND J. MAZÓN, *Qualitative Properties for the Total Variation Flow*, prepublicación, Departamento de Matematica Aplicada de la Universidad Complutense de Madrid, December 2000.

[7] D. ANDREUCCI AND A. F. TEDEEV, *Sharp estimates and finite speed of propagation for a Neumann problem in domains narrowing at infinity*, Adv. Differential Equations, 5 (2000), pp. 833–860.

[8] G. ANDREWS, *On the existence of solutions to the equation* $u_{tt} = u_{xxt} + \sigma(u_x)_x$, J. Differential Equations, 35 (1980), pp. 200–231.

[9] S. N. Antontsev, *Axially symmetric problems of gas dynamics with free boundaries*, Dokl. Akad. Nauk SSSR, 216 (1974), pp. 473–476 (English translation: Soviet Math. Dokl., 15-3 (1974), pp. 803–807).

[10] ——, *Boundary value problems for certain degenerate equations of continuum mechanics, Part* 1, Novosibirsk State University, Novosibirsk, 1976 (in Russian; "Краевые задачи для некоторых вырождающихся уравнений механики сплошной среды: Часть 1").

[11] ——, *The character of perturbations described by solutions of multidimensional degenerate parabolic equations*, Dinamika Sploshn. Sredy, 40, Dinamika Zhidkosti so Svobod. Granits., (1979), pp. 114–122 (in Russian).

[12] ——, *Finite rate of propagation of perturbations in two-dimensional problems of two-phase filtration*, Dinamika Sploshn. Sredy, 39 (1979), pp. 23–29 (in Russian).

[13] ——, *Finite rate of propagation of perturbations in multidimensional problems of two-phase filtration*, Zap. Nauchn. Sem. Leningrad. Otdel. Mat. Inst. Steklov. (LOMI), Boundary value problems of mathematical physics and related questions in the theory of functions, 12, 96 (1980), pp. 3–12, 305 (English translation: J. Sov. Math., 21 (1983), pp. 637–644).

[14] ——, *On localization of disturbances for multi-dimensional degenerate elliptic and parabolic equations*, Uspekhi Matem. Nauk, 35 (1980), pp. 165–166 (in Russian).

[15] ——, *On the localization of solutions of nonlinear degenerate elliptic and parabolic equations*, Dokl. Akad. Nauk SSSR, 260 (1981), pp. 1289–1293 (translation from Dokl. Akad. Nauk SSSR, 260-6 (1981), pp. 1289–1293).

[16] ——, *On localization of solutions of nonlinear degenerate elliptic equations*, in Partial Differential Equations and Problems with Free Boundaries, Kiev, Naukova Dumka, 1983, pp. 15–19.

[17] ——, *Localization of solutions of degenerate equations of continuum mechanics*, Akad. Nauk SSSR Sibirsk. Otdel. Inst. Gidrodinamiki, Novosibirsk, 1986 (in Russian; "Локализация решений вырождающихся уравнений механики сплошной среды").

[18] ——, *Metastable localization of solutions of degenerate parabolic equations of general form*, Dinamika Sploshn. Sredy, 83 (1987), pp. 138–144.

[19] ——, *Localization of solutions for degenerate equations in continuum mechanics*, in Free Boundary Problems: Theory and Applications, Vol. II (Irsee, 1987), Longman, Harlow, UK, 1990, pp. 725–739.

[20] ——, *Localization of solutions of a problem of mass transport in a porous medium*, Dokl. Akad. Nauk, 326 (1992), pp. 268–271.

[21] ——, *Quasilinear parabolic equations with non-isotropic nonlinearities: space and time localization*, in Energy Methods in Continuum Mechanics (Oviedo, 1994), Kluwer Academic Publishers, Dordrecht, the Netherlands, 1996, pp. 1–12.

[22] S. N. ANTONTSEV AND M. CHIPOT, *Existence, stability and blowup of the solution of the thermistor problem*, Dokl. Akad. Nauk, 324 (1992), pp. 309–313 (English translation: Soviet Math. Dokl., 37-5 (1992), pp. 229–231).

[23] ——, *The thermistor problem: existence, smoothness, uniqueness, blowup*, SIAM J. Math. Anal., 25 (1994), pp. 1128–1156.

[24] ——, *Analysis of blow up for the thermistor problem*, Siberian Math. J., 38 (1997), pp. 827–841.

[25] S. N. ANTONTSEV AND J. I. DÍAZ, *Application of the energy method in the localization of solutions of equations in continuum mechanics*, Dokl. Akad. Nauk SSSR, 303 (1988), pp. 320–325 (English translation: Soviet Phys. Dokl., 33-11 (1988), pp. 813–816).

[26] ——, *New results on the localization of solutions of nonlinear elliptic and parabolic equations that are obtained by the energy method*, Dokl. Akad. Nauk SSSR, 303 (1988), pp. 524–529 (English translation: Soviet Math. Dokl., 38-3 (1989), pp. 535–539).

[27] ——, *The energy method and the localization of solutions of equations of continuum mechanics*, Zh. Prikl. Mekh. i Tekhn. Fiz., (1989), pp. 18–25 (English translation: J. Appl. Mech. Tech. Phys., 30-2 (1989), pp. 182–189).

[28] S. N. ANTONTSEV, J. I. DÍAZ, AND A. V. DOMANSKIĬ, *Stability and stabilization of generalized solutions of degenerate problems of two-phase filtration*, Dokl. Akad. Nauk, 325 (1992), pp. 1151–1155 (English translation: Soviet Phys. Dokl., 37-8 (1993), pp. 411–413).

[29] ——, *Continuous dependence and stabilization of solutions of the degenerate system in two-phase filtration*, Dinamika Sploshn. Sredy, 107 (1993), pp. 11–25, 203–204.

[30] S. N. ANTONTSEV, J. I. DÍAZ, AND S. I. SHMAREV, *New applications of energy methods to parabolic free boundary problems*, Uspekhi. Mat. Nauk, 46 (1991), pp. 181–182 (English translation: Russian Math. Surveys, 46-6 (1991), pp. 193–194).

[31] ——, *New results on the character of localization of solutions for parabolic equations*, in Abstracts of the International conference "Free-Boundary Problems in Continuum Mechanics," Novosibirsk, July 15–19, 1991, pp. 11–12.

[32] ——, *New applications of energy methods to parabolic and elliptic free boundary problems*, in Free boundary problems in continuum mechanics (Novosibirsk, 1991), Birkhäuser, Basel, 1992, pp. 59–65.

[33] ——, *The support shrinking in solutions of parabolic equations with nonhomogeneous absorption terms*, in Elliptic and Parabolic Problems (Pont-à-Mousson, 1994), Longman, Harlow, UK, 1995, pp. 24–39.

[34] ——, *The support shrinking properties for solutions of quasilinear parabolic equations with strong absorption terms*, Ann. Fac. Sci. Toulouse Math. (6), 4 (1995), pp. 5–30.

[35] ——, *On the boundary layer for dilatant fluids*, in Energy Methods in Continuum Mechanics (Oviedo, 1994), Kluwer Academic Publishers, Dordrecht, the Netherlands, 1996, pp. 13–21.

[36] S. N. Antontsev, A. V. Domanskii, and V. I. Penkovskii, *Filtration in By-Well Zone of the Formation and the Well Productivity Stimulation Problems*, Lavrentyev Institute of Hydrodynamics, Siberian Branch of the USSR Academy of Science, Novosibirsk, 1989 (in Russian).

[37] S. N. Antontsev, G. P. Epikhov, and A. A. Kashevarov, *Mathematical system modelling of water exchange processes*, "Nauka" Sibirsk. Otdel., Novosibirsk, 1986 (in Russian; "Системное математическое моделирование процессов водообмена").

[38] S. N. Antontsev and A. A. Kashevarov, *Finite rate of propagation of perturbations in simultaneous flows of surface and ground water*, Dinamika Sploshn. Sredy, 57 (1982), pp. 21–27 (in Russian).

[39] ——, *Correctness of the mathematical model of joint motion of surface and ground waters*, Uspekhi Mat. Nauk, 4 (1984), p. 116 (in Russian).

[40] ——, *Splitting according to physical processes in the problem of interaction between surface and underground water*, Dokl. Akad. Nauk SSSR, 288 (1986), pp. 86–90 (English translation: Soviet Phys. Dokl, 31-5 (1986), pp. 381–383).

[41] ——, *Localization of solutions of nonlinear parabolic equations that are degenerate on a surface*, Dinamika Sploshn. Sredy, 111 (1996), pp. 7–14 (in Russian).

[42] ——, *Solutions localization of nonlinear parabolic equations with degeneracy on a surface*, in Abstracts of the International Conference "Nonlinear Partial Differential Equations," Kiev, August 26–30 1997, National Academy of Science of Ukraine.

[43] ——, *Mathematical models of mass transfer in interconnected processes of surface, soil and ground waters*, in Abstracts of the International Conference "Modern Approaches to Flows in Porous Media," September 6–8, Moscow, 1999, pp. 165–166.

[44] S. N. ANTONTSEV, A. A. KASHEVAROV, AND A. M. MEĬRMANOV, *Numerical modelling of simultaneous motion of surface channel and ground waters*, in Abstracts of the International conference on numerical modelling of river, channel overland flow for water resources and environmental applications, May 4–8, Bratislava, Czechoslovakia, 1981, pp. 1–11.

[45] S. N. ANTONTSEV, A. V. KAZHIKHOV, AND V. N. MONAKHOV, *Solvability of boundary value problems for some models of inhomogeneous fluids*, in Partial Differential Equations: Proceedings of the International Conference Dedicated to the Memory of I. G. Petrovskii, 1976, Moscow State University, Moscow, 1978, pp. 30–33.

[46] ——, *Boundary Value Problems in Mechanics of Nonhomogeneous Fluids*, North-Holland, Amsterdam, 1990 (translated from the original Russian edition, Nauka, Novosibirsk, 1993).

[47] S. N. ANTONTSEV AND A. M. MEĬRMANOV, *Questions of correctness of a model of the simultaneous motion of surface and ground waters*, Dokl. Akad. Nauk SSSR, 242 (1978), pp. 505–508.

[48] ——, *Mathematical models of simultaneous motions of surface and ground waters*, Novosibirsk State University, Lecture Notes, 1979.

[49] S. N. ANTONTSEV AND V. N. MONAKHOV, *Three-dimensional problems of time-dependent two-phase filtration in inhomogeneous anisotropic porous media*, Dokl. Akad. Nauk SSSR, 243 (1978), pp. 553–556 (in Russian; English translation: Soviet Math. Dokl., 19-6 (1978), pp. 1354–1358).

[50] S. N. ANTONTSEV AND A. A. PAPIN, *Localization of solutions of equations of a viscous gas, with viscosity depending on the density*, Dinamika Sploshn. Sredy, 86 (1988), pp. 24–40 (in Russian).

[51] S. N. ANTONTSEV AND P. REBELO, *Localização de soluções fracas para correntes de fluidos não-newtonianos com densidade inicial infinita ou nula*, Pré-Publicação 1, Centro de Matemática, Universidade da Beira Interior, Covilhã, Portugal, 1999.

[52] S. N. ANTONTSEV AND S. I. SHMAREV, *Localization of solutions of nonlinear parabolic equations with linear sources of general form*, Dinamika Sploshn. Sredy, 89 (1989), pp. 28–42 (in Russian).

[53] ——, *The local energy method and vanishing of weak solutions of nonlinear parabolic equations*, Dokl. Akad. Nauk SSSR, 318 (1991), pp. 777–781 (English translation: Soviet Math. Dokl., 43-3 (1991), pp. 738–742).

[54] ——, *Local energy method and vanishing properties of weak solutions of quasilinear parabolic equations*, in Free Boundary Problems Involving Solids (Montreal, 1990), Longman, Harlow, UK, 1993, pp. 2–6.

[55] D. ARCOYA AND M. CALAHORRANO, *Multivalued nonpositone problems*, Atti Accad. Naz. Lincei Cl. Sci. Fis. Mat. Natur. Rend. (9) Mat. Appl., 1 (1990), pp. 117–123.

[56] D. G. ARONSON, *The porous medium equation*, in Nonlinear Diffusion Problems (Montecatini Terme, 1985), Springer-Verlag, Berlin, 1986, pp. 1–46.

[57] C. ATKINSON AND J. E. BOUILLET, *Some qualitative properties of solutions of a generalised diffusion equation*, Math. Proc. Cambridge Philos. Soc., 86 (1979), pp. 495–510.

[58] H. ATTOUCH AND A. DAMLAMIAN, *Application des méthodes de convexité et monotonie à l'étude de certaines équations quasi linéaires*, Proc. Roy. Soc. Edinburgh Sect. A, 79 (1977/78), pp. 107–129.

[59] M. BALABANE, T. CAZENAVE, AND L. VÁZQUEZ, *Existence of standing waves for Dirac fields with singular nonlinearities*, Comm. Math. Phys., 133 (1990), pp. 53–74.

[60] A. BAMBERGER, *Étude d'une équation doublement non linéaire*, J. Functional Analysis, 24 (1977), pp. 148–155.

[61] C. BANDLE, T. NANBU, AND I. STAKGOLD, *Porous medium equation with absorption*, SIAM J. Math. Anal., 29 (1998), pp. 1268–1278.

[62] C. BANDLE AND I. STAKGOLD, *The formation of the dead core in parabolic reaction-diffusion problems*, Trans. Amer. Math. Soc., 286 (1984), pp. 275–293.

[63] V. BARBU, *Null controllability of first order quasilinear equations*, Differential Integral Equations, 4 (1991), pp. 673–681.

[64] G. I. BARENBLATT, *On some unsteady motions of a liquid and gas in a porous medium*, Akad. Nauk SSSR. Prikl. Mat. Meh., 16 (1952), pp. 67–78.

[65] G. BARLES, G. DÍAZ, AND J. I. DÍAZ, *Uniqueness and continuum of foliated solutions for a quasilinear elliptic equation with a non-Lipschitz nonlinearity*, Comm. Partial Differential Equations, 17 (1992), pp. 1037–1050.

[66] P. BENILAN, H. BREZIS, AND M. CRANDALL, *A semilinear equation in* $L^1(R^N)$, Ann. Scuola Norm. Sup. Pisa Cl. Sci. (4), 2 (1975), pp. 523–555.

[67] P. BÉNILAN AND M. CRANDALL, *The continuous dependence on* φ *of solutions of* $u_t - \delta\varphi(u) = 0$, Indiana Univ. Math. J., 30 (1981), pp. 161–177.

[68] A. Bensoussan and J.-L. Lions, *On the support of the solution of some variational inequalities of evolution*, J. Math. Soc. Japan, 28 (1976), pp. 1–17.

[69] H. Berestycki, L. A. Caffarelli, and L. Nirenberg, *Monotonicity for elliptic equations in unbounded Lipschitz domains*, Comm. Pure Appl. Math., 50 (1997), pp. 1089–1111.

[70] L. Berkovitz and H. Pollard, *A non-classical variational problem arising from an optimal filter problem*, Arch. Rational Mech. Anal., 26 (1967), pp. 281–304.

[71] F. Bernis, *Compactness of the support in convex and nonconvex fourth order elasticity problems*, Nonlinear Anal., 6 (1982), pp. 1221–1243.

[72] ——, *Asymptotic rates of decay for some nonlinear ordinary differential equations and variational problems of arbitrary order*, Ann. Fac. Sci. Toulouse Math. (5), 6 (1984), pp. 121–151.

[73] ——, *Compactness of the support for some nonlinear elliptic problems of arbitrary order in dimension n*, Comm. Partial Differential Equations, 9 (1984), pp. 271–312.

[74] ——, *Extinction of the solutions of some quasilinear elliptic problems of arbitrary order*, vol. 45 of Proceedings of Symposia in Pure Mathematics, Part 1, AMS, 1986, pp. 125–132.

[75] ——, *Finite speed of propagation and asymptotic rates for some nonlinear higher order parabolic equations with absorption*, Proc. Roy. Soc. Edinburgh Sect. A, 104 (1986), pp. 1–19.

[76] ——, *Existence results for doubly nonlinear higher order parabolic equations on unbounded domains*, Math. Ann., 279 (1988), pp. 373–394.

[77] ——, *Qualitative properties for some nonlinear higher order degenerate parabolic equations*, Houston J. Math., 14 (1988), pp. 319–352.

[78] ——, *Elliptic and parabolic semilinear problems without conditions at infinity*, Arch. Rational Mech. Anal., 106 (1989), pp. 217–241.

[79] ——, *Nonlinear parabolic equations arising in semiconductor and viscous droplets models*, in Nonlinear Diffusion Equations and Their Equilibrium States 3 (Gregynog, 1989), Birkhäuser, Boston, MA, 1992, pp. 77–88.

[80] ——, *Energy methods for higher order elliptic and parabolic problems*, in Energy Methods in Continuum Mechanics (Oviedo, 1994), Kluwer Academic Publishers, Dordrecht, the Netherlands, 1996, pp. 31–37.

[81] J. G. Berryman and C. J. Holland, *Stability of the separable solution for fast diffusion*, Arch. Rational Mech. Anal., 74 (1980), pp. 379–388.

[82] ——, *Asymptotic behavior of the nonlinear diffusion equation* $n_t = (n^{-1}n_x)_x$, J. Math. Phys., 23 (1982), pp. 983–987.

[83] L. Bers, *Mathematical Aspects of Subsonic and Transonic Gas Dynamics*, Surveys in Applied Mathematics, Vol. 3, John Wiley, New York, 1958.

[84] M.-F. Bidaut-Véron, *Propriété de support compact de la solution d'une équation aux dérivées partielles non linéaire d'ordre* 4, C. R. Acad. Sci. Paris Sér. A-B, 287 (1978), pp. A1005–A1008.

[85] ——, *Variational inequalities of order* 2m *in unbounded domains*, Nonlinear Anal., 6 (1982), pp. 253–269.

[86] ——, *Principe de maximum et support compact pour une classe d'équations elliptiques non linéaires d'ordre 4*, tech. rep., Publ. Math. Univ. Pau, 1988.

[87] G. Birkhoff and E. H. Zarantonello, *Jets, Wakes, and Cavities*, Academic Press, New York, 1957.

[88] D. Blanchard and F. Murat, *Renormalised solutions of nonlinear parabolic problems with* L^1 *data: existence and uniqueness*, Proc. Roy. Soc. Edinburgh Sect. A, 127 (1997), pp. 1137–1152.

[89] M. Böhm, *On a nonhomogeneous Bingham fluid*, J. Differential Equations, 60 (1985), pp. 259–284.

[90] S. Bonafede, G. R. Cirmi, and A. F. Tedeev, *Finite speed of propagation for the porous media equation*, SIAM J. Math. Anal., 29 (1998), pp. 1381–1398 (electronic).

[91] ——, *Finite speed of propagation for the porous media equation with lower order terms*, Discrete Contin. Dynam. Systems, 6 (2000), pp. 305–314.

[92] M. Borelli and M. Ughi, *The fast diffusion equation with strong absorption: the instantaneous shrinking phenomenon*, Rend. Istit. Mat. Univ. Trieste, 26 (1994), pp. 109–140 (1995).

[93] J. Boussinesq, *Théorie analytique de la chaleur*, Vol. 2, Gauthier-Villars, Paris, 1903.

[94] H. Brezis, *Monotone operators, nonlinear semigroupes and applications*, in Proceedings of the International Congress Math., Vancouver, 1974.

[95] H. Brézis, *Solutions à support compact d'inéquations variationnelles*, in Séminaire sur les Équations aux Dérivées Partielles (1973–1974), I, Exp. no. 3, Collège de France, Paris, 1974, p. 6.

[96] H. Brezis, *Solutions of variational inequalities, with compact support*, Uspehi Mat. Nauk, 29 (1974), pp. 103–108 (collection of articles dedicated to the memory of Ivan Georgievič Petrovskiĭ (1901–1973) I; translated from the English by Ju. A. Dubinskiĭ).

[97] H. Brezis, *A new method in the study of subsonic flows*, in Partial Differential Equations and Related Topics (Program, Tulane Univ., New Orleans, LA, 1974), Lecture Notes in Math., Vol. 446, Springer-Verlag, Berlin, 1975, pp. 50–64.

[98] H. Brézis and G. Duvaut, *Écoulements avec sillages autour d'un profil symétrique sans incidence*, C. R. Acad. Sci. Paris Sér. A–B, 276 (1973), pp. A875–A878.

[99] H. Brezis and A. Friedman, *Estimates on the support of solutions of parabolic variational inequalities*, Illinois J. Math., 20 (1976), pp. 82–97.

[100] H. Brezis and E. Lieb, *Minimum action solutions of some vector field equations*, Comm. Math. Phys., 96 (1984), pp. 97–113.

[101] H. Brezis and L. Nirenberg, *Removable singularities for nonlinear elliptic equations*, Topol. Methods Nonlinear Anal., 9 (1997), pp. 201–219.

[102] H. Brézis and G. Stampacchia, *Une nouvelle méthode pour l'étude d'écoulements stationnaires*, C. R. Acad. Sci. Paris Sér. A–B, 276 (1973), pp. A129–A132.

[103] ——, *The hodograph method in fluid-dynamics in the light of variational inequalities*, Arch. Rational Mech. Anal., 61 (1976), pp. 1–18.

[104] L. A. Caffarelli and A. Friedman, *Regularity of the free boundary of a gas flow in an n-dimensional porous medium*, Indiana Univ. Math. J., 29 (1980), pp. 361–391.

[105] J. Carrillo, *Solutions entropiques de problèmes non linéaires dégénérés*, C. R. Acad. Sci. Paris Sér. I Math., 327 (1998), pp. 155–160.

[106] J. Carrillo, *Entropy solutions for nonlinear degenerate problems*, Arch. Ration. Mech. Anal., 147 (1999), pp. 269–361.

[107] C. Y. Chan and H. G. Kaper, *Quenching for semilinear singular parabolic problems*, SIAM J. Math. Anal., 20 (1989), pp. 558–566.

[108] C. Y. Chan and M. K. Kwong, *Quenching phenomena for singular nonlinear parabolic equations*, Nonlinear Anal., 12 (1988), pp. 1377–1383.

[109] P. H. Chang and H. A. Levine, *The quenching of solutions of semilinear hyperbolic equations*, SIAM J. Math. Anal., 12 (1981), pp. 893–903.

[110] G. Chavent and J. Jaffre, *Mathematical Models and Finite Elements for Reservoir Simulation: Single Phase, Multiphase and Multicomponent Flows Through Porous Media*, Studies in Mathematics and Its Applications, Vol. 17, North-Holland, Amsterdam, 1986.

[111] X. Chen and A. Friedman, *The thermistor problem for conductivity which vanishes at large temperature*, Quart. Appl. Math., 51 (1993), pp. 101–115.

[112] M. Chipot and G. Cimatti, *A uniqueness result for the thermistor problem*, European J. Appl. Math., 2 (1991), pp. 97–103.

[113] G. Cimatti, *A bound for the temperature in the thermistor problem*, IMA J. Appl. Math., 40 (1988), pp. 15–22.

[114] ——, *Remark on existence and uniqueness for the thermistor problem under mixed boundary conditions*, Quart. Appl. Math., 47 (1989), pp. 117–121.

[115] G. Cimatti and G. Prodi, *Existence results for a nonlinear elliptic system modelling a temperature dependent electrical resistor*, Ann. Mat. Pura Appl. (4), 152 (1988), pp. 227–236.

[116] R. Courant and K. Friedrichs, *Supersonic flow and shock waves*, Interscience, 1967.

[117] M. G. Crandall, K. Fok, M. Kocan, and A. Święch, *Remarks on nonlinear uniformly parabolic equations*, Indiana Univ. Math. J., 47 (1998), pp. 1293–1326.

[118] P. M. Ogibalov and A. Kh. Mirzadžhanzade, *Nonstationary Motions of Visco-Plastic Media*, revised ed., Izdat. Moskov. Univ., Moscow, Moscow, 1977 (in Russian; П. М. Огибалов and А. Х. Мирзаджанзаде, Нестационарные движения вязкопластичных сред).

[119] C. Dafermos, *The mixed initial-boundary value problem for the equations of nonlinear one-dimensional viscoelasticity*, J. Differential Equations, 6 (1969), pp. 71–86.

[120] K. Deng, *The quenching of solutions of semilinear parabolic equations*, J. Huazhong Univ. Sci. Tech. (English Ed.), 7 (1985), pp. 1–6 (translated from the Chinese).

[121] ——, *Quenching for solutions of a plasma type equation*, Nonlinear Anal., 18 (1992), pp. 731–742.

[122] G. Díaz and J. I. Díaz, *Finite extinction time for a class of nonlinear parabolic equations*, Comm. Partial Differential Equations, 4 (1979), pp. 1213–1231.

[123] G. Díaz and R. Letelier, *Unbounded solutions of one-dimensional quasilinear elliptic equations*, Appl. Anal., 48 (1993), pp. 173–203.

[124] J. I. Díaz, *Solutions with compact support for some degenerate parabolic problems*, Nonlinear Anal., 3 (1979), pp. 831–847.

[125] ——, *Anulación de soluciones para operadores acretivos en espacios de banach. aplicaciones a ciertos problemas parabólicos no lineales*, Rev. Real. Acad. Ciencias Exactas, Físicas y Naturales de Madrid, LXXIV (1980), pp. 865–880.

[126] ——, *Results and methods concerning the finite extinction property for evolution equations*, in Proceedings of the Second Conference on Differential Equations and Their Applications, II (Valldoreix, 1979), no. 19, 1980, pp. 93–115.

[127] ——, *Técnica de supersoluciones locales para problemas estacionarios no lineales*, Memorias de la Real Academia de Ciencias Exactas, Fisicas y Naturales, Serie de Ciencias Exactas, XVI (1982).

[128] ——, *Nonlinear Partial Differential Equations and Free Boundaries* Vol. I, Pitman (Advanced Publishing Program), Boston, 1985.

[129] ——, *Mathematical analysis of some diffusive energy balance models in climatology*, in Mathematics, Climate and Environment (Madrid, 1991), Masson, Paris, 1993, pp. 28–56.

[130] J. I. Díaz and A. Dou, *On subsonic flow around a symmetric obstacle*, Collect. Math., 33 (1982), pp. 141–160.

[131] J. I. Díaz and F. de Thélin, *On a nonlinear parabolic problem arising in some models related to turbulent flows*, SIAM J. Math. Anal., 25 (1994), pp. 1085–1111.

[132] J. I. Díaz and G. Galiano, *Existence and uniqueness of solutions of the Boussinesq system with nonlinear thermal diffusion*, Topol. Methods Nonlinear Anal., 11 (1998), pp. 59–82.

[133] J. I. Díaz, G. Galiano, and A. Jüngel, *Space localization and uniqueness of vacuum solutions to a degenerate parabolic system in semiconductor theory*, C. R. Acad. Sci. Paris, Sér. I Math., 325 (1997), pp. 267–272.

[134] J. I. Díaz and J. Hernández, *On the existence of a free boundary for a class of reaction-diffusion systems*, SIAM J. Math. Anal., 15 (1984), pp. 670–685.

[135] ——, *Qualitative properties of free boundaries for some nonlinear degenerate parabolic equations*, in Nonlinear Parabolic Equations: Qualitative Properties of Solutions (Rome, 1985), Longman, Harlow, UK, 1987, pp. 85–93.

[136] ——, *Global bifurcation and continua of nonnegative solutions for a quasilinear elliptic problem*, C. R. Acad. Sci. Paris Sér. I Math., 329 (1999), pp. 587–592.

[137] J. I. Díaz and M. A. Herrero, *Propriétés de support compact pour certaines équations elliptiques et paraboliques non linéaires*, C. R. Acad. Sci. Paris Sér. A–B, 286 (1978), pp. A815–A817.

[138] ——, *Estimates on the support of the solutions of some nonlinear elliptic and parabolic problems*, Proc. Roy. Soc. Edinburgh Sect. A, 89 (1981), pp. 249–258.

[139] J. I. Díaz and R. Jimenez, *Comportamiento en el contorno de la solución del problema de signorini*, Actas del V CEDYA, Univ. de Zaragoza, 1983, pp. 308–314.

[140] J. I. Díaz and A. Liñán, *The release of gas in long pipelines: modeling and study of a doubly nonlinear parabolic equation*, in Proceedings of the Mathematical Meeting in Honor of A. Dou (Spanish) (Madrid, 1988), Madrid, 1989, Editorial Univ. Complutense Madrid, pp. 95–119.

[141] J. I. Diaz and J. Mossino, *Isoperimetric inequalities in the parabolic obstacle problems*, J. Math. Pures Appl. (9), 71 (1992), pp. 233–266.

[142] J. I. Díaz and O. Oleinik, *Nonlinear elliptic boundary value problems in unbounded domains and the asymptotic behaviour of its solutions*, C. R. Acad. Sci. Paris Sér. I Math., 315 (1992), pp. 787–792.

[143] J. I. Díaz and R. Quintanilla, *Spatial and continuous dependence estimates in linear viscoelasticity*, preprint, Universidad Complutense de Madrid, Depto. de Matemática Aplicada, December 1999.

[144] J. I. Díaz and Á. M. Ramos, *Some results about the approximate controllability property for quasilinear diffusion equations*, C. R. Acad. Sci. Paris Sér. I Math., 324 (1997), pp. 1243–1248.

[145] J. I. Díaz, J. E. Saá, and U. Thiel, *On the equation of prescribed mean curvature and other elliptic quasilinear equations with locally vanishing solutions*, Rev. Un. Mat. Argentina, 35 (1989), pp. 175–206 (1991).

[146] J. I. Díaz and I. Stakgold, *Mathematical aspects of the combustion of a solid by a distributed isothermal gas reaction*, SIAM J. Math. Anal., 26 (1995), pp. 305–328.

[147] J. I. Díaz and L. Véron, *Compacité du support des solutions d'équations quasi linéaires elliptiques ou paraboliques*, C. R. Acad. Sci. Paris Sér. I Math., 297 (1983), pp. 149–152.

[148] ——, *Existence theory and qualitative properties of the solutions of some first order quasilinear variational inequalities*, Indiana Univ. Math. J., 32 (1983), pp. 319–361.

[149] ——, *Local vanishing properties of solutions of elliptic and parabolic quasilinear equations*, Trans. Amer. Math. Soc., 290 (1985), pp. 787–814.

[150] E. DiBenedetto, *Degenerate Parabolic Equations*, Springer-Verlag, New York, 1993.

[151] Y. Dubinskii, *Nonlinear elliptic and parabolic equations*, in Itogi nauki i tekhniki. Ser. Sovremennye problemy matematiki, vol. 9, VINITI, Moscow, 1976, pp. 5–126.

[152] G. Duvaut, *Mécanique des milieux continus*, Masson, Paris, 1990.

[153] J. Esquinas and M. A. Herrero, *Travelling wave solutions to a semilinear diffusion system*, SIAM J. Math. Anal., 21 (1990), pp. 123–136.

[154] J. R. Esteban, A. Rodríguez, and J. L. Vázquez, *A nonlinear heat equation with singular diffusivity*, Comm. Partial Differential Equations, 13 (1988), pp. 985–1039.

[155] L. C. Evans and B. F. Knerr, *Instantaneous shrinking of the support of nonnegative solutions to certain nonlinear parabolic equations and variational inequalities*, Illinois J. Math., 23 (1979), pp. 153–166.

[156] E. Fernández-Cara, F. Guillén, and R. R. Ortega, *Some theoretical results for visco-plastic and dilatant fluids with variable density*, Nonlinear Anal., Theory Methods Appl., 28 (1997), pp. 1079–1100.

[157] ——, *Some theoretical results concerning non-Newtonian fluids of the Oldroyd kind*, Ann. Scuola Norm. Sup. Pisa Cl. Sci. (4), 26 (1998), pp. 1–29.

[158] V. Ferone and F. Murat, *Nonlinear problems having natural growth in the gradient: an existence result when the source terms are small*, Nonlinear Anal., 42 (2000), pp. 1309–1326.

[159] P. C. Fife, *The Bénard problem for general fluid dynamical equations and remarks on the Boussinesq approximation*, Indiana Univ. Math. J., 20 (1970/1971), pp. 303–326.

[160] M. Fila, J. Hulshof, and P. Quittner, *The quenching problem on the N-dimensional ball*, in Nonlinear Diffusion Equations and Their Equilibrium States 3 (Gregynog, 1989), Birkhäuser, Boston, MA, 1992, pp. 183–196.

[161] M. Fila and B. Kawohl, *Is quenching in infinite time possible?*, Quart. Appl. Math., 48 (1990), pp. 531–534.

[162] M. Fila and H. A. Levine, *Quenching on the boundary*, Nonlinear Anal., 21 (1993), pp. 795–802.

[163] J. N. Flavin and S. Rionero, *Qualitative Estimates for Partial Differential Equations: An Introduction*, CRC Press, Boca Raton, FL, 1996.

[164] C. Foias, O. Manley, and R. Temam, *Attractors for the Benard problem: Existence and physical bounds for their fractal dimension*, Nonlinear Analysis, 11 (1987), pp. 939–967.

[165] A. Friedman, *Partial Differential Equations*, Holt, Rinehart and Winston, Inc., New York, 1969.

[166] A. Friedman and P. Knabner, *A transport model with micro- and macro-structure*, J. Differential Equations, 98 (1992), pp. 328–354.

[167] G. Gagneux, *Une approche analytique nouvelle des modèles de la récupération secondée en ingéniérie pétrolière*, J. Méc. Théor. Appl., 5 (1986), pp. 3–20.

[168] G. Gagneux and M. Madaune-Tort, *Analyse mathématique de modèles non linéaires de l'ingéniérie petrolière (Mathematical Analysis of Nonlinear Models of Petrol Engineering)*, Mathématiques et Applications (Paris), Vol. 22, Springer-Verlag, Berlin, 1995.

[169] V. A. Galaktionov, J. Hulshof, and J. L. Vazquez, *Extinction and focusing behaviour of spherical and annular flames described by a free boundary problem*, J. Math. Pures Appl. (9), 76 (1997), pp. 563–608.

[170] V. A. Galaktionov and R. Kersner, *On a discontinuous parabolic semigroup*, in Free boundary problems: theory and applications, II (Chiba, 1999), Gakkōtosho, Tokyo, 2000, pp. 135–145.

[171] V. A. Galaktionov, S. I. Shmarev, and J. L. Vazquez, *Regularity of interfaces in diffusion processes under the influence of strong absorption*, Arch. Ration. Mech. Anal., 149 (1999), pp. 183–212.

[172] ——, *Regularity of solutions and interfaces to degenerate parabolic equations. The intersection comparison method*, in Free Boundary Problems: Theory and Applications (Crete, 1997), Chapman and Hall/CRC, Boca Raton, FL, 1999, pp. 115–130.

[173] ——, *Second-order interface equations for nonlinear diffusion with very strong absorption*, Commun. Contemp. Math., 1 (1999), pp. 51–64.

[174] ——, *Behaviour of interfaces in a diffusion-absorption equation with critical exponents*, Interfaces and Free Boundaries, 2 (2000), pp. 425–448.

[175] V. A. Galaktionov and J. L. Vázquez, *Extinction for a quasilinear heat equation with absorption. II. A dynamical systems approach*, Comm. Partial Differential Equations, 19 (1994), pp. 1107–1137.

[176] ——, *Extinction for a quasilinear heat equation with absorption. I. Technique of intersection comparison*, Comm. Partial Differential Equations, 19 (1994), pp. 1075–1106.

[177] ——, *Geometrical properties of the solutions of one-dimensional nonlinear parabolic equations*, Math. Ann., 303 (1995), pp. 741–769.

[178] ——, *A dynamical systems approach for the asymptotic analysis of nonlinear heat equations*, in International Conference on Differential Equations (Lisboa, 1995), World Sci. Publishing, River Edge, NJ, 1998, pp. 82–106.

[179] G. P. Galdi and S. Rionero, *Weighted Energy Methods in Fluid Dynamics and Elasticity*, Springer-Verlag, Berlin, 1985.

[180] G. Galiano and M. A. Peletier, *Spatial localization for a general reaction-diffusion system*, Ann. Fac. Sci. Toulouse Math. (6), 7 (1998), pp. 419–441.

[181] J. García-Melián and J. Sabina de Lis, *Stationary patterns to diffusion problems*, Math. Methods Appl. Sci., 23 (2000), pp. 1467–1489.

[182] M. G. Garroni and M. A. Vivaldi, *Stability of free boundaries*, Nonlinear Anal., 12 (1988), pp. 1339–1347.

[183] D. Gilbarg and N. Trudinger, *Elliptic Partial Differential Equations of Second Order*, 2nd ed., Springer-Verlag, Berlin, 1983.

[184] B. H. Gilding and R. Kersner, *The characterization of reaction-convection-diffusion processes by travelling waves*, J. Differential Equations, 124 (1996), pp. 27–79.

[185] O. N. Goncharova, *Solvability of a nonstationary problem for equations of free convection with temperature-dependent viscosity*, Dinamika Sploshn. Sredy, (1990), pp. 35–58, 154.

[186] J. Greenberg, R. MacCamy, and V. Mizei, *On the existence, uniqueness and stability of the equation* $\sigma'(u_x)u_{xx} + \lambda u_{xtx} = \rho_0 u_{tt}$, J. Math. Mech., 17 (1968), pp. 707–728.

[187] M. Guedda and L. Veron, *Bifurcation phenomena associated to the p–Laplace operator*, Trans. Amer. Math. Soc., 310 (1988), pp. 419–431.

[188] M. E. Gurtin, *An Introduction to Continuum Mechanics*, Academic Press, New York, 1981.

[189] J. K. HALE, *Asymptotic Behavior of Dissipative Systems*, American Mathematical Society, Providence, RI, 1988.

[190] A. HARAUX, *Comportement à l'infini pour certains systèmes dissipatifs non linéaires*, Proc. Roy. Soc. Edinburgh Sect. A, 84 (1979), pp. 213–234.

[191] G. H. HARDY, J. E. LITTLEWOOD, AND G. PÓLYA, *Inequalities*, Second ed., Cambridge University Press, Cambridge, 1952.

[192] M. A. HERRERO, *A limit case in nonlinear diffusion*, Nonlinear Anal., 13 (1989), pp. 611–628.

[193] M. A. HERRERO AND J. L. VÁZQUEZ, *Asymptotic behaviour of the solutions of a strongly nonlinear parabolic problem*, Ann. Fac. Sci. Toulouse Math. (5), 3 (1981), pp. 113–127.

[194] M. HESTENES AND R. REDHEFFER, *On the minimization of certain quadratic functionals. I*, Arch. Rational Mech. Anal., 56 (1974), pp. 1–14.

[195] S. HOWISON, J. RODRIGUES, AND M. SHILLOR, *Stationary solutions to the thermistor problem*, J. Math. Anal. Appl., 174 (1993), pp. 573–588.

[196] A. V. IVANOV, *Quasilinear degenerate and nonuniformly elliptic and parabolic equations of second order*, Proc. Steklov Inst. Math., (1984) (translation of Trudy Mat. Inst. Steklov, 160 (1982); translated from the Russian by J. R. Schulenberger).

[197] D. D. JOSEPH, *Stability of Fluid Motions*, Vol. II, Springer-Verlag, Berlin, 1976.

[198] A. JÜNGEL, *On the existence and uniqueness of transient solutions of a degenerate nonlinear drift-diffusion model for semiconductors*, Math. Models Mech. Appl. Sci., 4 (1994), pp. 677–703.

[199] ——, *Qualitative behavior of solutions of a degenerate nonlinear drift-diffusion model for semiconductors*, Math. Models Mech. Appl. Sci., 5 (1995), pp. 497–518.

[200] ——, *Asymptotic analysis of a semiconductor model based on Fermi-Dirac statistics*, Math. Mach. Appl. Sci., 19 (1996), pp. 497–518.

[201] A. JÜNGEL, *A nonlinear drift-diffusion system with electric convection arising in semiconductor and electrophoretic modeling*, Tech. Rep. 400, Fachbereich Mathematik, TU Berlin, Germany, 1997.

[202] A. JÜNGEL AND P. PIETRA, *A discretization scheme of a quasi-hydrodynamic semiconductor model*, Math. Nachr., 185 (1997), pp. 85–110.

[203] J. KAČUR, *On a solution of degenerate elliptic-parabolic systems in Orlicz-Sobolev spaces. II*, Math. Z., 203 (1990), pp. 569–579.

[204] A. S. KALAŠHNIKOV, *Some problems of the qualitative theory of second-order nonlinear degenerate parabolic equations*, Uspekhi Mat. Nauk, 42 (1987), pp. 135–176, 287.

[205] ——, *Equations of the type of a nonstationary filtration with infinite rate of propagation of perturbations*, Vestnik Moskov. Univ. Ser. I Mat. Meh., 27 (1972), pp. 45–49.

[206] ——, *The nature of the propagation of perturbations in problems of nonlinear heat conduction with absorption*, Ž. Vyčisl. Mat. i Mat. Fiz., 14 (1974), pp. 891–905, 1075.

[207] ——, *A nonlinear equation arising in the theory of nonlinear filtration*, Trudy Sem. Petrovsk., (1978), pp. 137–146.

[208] ——, *On the concept of finite rate of propagation of perturbations*, Uspekhi Mat. Nauk, 34 (1979), pp. 199–200.

[209] E. A. KALITA, *Compactness of the support and the nonexistence of singularities for nonlinear elliptic systems of arbitrary order*, Sibirsk. Mat. Zh., 35 (1994), pp. 327–339, ii.

[210] S. KAMIN AND L. VERON, *Flat core properties associated to the p–Laplace operator*, Proceedings of the Amer. Math. Soc., 118 (1993), pp. 1079–1085.

[211] J. I. KANEL′, *A model system of equations for the one-dimensional motion of a gas*, Differencial′ nye Uravnenija, 4 (1968), pp. 721–734.

[212] H. KAWARADA, *On solutions of initial-boundary problem for* $u_t = u_{xx} + 1/(1-u)$, Publ. Res. Inst. Math. Sci., 10 (1974/75), pp. 729–736.

[213] B. KAWOHL AND R. KERSNER, *On degenerate diffusion with very strong absorption*, Math. Methods Appl. Sci., 15 (1992), pp. 469–477.

[214] B. KAWOHL AND L. A. PELETIER, *Observations on blow up and dead cores for nonlinear parabolic equations*, Math. Z., 202 (1989), pp. 207–217.

[215] A. KAZHIKHOV, *The equations of potential flow of compressible viscous fluid at low Reynolds number*, Acta. Appl. Math., 37 (1994), pp. 77–81.

[216] A. KAZHIKHOV AND V. WEIGANT, *On the existance of global solutions to two-dimensional Navier-Stokes equations of compressible viscous fluid*, Siberian Math. J., 36 (1995), pp. 1108–1141.

[217] R. KERSNER, *Nonlinear heat conduction with absorption: space localization and extinction in finite time*, SIAM J. Appl. Math., 43 (1983), pp. 1274–1285.

[218] ——, *Nonlinear heat conduction with absorption: space localization and extinction in finite time*, SIAM J. Appl. Math., 43 (1983), pp. 1274–1285.

[219] ——, *The behavior of temperature fronts in media with nonlinear heat conductivity under absorption*, Vestnik Moskov. Univ. Ser. I Mat. Mekh., (1987), pp. 44–51.

[220] R. Kersner and A. Shishkov, *Instantaneous shrinking of the support of energy solutions*, J. Math. Anal. Appl., 198 (1996), pp. 729–750.

[221] S. Kichenassamy and J. Smoller, *On the existence of radial solutions of quasi-linear elliptic equations*, Nonlinearity, 3 (1990), pp. 677–694.

[222] D. Kinderlehrer and P. Pedregal, *Weak convergence of integrands and the Young measure representation*, SIAM J. Math. Anal., 23 (1992), pp. 1–19.

[223] B. F. Knerr, *The porous medium equation in one dimension*, Trans. Amer. Math. Soc., 234 (1977), pp. 381–415.

[224] R. Knops and B. Straughan, *Decay and Nonexistence for Sublinearly Forced Systems in Continuum Mechanics*, Collège de France Seminar, Vol. II, H. Brezis and J. L. Lions, eds., Pitman, London, 1982.

[225] V. A. Kondratiev and O. A. Oleinik, *On Korn's inequalities*, C. R. Acad. Sci. Paris Sér. I Math., 308 (1989), pp. 483–487.

[226] A. N. Konovalov, *Problems of Multiphase Fluid Filtration*, World Scientific, Singapore, 1994.

[227] A. I. Kozhanov, N. A. Lar′kin, and N. N. Yanenko, *A mixed problem for a class of third-order equations*, Sibirsk. Mat. Zh., 22 (1981), pp. 81–86, 225.

[228] M. A. Krasnosel′skiĭ and J. B. Rutickiĭ, Выпуклые функции и пространства Орлича*; Convex Functions and Orlicz Spaces*, Gosudarstv. Izdat. Fiz.-Mat. Lit., Moscow (Noordhoff, Groningen), (Russian/English), 1958/1961.

[229] S. Kruzhkov and A. Korolev, *Towards a theory of embedding of anisotropic functional spaces*, Dokl. Acad. Nauk. SSSR, 285 (1985), pp. 1054–1057.

[230] N. V. Krylov, *Lectures on Elliptic and Parabolic Equations in Hölder Spaces*, American Mathematical Society, Providence, RI, 1996.

[231] A. Kufner, *Weighted Sobolev Spaces*, John Wiley, New York, 1985 (translated from the Czech).

[232] Y. C. Kwong, *Asymptotic behavior of a plasma type equation with finite extinction*, Arch. Rational Mech. Anal., 104 (1988), pp. 277–294.

[233] O. A. Ladyženskaja, V. A. Solonnikov, and N. N. Ural'tseva, *Linear and Quasilinear Equations of Parabolic Type*, American Mathematical Society, Providence, RI, 1967 (translation of Math. Monographs, Vol. 23; translated from the Russian by S. Smith).

[234] O. A. Ladyzhenskaya, *The Mathematical Theory of Viscous Incomprehensible Flow*, Second English ed., revised and enlarged, Mathematics and Its Applications, Vol. 2, Gordon and Breach, New York, 1969 (translated from the Russian by Richard A. Silverman and John Chu).

[235] O. A. Ladyzhenskaya and N. N. Ural'tseva, *Linear and Quasilinear Elliptic Equations*, Academic Press, New York, 1968 (translated from the Russian by Scripta Technica; Leon Ehrenpreis, translation ed.).

[236] A. V. Lair, *Finite extinction time for solutions of nonlinear parabolic equations*, Nonlinear Anal., 21 (1993), pp. 1–8.

[237] A. V. Lair and M. E. Oxley, *Extinction in finite time of solutions to nonlinear absorption-diffusion equations*, J. Math. Anal. Appl., 182 (1994), pp. 857–866.

[238] ——, *Finite extinction time for a nonlinear parabolic Neumann boundary value problem*, Appl. Anal., 55 (1994), pp. 41–50.

[239] M.-N. Le Roux, *Semi-discretization in time of a fast diffusion equation*, J. Math. Anal. Appl., 137 (1989), pp. 354–370.

[240] H. A. Levine, *Quenching, nonquenching, and beyond quenching for solution of some parabolic equations*, Ann. Mat. Pura Appl. (4), 155 (1989), pp. 243–260.

[241] H. A. Levine and R. Quintanilla, *Some remarks on Saint-Venant's principle*, Math. Methods Appl. Sci., 11 (1989), pp. 71–77.

[242] G. M. Lieberman, *Second order parabolic differential equations*, World Scientific Publishing Co. Inc., River Edge, NJ, 1996.

[243] J.-L. Lions, *Quelques méthodes de résolution des problèmes aux limites non linéaires*, Dunod, 1969.

[244] P.-L. Lions, *Mathematical Topics in Fluid Mechanics, Vol. 2: Compressible models*, Oxford Lecture Series in Mathematics and Its Applications, Vol. 10, Clarendon Press, Oxford, 1996.

[245] ——, *Mathematical Topics in Fluid Mechanics, Vol. 2: Incompressible Models*, Oxford Lecture Series in Mathematics and Its Applications, Vol. 3, Clarendon Press, Oxford, 1996.

[246] G. LUMER, R. REDHEFFER, AND W. WALTER, *Estimates for solutions of degenerate second-order differential equations and inequalities with applications to diffusion*, Nonlinear Anal., Theory Methods Appl., 12 (1988), pp. 1105–1121.

[247] A. MAJDA, *Disappearing solutions for the dissipative wave equation*, Indiana Univ. Math. J., 24 (1974/75), pp. 1119–1133.

[248] J. MÁLEK, J. NEČAS, M. ROKYTA, AND M. RUŽIČKA, *Weak and Measure-Valued Solutions to Evolutionary PDEs*, Chapman and Hall, London, 1996.

[249] P. MARKOVICH, C. RINGHOFER, AND C. SCHMEISER, *Semiconductor Equations*, Springer-Verlag, Vienna, 1986.

[250] L. K. MARTINSON AND K. B. PAVLOV, *The problem of the three-dimensional localization of thermal pertubations in the theory of nonlinear heat conduction*, USSR Comput. Math. and Math. Phys., (1972), pp. 261–268 (translated from the Russian).

[251] V. G. MAZ'YA, *Sobolev Spaces*, Springer-Verlag, Berlin, 1985 (translated from the Russian by T. O. Shaposhnikova).

[252] J. MILHAJAN, *A rigorous exposition of the Boussinesq approximations applicable to a thin layer of fluid*, Astrophysics J., 136 (1962).

[253] S. MIZOHATA, *Unicité du prolongement des solutions pour quelques opérateurs différentiels paraboliques*, Mem. Coll. Sci. Univ. Kyoto. Ser. A. Math., 31 (1958), pp. 219–239.

[254] M. MOCK, *On equations describing steady state carrier distributions in a semiconductor device*, Comm. Pure. Appl. Math., 25 (1972), pp. 781–792.

[255] V. MONAKHOV, *Boundary-Value Problems with Free Boundaries for Elliptic Systems of Equations*, Translations of Mathematical Monographs, Vol. 57, American Mathematical Society, Providence, RI, 1983 (translated from the Russian).

[256] P. P. MOSOLOV AND V. P. MJASNIKOV, *A proof of Korn's inequality*, Dokl. Akad. Nauk SSSR, 201 (1971), pp. 36–39.

[257] J. D. MURRAY, *Perturbation effects on the decay of discontinuous solutions of nonlinear first order wave equations*, SIAM J. Appl. Math., 19 (1970), pp. 273–298.

[258] T. NAKAKI, *Numerical interfaces in nonlinear diffusion equations with finite extinction phenomena*, Hiroshima Math. J., 18 (1988), pp. 373–397.

[259] T. NAKAKI AND K. TOMOEDA, *A finite difference approach to the interface equation for some nonlinear diffusion equations with absorption*, Proc. Japan Acad. Ser. A Math. Sci., 77 (2001), pp. 32–37.

[260] R. NATALINI, *The bipolar hydrodynamic model for semiconductors and the drift-diffusion equations*, J. Math. Anal. Appl., 198 (1996), pp. 262–281.

[261] I. NATANSON, *Theorie der Funktionen einer reellen Veraenderlichen. Hrsg. von Karl Boegel. Uebers. aus dem Russischen. 5 Aufl.*, Mathematische Lehrbücher und Monographien I: Abteilung: Mathematische Lehrbücher, Bd. VI, Akademie-Verlag, Berlin, 1981.

[262] J. NEČAS, *Écoulements de Fluide: Compacité par Entropie*, Recherches en Mathématiques Appliquées, Masson, 1989.

[263] L. NIRENBERG, *An extended interpolation inequality*, Ann. Scuola Norm. Sup. Pisa (3), 20 (1966), pp. 733–737.

[264] O. A. OLEĬNIK, A. S. KALAŠINKOV, AND Y.-L. ČŽOU, *The Cauchy problem and boundary problems for equations of the type of non-stationary filtration*, Izv. Akad. Nauk SSSR. Ser. Mat., 22 (1958), pp. 667–704.

[265] O. A. OLEĬNIK AND G. A. YOSIFIAN, *On the asymptotic behaviour at infinity of solutions in linear elasticity*, Arch. Rational Mech. Anal., 78 (1982), pp. 29–53.

[266] L. V. OVSYANNIKOV, *On a gas flow with a straight transition line*, Akad. Nauk SSSR. Prikl. Mat. Meh., 13 (1949), pp. 537–542.

[267] R. E. PATTLE, *Diffusion from an instantaneous point source with a concentration-dependent coefficient*, Quart. J. Mech. Appl. Math., 12 (1959), pp. 407–409.

[268] A. PAZY, *Strong convergence of semigroups of nonlinear contractions in Hilbert space*, J. Analyse Math., 34 (1978), pp. 1–35 (1979).

[269] D. PHILLIPS, *Existence of solutions of quenching problems*, Appl. Anal., 24 (1987), pp. 253–264.

[270] P. PIETRA AND C. VERDI, *On the convergence of the approximate free boundary for the parabolic obstacle problem*, Atti Accad. Naz. Lincei Rend. Cl. Sci. Fis. Mat. Natur. (8), 79 (1985), pp. 159–171 (1986).

[271] M. A. POZIO AND A. TESEI, *On a strongly coupled parabolic system in population dynamics*, in Biomathematics and Related Computational Problems (Naples, 1987), Kluwer Academic Publishers, Dordrecht, the Netherlands, 1988, pp. 559–565.

[272] M. H. PROTTER AND H. F. WEINBERGER, *Maximum Principles in Differential Equations*, Prentice–Hall, Englewood Cliffs, NJ, 1967.

[273] P. PUCCI, J. SERRIN, AND H. ZOU, *A strong maximum principle and a compact support principle for singular elliptic inequalities*, J. Math. Pures Appl. (9), 78 (1999), pp. 769–789.

[274] R. REDHEFFER, *On a nonlinear functional of Berkovitz and Pollard*, Arch. Rational Mech. Anal., 50 (1973), pp. 1–9.

[275] ——, *Nonlinear differential inequalities and functions of compact support*, Trans. Amer. Math. Soc., 220 (1976), pp. 133–157.

[276] R. RICCI, *Large time behavior of the solution of the heat equation with nonlinear strong absorption*, J. Differential Equations, 79 (1989), pp. 1–13.

[277] R. RICCI AND D. A. TARZIA, *Asymptotic behavior of the solutions of the dead-core problem*, Nonlinear Anal., 13 (1989), pp. 405–411.

[278] J.-F. RODRIGUES, *Weak solutions for thermoconvective flows of Boussinesq-Stefan type*, in Mathematical Topics in Fluid Mechanics (Lisbon, 1991), Longman, Harlow, UK, 1992, pp. 93–116.

[279] A. RODRÍGUEZ AND J. L. VÁZQUEZ, *A well-posed problem in singular Fickian diffusion*, Arch. Rational Mech. Anal., 110 (1990), pp. 141–163.

[280] J. RUBIN, *Transport of reacting solutes in porous media: relation between mathematical nature of problem formulation and chemical nature of reactions*, Water Resour. Res., 19 (1983), pp. 1231–1252.

[281] Y. G. RYKOV, *The behavior, with increasing time, of the generalized solutions of degenerate parabolic equations with nonisotropic nonlinearities not of power type*, in Differential Equations and Their Applications, Moskov. Gos. Univ., Moscow, 1984, pp. 160–164 (in Russian).

[282] ——, *The finite propagation rate of perturbations and momentum, for the generalized solution of quasilinear degenerate parabolic equations*, in Differential Equations and Their Applications (Russian), Moskov. Gos. Univ., Moscow, 1984, pp. 101–105.

[283] E. S. SABININA, *On a class of non-linear degenerate parabolic equations*, Dokl. Akad. Nauk SSSR, 143 (1962), pp. 794–797.

[284] A. A. SAMARSKII, V. A. GALAKTIONOV, S. P. KURDYUMOV, AND A. P. MIKHAILOV, *Blow-Up in Quasilinear Parabolic Equations*, Walter de Gruyter, Berlin, 1995 (translated from the 1987 Russian original by Michael Grinfeld and revised by the authors).

[285] V. N. SAMOKHIN, *A system of equations of a stationary boundary layer of a dilatant medium*, Trudy Sem. Petrovsk., (1989), pp. 89–108, 266.

[286] L. SANTOS, *Variational limit of compressible to incompressible fluid*, in Energy Methods in Continuum Mechanics (Oviedo, 1994), Kluwer Academic Publishers, Dordrecht, the Netherlands, 1996, pp. 126–144.

[287] L. A. SEGEL, *Mathematics Applied to Continuum Mechanics*, MacMillan, New York, 1977 (with material on elasticity by G. H. Handelman).

[288] A. E. SHISHKOV, *Estimates for the rate of propagation of perturbations in quasilinear degenerate higher-order parabolic equations in divergence form*, Ukraïn. Mat. Zh., 44 (1992), pp. 1451–1456.

[289] ——, *Dynamics of the geometry of the support of the generalized solution of a higher-order quasilinear parabolic equation in divergence form*, Differentsial′nye Uravneniya, 29 (1993), pp. 537–547, 552.

[290] S. I. SHMAREV, *Local properties of solutions of a nonhomogeneous nonlinear heat equation with absorption*, Din. Sploshn. Sredy, (1990), pp. 131–147, 164.

[291] ——, *The diminution effect for the support of a solution of a degenerate parabolic equation*, Din. Splosh. Sredy, 97 (1990), pp. 194–200.

[292] R. SHOWALTER, *Mechanics of Non-Newtonian Fluids*, Pergamon Press, Oxford, 1978.

[293] J. SIMON, *Nonhomogeneous viscous incompressible fluids: Existence of velocity, density, and pressure*, SIAM J. Math. Anal., 21 (1990), pp. 1093–1117.

[294] M. SLEMROD, *Stabilization of bilinear control systems with applications to nonconservative problems in elasticity*, SIAM J. Control Optimization, 16 (1978), pp. 131–141.

[295] T. B. SOLOMJAK, *An application of the theory of degenerate equations to inverse problems for subsonic flows of a compressible fluid*, in Proceedings of a Conference on Boundary Value Problems (Kazan, 1969), Kazan, 1970, Izdat. Kazan. Univ., pp. 238–243 (in Russian).

[296] I. STAKGOLD, *Partial extinction in reaction-diffusion*, Confer. Sem. Mat. Univ. Bari, 1987, p. 21.

[297] G. STAMPACCHIA, *Équations elliptiques du second ordre à coefficients discontinus*, Séminaire de Mathématiques Supérieures, 16 (Été, 1965), Les Presses de l'Université de Montréal, Montreal, 1966.

[298] B. STRAUGHAN, *The Energy Method, Stability, and Nonlinear Convection*, Springer-Verlag, New York, 1992.

[299] R. TEMAM, *Navier-Stokes Equations: Theory and Numerical Analysis*, Studies in Mathematics and Its Applications, Vol. 2, North-Holland, Amsterdam, 1977.

[300] ——, *Infinite-Dimensional Dynamical Systems in Mechanics and Physics*, Springer-Verlag, New York, 1988.

[301] M. TSUTSUMI, *On solutions of some doubly nonlinear degenerate parabolic equations with absorption*, J. Math. Anal. Appl., 132 (1988), pp. 187–212.

[302] C. J. VAN DUIJN, G. GALIANO, AND M. PELETIER, *How mangroves salinize the soil*, tech. rep., 2000.

[303] C. J. VAN DUIJN AND P. KNABNER, *Solute transport through porous media with slow adsorption*, in Free Boundary Problems: Theory and Applications, Vol. I (Irsee, 1987), Longman, Harlow, UK, 1990, pp. 375–388.

[304] J. L. VÁZQUEZ, *An introduction to the mathematical theory of the porous medium equation*, in Shape Optimization and Free Boundaries, M. Delfour, ed., Mathematical and Physical Sciences, Kluwer Academic Publishers, Dordrecht, the Netherlands, 1992, pp. 347–389.

[305] ——, *A strong maximum principle for some quasilinear elliptic equations*, Appl. Math. Optim., 12 (1984), pp. 191–202.

[306] L. VÉRON, *Effets régularisants de semi-groupes non linéaires dans des espaces de Banach*, Ann. Fac. Sci. Toulouse Math. (5), 1 (1979), pp. 171–200.

[307] J. VIDEMAN, *Mathematical Analysis of Viscoelastic Non-Newtonian Fluids*, Ph.D. thesis, Instituto Técnico Superior, Lisboa, 1997.

[308] W. WALTER, *Parabolic differential equations with a singular nonlinear term*, Funkcial. Ekvac., 19 (1976), pp. 271–277.

[309] J. WU, *Theory and Applications of Partial Functional-Differential Equations*, Springer-Verlag, New York, 1996.

[310] N. YAMADA, *Estimates on the support of solutions of parabolic variational inequalities in bounded cylindrical domains*, Hiroshima Math. J., 10 (1980), pp. 337–349.

[311] Y. B. ZEL′DOVIČ AND A. S. KOMPANEEC, *On the theory of propagation of heat with the heat conductivity depending upon the temperature*, in Collection in Honor of the Seventieth Birthday of Academician A. F. Ioffe, Izdat. Akad. Nauk SSSR, Moscow, 1950, pp. 61–71.

[312] H. F. ZHANG, *Large time behavior of the maximal solution of equation* $u_t = (u^{m-1}u_x)_x$ *with* $-1 < m \leq 0$, Differential Integral Equations, 6 (1993), pp. 613–626.

Index

Progress in Nonlinear Differential Equations and Their Applications

Editor
Haim Brezis
Département de Mathématiques
Université P. et M. Curie
4, Place Jussieu
75252 Paris Cedex 05
France
and
Department of Mathematics
Rutgers University
Piscataway, NJ 08854-8019
U.S.A.

Progress in Nonlinear Differential Equations and Their Applications is a book series that lies at the interface of pure and applied mathematics. Many differential equations are motivated by problems arising in such diversified fields as Mechanics, Physics, Differential Geometry, Engineering, Control Theory, Biology, and Economics. This series is open to both the theoretical and applied aspects, hopefully stimulating a fruitful interaction between the two sides. It will publish monographs, polished notes arising from lectures and seminars, graduate level texts, and proceedings of focused and refereed conferences.

We encourage preparation of manuscripts in some form of TeX for delivery in camera-ready copy, which leads to rapid publication, or in electronic form for interfacing with laser printers or typesetters.

Proposals should be sent directly to the editor or to: Birkhäuser Boston, 675 Massachusetts Avenue, Cambridge, MA 02139

PNLDE 30 Parametrized Measures and Variational Principles
Pablo Pedregal

PNLDE 31 Topics in Mathematical Modelling of Composite Materials
Andrej Cherkaev and Robert Kohn, editors

PNLDE 32 Geometrical Optics and Related Topics
Ferruccio Colombini and Nicolas Lerner, editors

PNLDE 33 Entire Solutions of Semilinear Elliptic Equations
I. Kuzin and S. Pohozaev

PNLDE 34 Stability of Functional Equations in Several Variables
Donald H. Hyers†, George Isac and Themistocles M. Rassias

PNLDE 35 Topics in Nonlinear Analysis: The Herbert Amann Anniversary Volume
Joachim Escher and Gieri Simonett, editors

PNLDE 36 Variational Problems with Concentration
Martin Flucher

PNLDE 37 Implicit Partial Differential Equations
Bernard Dacorogna and Paolo Marcellini

PNLDE 38 Systems of Conservation Laws: Two-Dimensional Riemann Problems
Yuxi Zheng

PNLDE 39 Linear and Nonlinear Aspects of Vortices: The Ginzburg–Landau Model
Frank Pacard and Tristan Rivière

PNLDE 40 Recent Trends in Nonlinear Analysis: Festschrift Dedicated to Alfonso Vignoli on the Occasion of his Sixtieth Birthday
Jürgen Appell, editor

PNLDE 41 Quasi-hydrodynamic Semiconductor Equations
Ansgar Jüngel

PNLDE 42 Semigroups of Operators: Theory and Applications
A.V. Balakrishnan, editor

PNLDE 43 Nonlinear Analysis and its Applications to Differential Equations
M.R. Grosshino, M. Ramos, C. Rebelo, and L. Sanchez, editors

PNLDE 44 The Monge–Ampère Equation
Cristian E. Gutiérrez

PNLDE 45 Spatial Patterns: Higher Order Models in Physics and Mechanics
L.A. Peletier and W.C. Troy

PNLDE 46 Carleman Estimates and Applications to Uniqueness and Control Theory
Ferrucio Colombini and Claude Zuily, editors

PNLDE 47 Recent Advances in the Theory of Shock Waves
Heinrich Freistühler and Anders Szepessy, editors

PNLDE 48 Energy Methods for Free Boundary Problems: Applications to Nonlinear PDEs and Fluid Mechanics
S.N. Antontsev, J.I. Díaz, and S. Shmarev